Copyright Page

AF880049

Title:

Spatial Relativity: A Revision and Reconstruction of the Fundamental Theoretical Framework of Physics Based on the Objective Reality of Space

Authors:

Su Enjie

Fang Wenxiang

Publisher:

Hope Grace Publishing

ISBN:

ISBN (Chinese edition, ebook): 978-1-966423-37-9

ISBN (English edition, paperback): 978-1-966423-38-6

LCCN:

Library of Congress Control Number: 2025928021

Printed in the United States of America

There is no new thing in the universe, nor outside of it.

Space Relativity: A Revision and Reconstruction of the Fundamental Theoretical Framework of Physics Based on the Objective Reality of Space

Su Enjie [1] , Fang Wenxiang [2]

(1. School of Automation, Northwestern Polytechnical University, Shanghai 201800)
(2. Founder of Datable Wuli Sheng Technology Co., Ltd., Shanxi, Datong)

[Abstract] Based on the objective reality of the void space and the principle of dynamic characteristics, a superfluid model of void space is established. This model is used to analyze the principles of the universe's spacetime, the origin of physical forces, and the mechanisms of quantum behavior. A new unified concept of spacetime at both macroscopic and microscopic levels is proposed. The unified origin and formation mechanism of all physical forces, including electric field force, magnetic field force, and gravitational force, are discussed. A new prediction of the range of gravitational fields is given, which is consistent with the observed results of cosmic gravitational phenomena and astronomical events. The core issues in quantum theory, such as probability, wave-particle duality, and double-slit interference, are explored and explained, forming a complete foundational theoretical framework of physics based on the objective reality of void space. Based on this framework, the special and general theories of relativity, quantum mechanics, classical mechanics, electromagnetism, and particle mechanics are revised and reconstructed, forming a unified physical theory framework: Spatial Relativity.

[Keywords] Void superfluid, origin of force, gravity model, wave-particle duality, electromagnetic field, origin of particles, origin of the universe

Spatial Relativity: Revision and Reconstruction of the Fundamental Theoretical Framework of Physics Based on the Objective Reality of Space

Su Enjie [1] , Fang Wen Xiang [2]
(1.College of Automation, Northwestern Polytechnical University shanghai 201800)
(2.Founder of Datong WanWuSheng Technology Co., Ltd., Datong, Shanxi)

Abstract

Based on the objective reality of void space and its dynamic mechanical charac teristics, a void superfluid model is established to analyze the principles of cosmic spacetime, the origin of physical forces, and quantum behavior mechanisms. This framework proposes a unified spatiotemporal concept integrating macroscopic and microscopic scales, explores the unified ori gins and formation mechanisms of fundamental forces (electric, magnetic, and gravitational), and predicts novel gravitational field ranges consistent with galactic-scale gravitational observations and astronomical phenomena. It further addresses core quantum mechanical issues such as prob ability, wave-particle duality, and double-slit interference, forming a comprehensive physics unification theory grounded in the objective reality of void space. Within this framework, revisions and reconstructions are proposed for special/general relativity, quantum mechanics, classical me chanics, electromagnetics, and particle physics, culminating in a cohesive theoretical framework: Spatial Relativity.

Keywords: Void Superfluid, Origin of Forces, Gravitational Model, Wave-Particle Duality, Electromagnetic Field, Particle Origin, Cosmic Origin

Preface

Since I accidentally derived the mass-energy equation from the textbook's de Broglie matter wave formula in junior high school, I have felt an indissoluble connection with relativity. I still remember when my teacher asked me if I knew about relativity after seeing the equation, I was completely confused at the time. The name seemed very strange to me. I knew each character, but when put together, I had no idea what it meant.

Later, I continued my studies at Northwestern Polytechnical University, earned my bachelor's and master's degrees, and then joined Huawei to start my career. Over the past decade, I have always paid attention to relativity and felt that there were many aspects of the theory that left me unsatisfied. After leaving Huawei, I wanted to study it thoroughly and write down my thoughts and questions, but I never had the time. Until now, I feel it's time to sit down seriously and conclude this matter, no longer postponing it. It's also a way to give myself an answer. After several months of work, this article was finally completed.

At first, I only focused on the content of special relativity that I thought had problems, such as the principle of the constancy of the speed of light, the principle of relativity, and the determination of simultaneity. I wrote dozens of chapters on these topics (a summary of the relevant content can be found in "Extended Reading 12.1: 8 Internal Issues in the Theories of Special and General Relativity"). I intended to stop writing there, without planning to go further. However, through many readers' feedback and discussions, I discovered some new areas and controversies, including frequent mentions of general relativity in the discussions. I then paid more attention to it and further found that there are many contradictions between general relativity and special relativity (see "Extended Reading 12.2: Einstein's Boomerang: Internal Contradictions in General and Special Relativity"). In addition, general relativity itself also has significant controversies. As a result, the work greatly exceeded my original plan and ended up being written over thirty chapters.

I thought it could be concluded here, but later, through discussions with more people, I found that even if so many problems are pointed out, scholars in the relevant fields can provide many explanations. This made me realize that although the special theory of relativity is a theory with serious defects or even errors, it has become a consistent system mathematically after many years of development. It is very difficult to break this consistent system from within.

The only solution is to provide an alternative theory. Therefore, since we are aware of so many problems in this theory, can we correct these issues and offer a new theoretical explanation based on the correct approach?

Thus, the author began to start over, continuing the writing according to this line of thought, including the basic principles on which the correct theory is based, summarized seven basic principles (see Chapter 3: "Basic Principles of Spatial Relativity"), and then began to reinterpret some related experiments and theories of special relativity based on these principles. After completing this part, during the process of reviewing literature, the author learned that there are also similar theories and discussions in the academic community, such as the theory of spatial superfluidity, which made the author realize that the "void space" entity always mentioned in "Spatial Relativity" is actually similar to such a superfluid space.

Fluids, including theoretical circles, also have similar expressions such as "superfluid universe." The author further thinks that according to the current basic principles and theories of "Space Relativity," it is possible to extend the analysis to the cosmic level, using the theoretical approach of "vacuum space superfluidity" to continue exploring the origins of various fundamental forces, including the mystery of gravity in cosmology. It seems to fit very well with various astronomical phenomena in the universe.

Thus, following this line of thought, the author began to continue the analysis and writing. However, as the number of content chapters increased, the author thought about how to distinguish them, and eventually separated the earlier part into "Space Electrodynamics" and the latter part as "Space Wave Dynamics". Of course, since the content was mainly framework-based analysis and discussion, they could only be considered as "General Conceptions".

In the part of "Space Wave Dynamics", after analyzing the causes and origins of gravity, the author felt somewhat excited, as it seemed that the author had discovered something, possibly touching upon the truth of gravity—the great physics mystery that has puzzled humanity for over a hundred years. The author even felt a bit proud. When analyzing parts such as magnetic fields and Lorentz force, the author considered magnetic fields as vortex flow fields, which felt very logical and consistent with experimental phenomena. Later, upon偶然ly discovering that Maxwell had already made such an analysis and assumption a hundred years ago, the author was both happy and regretful. The author felt that many scholars, such as Maxwell and Fresnel, had made tremendous efforts and achieved significant progress on this theoretical path over a hundred years ago, but it was later defeated and completely discarded by the special theory of relativity due to just one Michelson-Morley experiment and a minor, easily fixable flaw in the "aether" theory. This delay has lasted over a century.

The author often thought that if the theoretical community at that time had chosen the special theory of relativity, it was not the result of natural scientific selection and evolution, but rather manipulated by some force, intentionally steering the direction. If that were the case, the implications would be far more than just regret; it would mean that our scientific development has been deliberately distorted and suppressed for over a hundred years.

Subsequently, the author provided reasonable explanations for phenomena such as black holes, dark energy, and the big bang theory based on the basic principles and framework of "Space Relativity". When writing this part, the author felt somewhat uneasy. Such important areas as gravity and the big bang theory, especially the latter, which has gained widespread recognition, were being negated by the author? However, later the author discovered that the theory of energy loss in starlight redshift proposed by "Space Relativity" was actually similar to an existing academic theory called the "fatigue light theory", which was proposed by Zwicky as an alternative explanation for redshift. This was in perfect alignment with the explanation provided in the wave dynamics section of "Space Relativity", which gave the author renewed confidence. It showed that the analysis in "Space Relativity" was reasonable, and thus the author continued writing.

During the writing process, aspects that I had previously found confusing, such as quantum entanglement and the double-slit interference, became clear and could be naturally explained with reasonable and classical interpretations based on the theory of Spatial Relativity. I then continued to update and revise the content related to quantum mechanics, and this section was eventually established as Spatial Quantum Dynamics. After completing it, I learned that there have been many efforts in the theoretical community to revise quantum mechanics, among which Bohm's mechanics and other pilot-wave theories provide perfect explanations for all content of quantum mechanics from a classical perspective, which made me very excited and felt that this could serve as a foundation for spatial..

A reference to quantum mechanics, but upon careful study, it was found that Bohmian mechanics also has its limitations, failing to provide a reasonable explanation for "pilot waves" and not offering an interpretation of "entanglement" that aligns with the "new locality" (see Chapter 7.4: "(1) New Locality Principle"). However, its overall approach is consistent with quantum space dynamics, which has increased the author's confidence.

After completing the quantum dynamics part, the classical mechanics aspect was also thoroughly described, forming the space classical dynamics. In this way, the four parts of space relativity seem to be complete.

I thought the work was about to be completed, but some readers raised questions about how antimatter is explained. During this time, I also learned some information about "cosmic superfluid," realizing there was another important particle mechanics module that needed analysis and organization. Particle mechanics is considered the last module in the field of fundamental physics, so I began working again and incorporated this content into the framework of "Space Relativity."

In the section on spatial particle dynamics, a more fundamental analysis of the principles of electromagnetic fields is provided, offering a simpler and more reasonable model, including the principles of electromagnetic wave generation and propagation. It also gives consistent explanations and even provides corrections and improvements to the Maxwell equations, the origin of the universe, and related models. Because these aspects are widely involved, the Maxwell equations are considered one of the most core equations in theoretical physics, and the Big Bang model has been widely accepted by the physics community. According to the theoretical analysis of spatial relativity, there are issues with these two aspects, which have been corrected, causing the author to have some self-doubt. However, later it was found that the academic community has actually discussed and questioned the issue of displacement current in the Maxwell equations, with many scholars having published papers on this topic, consistent with the analysis in "Spatial Relativity" (see Chapter 8.4, "(4) Correction of the Maxwell Equations"). Meanwhile, recently some scholars have proposed the theory that the universe may have an overall rotation to address the anomaly in the measurement of the Hubble constant, which also aligns with the analysis in "Spatial Relativity" (see Chapter 8.12, "(3) The Five Major Problems in Current Cosmology"). These findings have further increased the author's confidence in spatial relativity.

Finally, based on the author's actual experiences and thoughts, a section on spatial solitary dynamics was added. All particles, matter, life forms, as well as stars, galaxies, and even the universe, can be considered as quantum solitons in the void space. The different interactions between quantum solitons manifest as strong force, weak force, electric field force, magnetic field force, and gravitational force. It also introduces a new type of force: entanglement force, which enables remote, precise, and real-time interaction between quantum solitons. This is key to the subsequent realization of a mind internet and a quantum internet. The mind internet and quantum internet are important means for the next leap and evolution of human civilization, and their significance is self-evident.

By this point, all parts of the theory of spatial relativity have been basically completed. Throughout the process, the more I have studied the theories and knowledge from various fields, the more I have felt a deep respect for science. Science does not reject innovation, but innovation must be carried out on the basis of respect and compatibility with the current scientific system, and it cannot be blind. The author has extensively reviewed the theories and writings of previous scholars, feeling as if engaging in cross-temporal communication with these historical scientists and scholars. One can sense their thoughts and their efforts. These materials and writings are like weapons left behind by these warriors on their journey, paving the way for future generations. The author has been on this path...

Pick up these weapons and forge your own new ones with them, continuing to explore the path of science. One day, I will also fall, but I hope that the weapon I have created can be seen by other fellow travelers or future generations, to be used by them in cutting through the unknown thorns and fog, and continuing along the path of science.

This book, as an exploration and construction of a unified theoretical framework for physics, can serve as a reference for theoretical researchers in related fields for further study, also provide insights for applied physics researchers, and may be used as a popular science book for general readers.

Due to the urgency of time and limited expertise, there are inevitably many omissions and errors in the text. I hope this work can serve as a starting point to attract more people's attention and participation, jointly promoting the development of theory.

Space Relativity, Su Enjie
September 2024, started writing in Beijing
May 2025, written in Shanghai

Table of Contents

1 Introduction

Relativity theory has been around for over a century, during which debates and doubts about it have never ceased. The fundamental reason lies in the fact that special relativity contains several important defects and errors within its theoretical framework. On the other hand, general relativity's geometric description of spacetime curvature does not provide a suitable physical essence explanation, making it impossible to quantize and leading to structural contradictions with quantum mechanics. Quantum mechanics itself also faces difficulties in understanding and explaining phenomena such as wave-particle duality and the double-slit interference. Meanwhile, particle physics has not provided a fundamental physical image of electromagnetism and electric charge, and there is a misunderstanding of electromagnetic wave propagation. Cosmology also lacks clear theoretical explanations for issues such as the singularity problem and the anomalous Hubble constant, among others. These are all problems that physics development must address.

Modern scientific research has confirmed that empty space is an objective physical reality and exists in relative motion. As stated in quantum mechanics [2] , empty space has endless quantum fluctuations, and general relativity [3] suggests that empty space can be distorted, all of which demonstrate the characteristics of the objective reality and motion of empty space.

We can further infer that it possesses properties such as mass and density, and has mechanical characteristics. We do not consider the composition of matter within the void space, treating it as a whole. Overall, it can be inferred to resemble an invisible ocean of superfluid, where all stars, matter, and particles interact and move within the void space. Therefore, when considering the motion of matter and the formation mechanism of related forces, we must take into account the role of void space and its interaction with matter. The failure to consider this factor is an important reason for the inconsistencies or difficulties in fundamental theories such as special relativity, general relativity, quantum mechanics, and classical mechanics.

Several Issues with Current Theories

2.1 Misuse of the Principle of the Invariance of the Speed of Light

Special relativity's constancy of the speed of light hypothesis includes three meanings: 1) the speed of light in a uniform flat vacuum space is constant at C, which is approximately 300,000 kilometers per second; 2) the speed of light is independent of the motion of the light source; 3) the speed of light is independent of the motion of the observer.

The main error lies in the third point: indeed, the speed of light is constant from the perspective of a reference frame in relative vacuum space and is independent of the observer's motion, but this does not mean that the relative speed of light observed directly is also constant at C.

When observers move differently in the void space, the light propagating within this void space will be observed to have different relative velocities, not a constant value of C.

Because the propagation of light relies on the void space as a background reference frame, its speed is only constant relative to the void space, not relative to the observer.

This misunderstanding has led to many paradoxes in special relativity, making it inconsistent, and even caused direct contradictions between special relativity and general relativity.

For example, the hypothetical experiment observing the motion of light in special relativity.
Suppose a spaceship is moving at a constant velocity V relative to Earth, and there are two parallel mirrors inside the spaceship aligned with the direction of motion. As a light pulse is emitted perpendicularly, observer A inside the spaceship observes the light reflecting vertically between the two mirrors, with the speed of light being C, as shown in Figure 2-1.

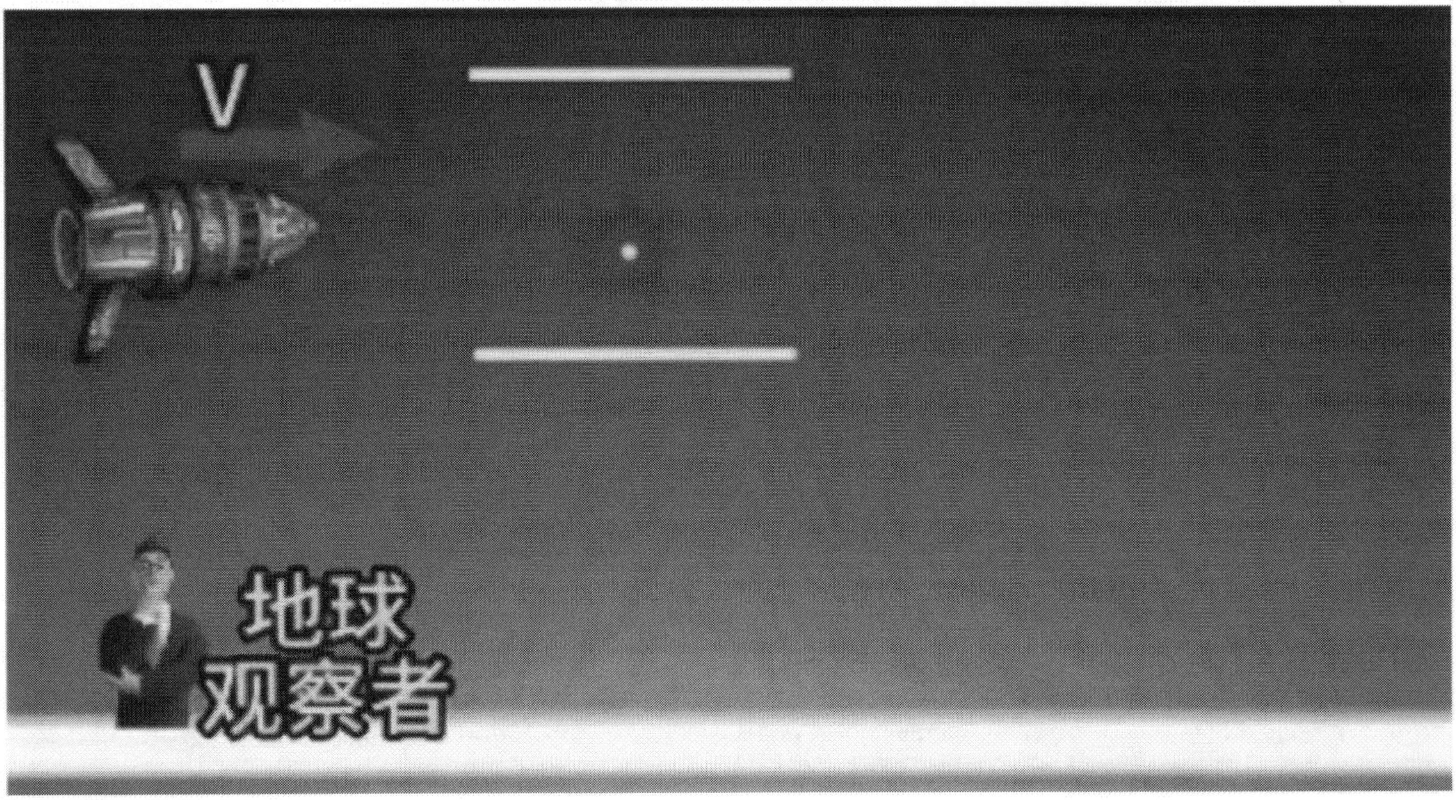

Figure 2-1 Vertical Reflection of Light Pulses in a Spaceship

However, due to the movement of the spaceship itself, to an observer B on Earth who is stationary, the path of the light appears as a slanted line, no longer a vertical motion.

The light paths observed by both parties can be considered as a right-angled triangle, as shown in Figure 2-2.
If we assume the speed of light is constant here, meaning the speed of light observed by B is also equal to C, then in order to satisfy the mathematical formula of the Pythagorean theorem, the time parameter t observed by A and B must change.

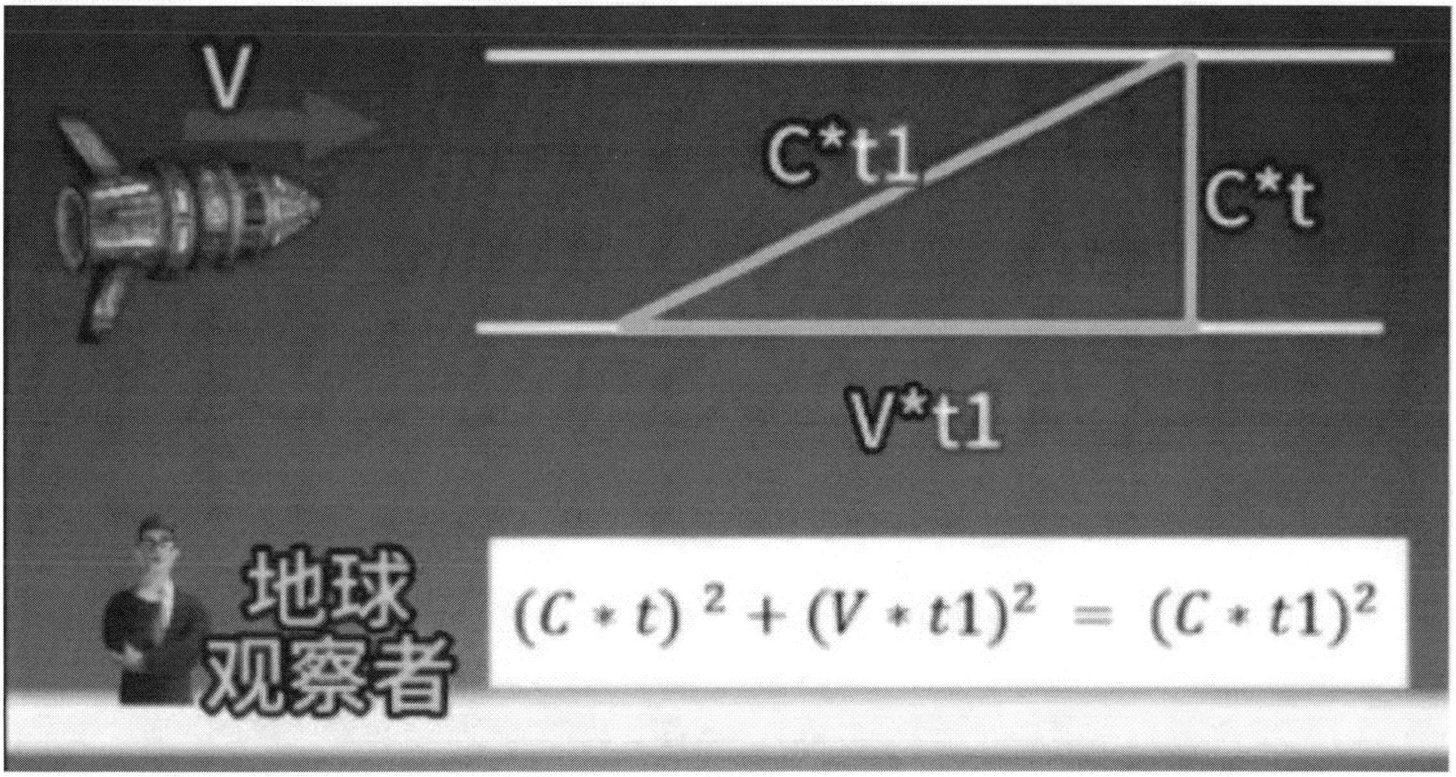

Figure 2-2 Light trajectory observed by a stationary observer

For example, suppose observer A on the spaceship observes time t, and observer B on Earth, who is stationary, observes time t_1 , then at this moment, we can simply calculate using the Pythagorean theorem mathematical formula:

$$(C^*t)^2 + (V^*t_1)^2 = (C^*t_1)^2$$

The calculation can yield:

$$t_1 = \frac{t}{\sqrt{1 - (\frac{v}{c})^2}}$$

This is the time transformation in special relativity, as shown in Figure 2-3.

The $\frac{1}{\sqrt{1-(\frac{v}{c})^2}}$ is the Lorentz factor, and this transformation is known as the Lorentz transformation.

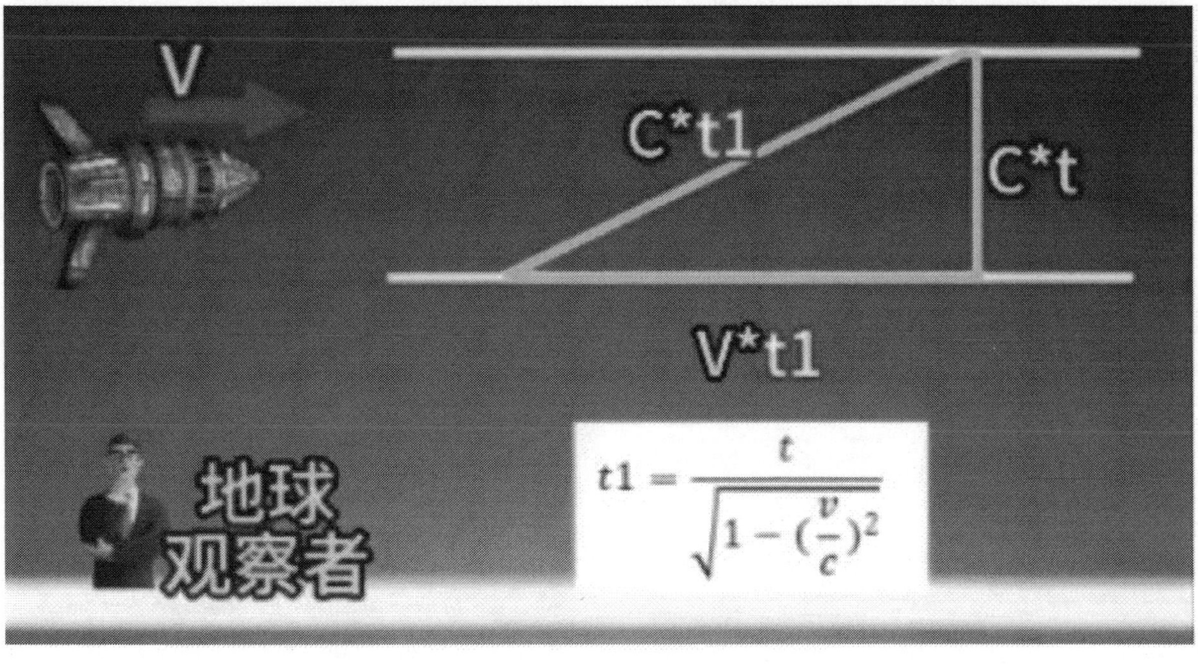

Figure 1: Figure 2-3 Lorentz transformation between different inertial frames

In this transformation, t_1 is greater than t, meaning that the time experienced by the spaceship's observer A is shorter than the time observed by the ground-based observer B. It seems as if the spaceship's observer's time is running slower, which is the phenomenon known as time dilation.

However, the situation changed when Einstein introduced the general theory of relativity.
To illustrate the equivalence between accelerated reference frames and gravitational fields, Einstein imagined a thought experiment involving an elevator.

In this experiment, Einstein assumed there was a closed elevator in uniform motion upward. At this moment, a beam of light was horizontally shone through a small hole on one side of the elevator. Because of the elevator's upward movement, this beam of light

As the light beam travels toward the other side of the elevator, observer A inside the elevator will observe it as a downward-slanted diagonal line.

However, if at this moment there is a stationary observer B outside, he would observe the light beam as still moving horizontally in a straight line, as shown in Figure 2-4.

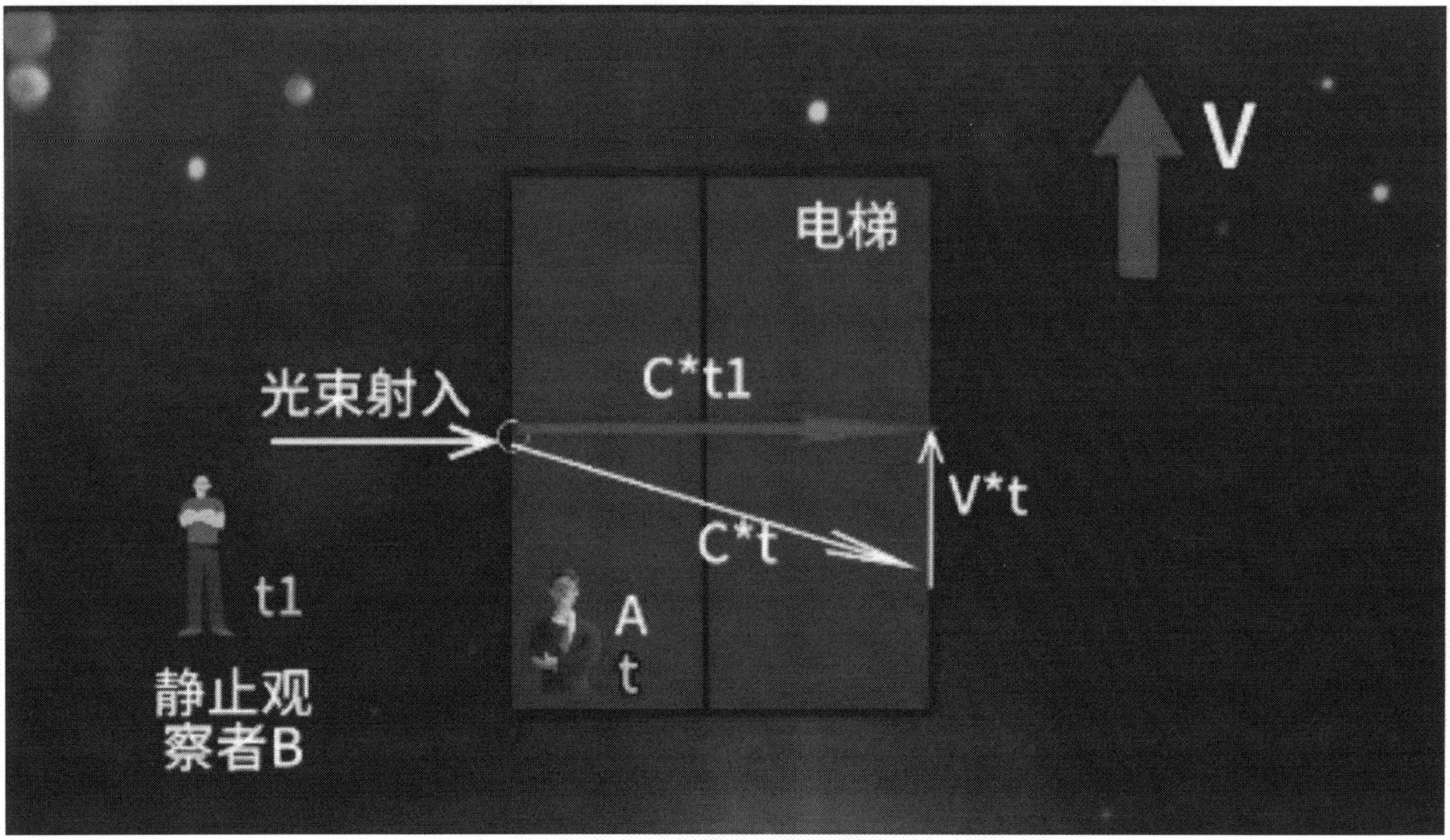

Figure 2-4: Observation of the light beam entering the elevator

Thus, the light paths observed by A and B observers, together with the distance the elevator moves upward, form a right triangle.

We also assume that the speed of light observed by A、 B is C, the time observed by A is t, B , and the time observed by t_1 is the time experienced by A, then according to the Pythagorean theorem, the calculation can also be simply performed:

$$(C^*t_1)^2 + (V^*t1)^2 = (C^*t)^2$$

It can be calculated that:

$$t = \frac{t_1}{\sqrt{1 - \left(\frac{v}{c}\right)^2}}$$

In this transformation, t is greater than t_1 , and the time experienced by the spaceship observer A is shorter than the time observed by the ground observer B, as if time inside the spaceship is moving faster.

Therefore, this contradicts the time dilation effect in special relativity, as shown in Figure 2-5.

A: Moving observer, observed time t B: Stationary observer, observed time t1

Special Relativity:

$$t1 = \frac{t}{\sqrt{1-\left(\frac{v}{c}\right)^2}}$$

General Relativity:

$$t = \frac{t_1}{\sqrt{1-\left(\frac{v}{c}\right)^2}}$$

Clock runs fast

Figure 2-5 The difference between time dilation in special and general relativity

In special relativity, clocks on moving objects run slower, but according to the equivalence principle in general relativity, clocks on moving objects actually run longer.

The boomerang Einstein threw in special relativity came back to him in general relativity

On the body.

The fundamental issue here lies in the constancy of the speed of light hypothesis, which forcibly assumes that the speed of light is always constant at C for all different observers. This assumption causes relativity to distort the description of objective physical reality, turning it into a twisted mathematical system and leading to numerous distorted, counterintuitive conclusions, which are the famous various paradoxes and contradictions.

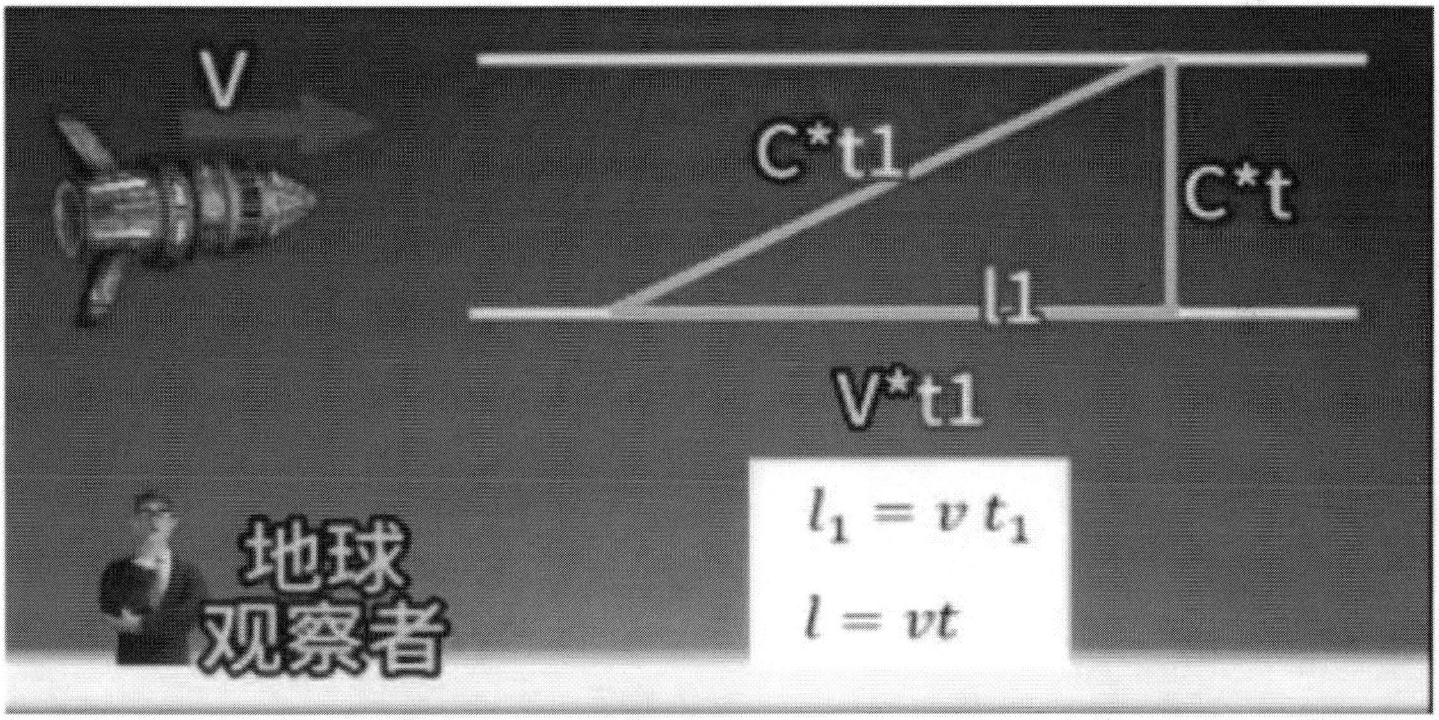

Time Dilation and Length Contraction under Constant Speed of Light

In fact, even according to the calculation framework of special relativity, we can still deduce in the above thought experiment of the spaceship that a moving object should experience time dilation and length contraction, not time dilation and length contraction (as shown in Figure 2-6). This error in special relativity leads to incompatibility with gravitational formulas, forcing the development of general relativity and moving further away, as shown in Figure 2-7.

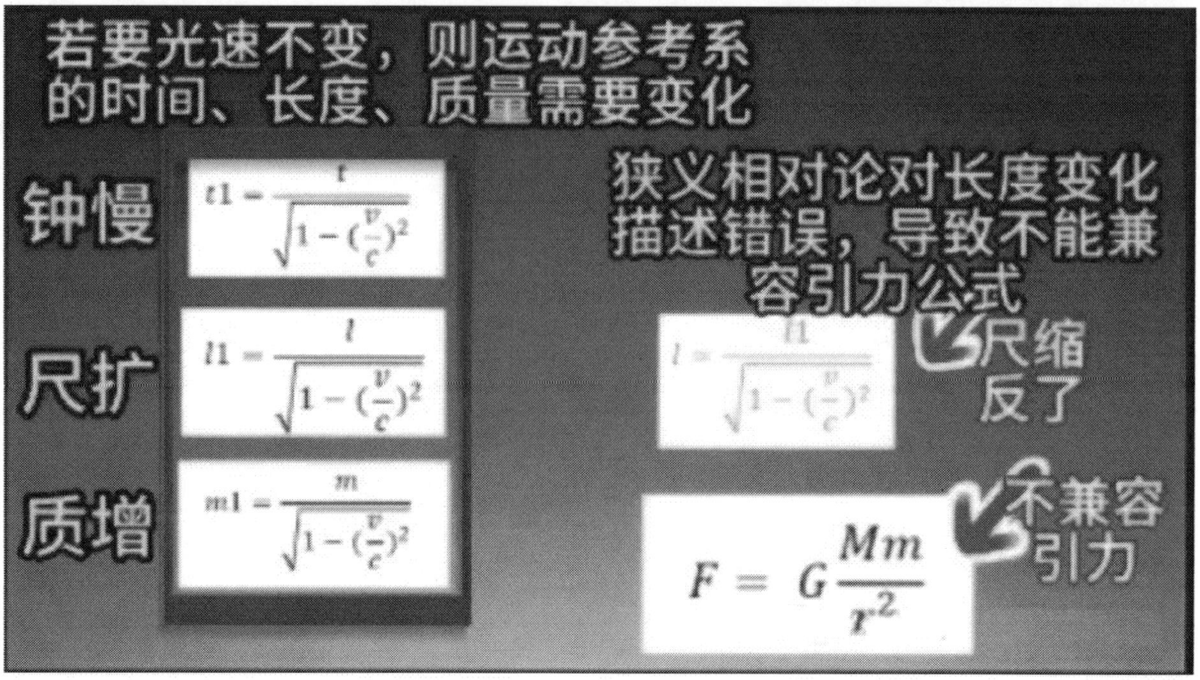

Figure 2: Figure 2-7 Incompatibility of Time Dilation and Length Contraction in Special Relativity with Gravitational Formulas

2.2 Misuse of the Principle of Relativity

The understanding of the principle of relativity in special relativity is that all motion is relative. Two objects in relative motion can have their actual velocities arbitrarily assumed; one can consider A moving and B stationary, or B moving and A stationary, or both moving, as long as the total relative velocity remains consistent.

In reality, all motion occurs within the surrounding vacuum space, so their motion speed relative to the vacuum space is unique and cannot be arbitrarily assumed. The sum of their respective motion speeds relative to the vacuum space equals their relative motion speed.

Especially when it comes to situations involving light, electromagnetic forces, and gravity, this point must be taken into consideration, because the propagation of electromagnetic waves and gravity relies on the vacuum space, and their motion speeds are all relative to the vacuum space. Deviating from this principle inevitably leads to errors.

For example, we assume a hypothetical experiment where two inertial systems in relative motion measure each other. Since each inertial system can have multiple observers, we assume there are two inertial reference frames A and B, moving relative to each other at 0.9c. Each inertial system has two observers, A_1、A_2 and B_1、B_2 .

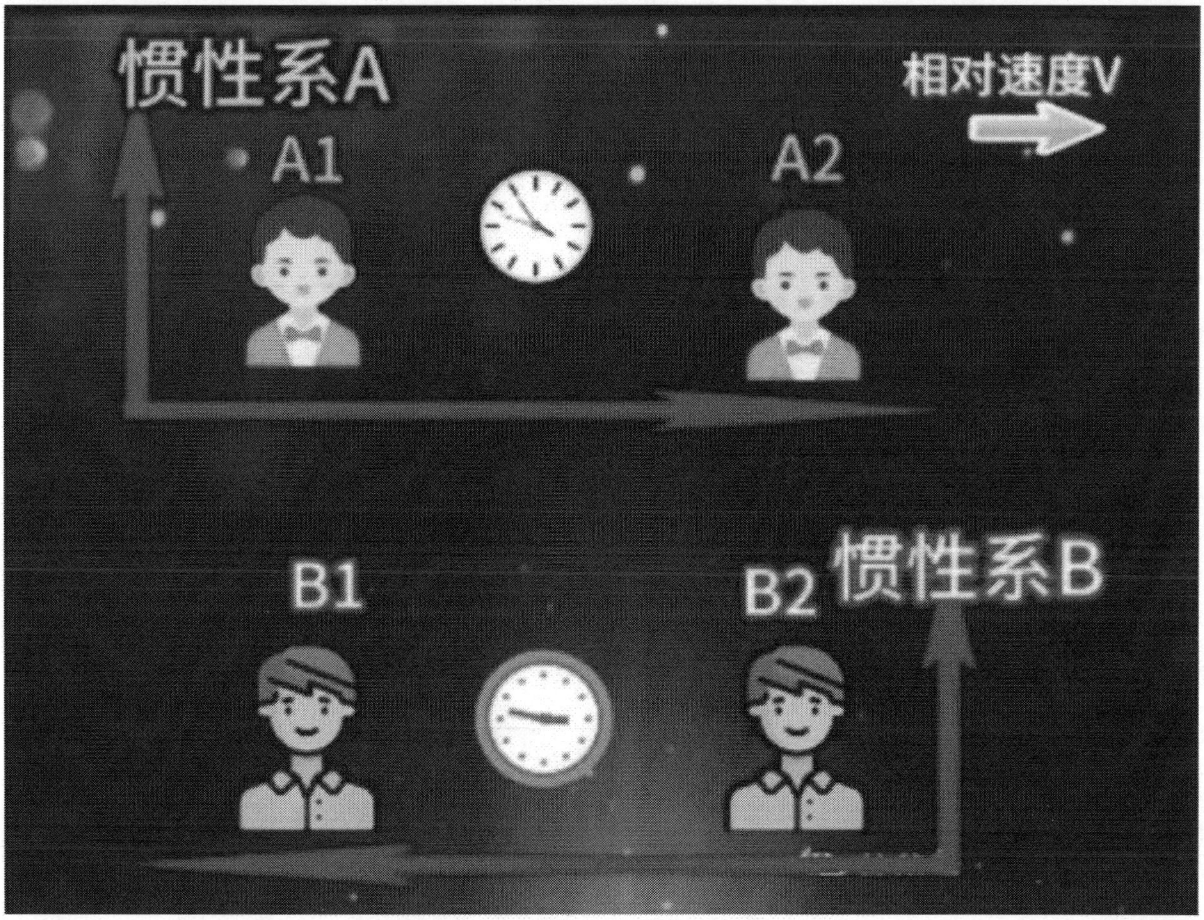

Figure 3: Figure 2-8 shows the clocks of the observers in the two inertial systems are aligned respectively.

Because they are each in the same inertial system, their clocks are aligned respectively, that is, A_1 and A_2 have aligned clocks, and B_1 and B_2 have aligned clocks, as shown in Figure 2-8.

At a certain moment when A_1 and B_1 meet, A_2 also meets B_2 at the same time, because clocks in the same reference frame are synchronized, so A_1 and A_2、 B_1 and B_2 can respectively confirm that their meetings are simultaneous.

Assuming two inertial frames, the distance between A_1 and A_2 is L_A, B_1 , and the distance between B_2 and L_B is B_1、 B_2 , then at this time, what is the distance measured by A inertial frame for A_1、 A_2 ? What is the distance measured by B inertial frame for @7#?

As shown in Figure 2-9, is L_A equal to L_B at this moment?

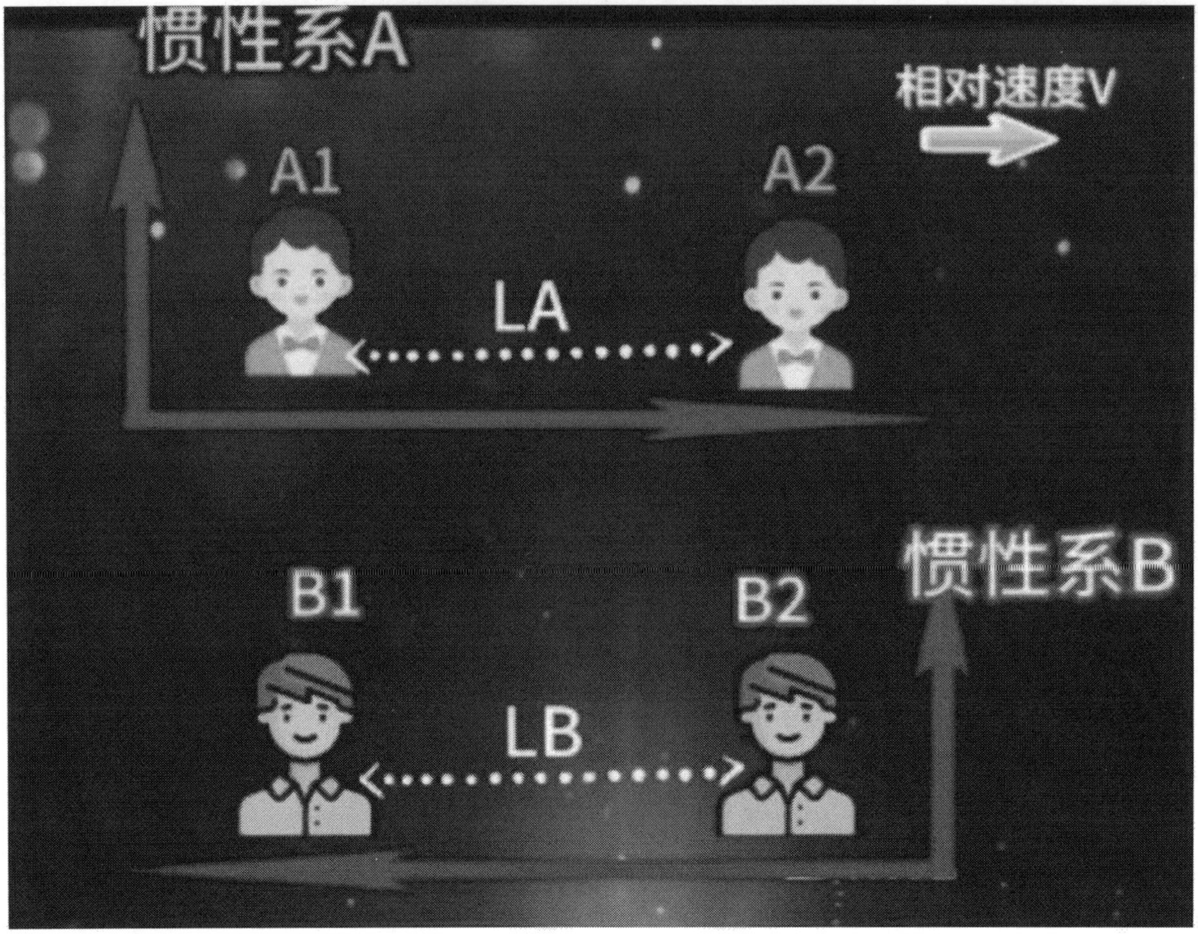

Figure 4: Figure 2-9 Two inertial frames observing each other's distance length

Due to the relativity of motion, it can be assumed that two inertial frames have various different velocities, such as assuming frame A is stationary and frame B is moving, or assuming frame B is stationary and frame A is moving, or both are moving, etc. In these cases, according to the length contraction effect in special relativity, should the data they measure also have multiple possibilities?

The mass will be different at different speeds, so will the measured mass of each other also vary?

Suppose at this moment, they measured each other's ages. Then, when the two inertial frames continue moving towards each other and eventually meet at A_1 , they observe that each other's ages have increased by T_A and T_B years respectively. Here, is T_A equal to T_B ?

Similarly, assuming various different motion speeds for them while keeping their relative velocity unchanged, would the observed T_A、T_B change according to the time dilation and length contraction effects in special relativity?

This is obviously impossible.
Their measurement results should have and only have one. The problem lies in the misunderstanding of the principle of relativity in special relativity.

The motion states of relatively moving objects cannot be arbitrarily assumed, especially when involving factors such as electricity, magnetism, light, and gravity, where it is necessary to consider their respective "absolute" motion speeds relative to the void space, which is unique.

2.3 Misunderstanding of Void Space

Special relativity treats empty space as an illusory, conceptual reference frame rather than an objective reality, which is also the fundamental reason for its misapplication of the constancy of the speed of light and the principle of relativity.

In reality, empty space is an objective reality, a real existing background reference frame, and it itself exists in a state of relative motion. The propagation and movement of electromagnetic waves and gravity all rely on empty space, and the so-called constant propagation speed C they refer to is also relative to the particular empty space they are in.

When there is motion in the void space, or when an observer is in motion relative to the void space, the speed of light observed by him will inevitably be affected, showing a relative change.

The hypothesis of the constancy of the speed of light did not take this into account, leading to the mistaken belief that "the speed of light is constant relative to different observers in motion."

The principle of relativity also failed to consider this, resulting in the erroneous notion that "the relative velocity of an object is absolutely relative and can be arbitrarily assumed."

Einstein had already recognized this in general relativity. In a lecture he gave at the National University of Leiden in the Netherlands in 1920, he mentioned:
According to general relativity, space has been given physical properties, so in this sense, there exists an ether. According to general relativity, a space without ether is inconceivable. Because in such a space, not only can light not propagate, but measuring rods and clocks would also be impossible, and thus there would be no physical space-time interval.

here, einstein acknowledged the objective reality of space, which is no longer just a concept, but

It is objectively existing and has physical properties.
Quantum theory has already demonstrated that electric fields, magnetic fields, gravitational fields, and even endless quantum fluctuations exist in space, and they themselves are forms of objective reality, even existing in motion.

In general relativity, space can be distorted by massive objects, which itself demonstrates that space is an objective reality and even exists in motion.

Light is deflected due to the distortion of space, indicating that the motion of light is relative to space, directly contradicting his conclusion in special relativity that "motion can never be considered as 'motion relative to space'."

However, this error cannot be corrected anymore, because the constancy of the speed of light and the principle of relativity are the two fundamental pillars of special relativity. The errors in these two principles have caused special relativity to lose its rational foundation.

2.4 Misunderstanding the Logic of Determining Simultaneity

There is an obvious logical problem in Einstein's determination of simultaneity. He believed that different inertial frames would judge the simultaneity of the same event differently due to their own motion, which is the fundamental logical basis of his discussion in special relativity.

But here he confuses an inertial frame with an observer within that inertial frame.
We know that an inertial frame is a complete reference frame, whose scope can be infinitely expanded, even encompassing the entire universe. Therefore, there can be not just one, but an infinite number of observers within this reference frame. We can consider that at every spatial point in the universe, there exists an observer of this reference frame, whose clocks are synchronized, as shown in Figure 2-10.

Thus, whenever two events occur, we can ensure that there exists an observer at zero distance from both locations who can instantly determine the time of each event and thus judge whether they are simultaneous.

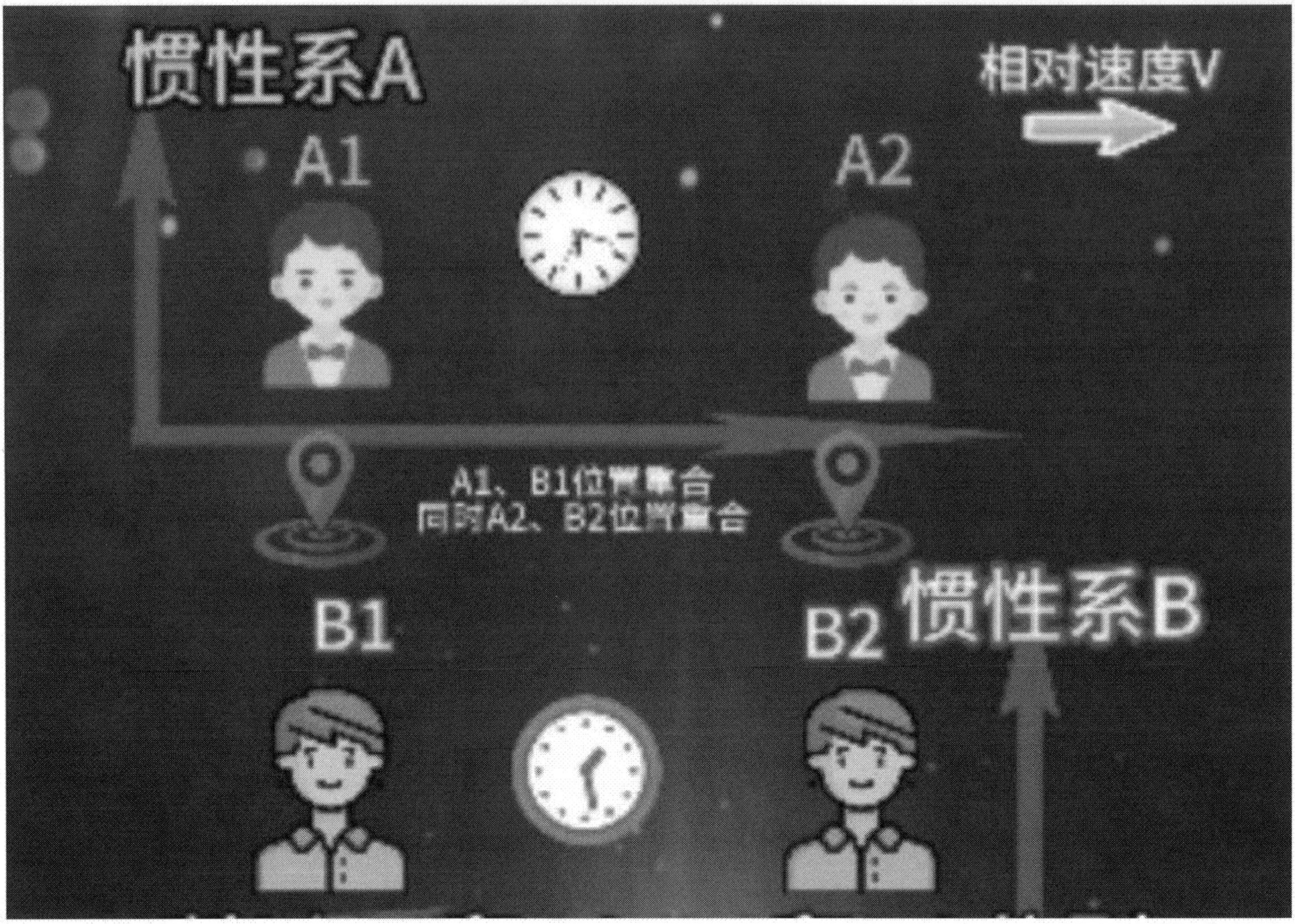

Figure 2-10 Each inertial frame can have multiple observers with synchronized clocks

In this case, regardless of how many inertial frames there are and regardless of their motion, they will all be able to confirm the simultaneity of these two events, and their judgments will remain consistent without any differences due to the different inertial frames.

Step back, even if each inertial frame has only one human observer, we do not rely solely on our own eyes to observe things. We can have countless sensors and detectors, which can also be deployed anywhere within an inertial frame. Therefore, we can also accurately determine the time when these two events occur and confirm their simultaneity.

Therefore, Einstein's belief that different inertial frames moving at different velocities would judge the simultaneity of the same event differently is a logical fallacy.

The constancy of the speed of light and the principle of relativity are the two fundamental pillars of special relativity, while the determination of simultaneity forms the basic logic of special relativity. The errors in all three have led to the complete loss of a reasonable foundation for the theory.

Although various observations and experimental results are considered as evidence verifying relativity, most of these experiments and

observational phenomena are not unique to relativity and can be derived and explained by many other theories. Despite these results being placed within the framework of relativity, they do not change the fundamental fact that relativity has significant defects and errors.

The theory of special relativity, including the effects it argues for such as time dilation and length contraction, is merely a mathematical calculation system, not a true reflection of objective reality, nor is it所谓的 observation effect. This mathematical system does not reflect the objective physical world and distorts the understanding and description of physical reality, leading to distorted and erroneous conclusions.

2.5 Contradictions Between General Relativity and Quantum Mechanics

Since its inception, general relativity has successfully explained gravitational effects through the principle of spacetime distortion, but this geometric theory system has structural contradictions with quantum mechanics, making it difficult to quantize gravity and achieve a unified physical theory.

For more analysis on related issues of special and general relativity, see Extended Reading 12.1: 8 Intrinsic Problems of Special and General Relativity Theories.

2.6 The Probabilistic Nature of Quantum Mechanics and the Difficulty in Interpreting Quantum Phenomena

The development of quantum mechanics has driven humanity to make significant leaps in the field of microscopic particle physics, but the probabilistic descriptions inherent in its theory have always been controversial. It also faces challenges in explaining various quantum phenomena such as wave-particle duality and the double-slit interference, leading people to question whether there are deeper underlying principles yet to be discovered behind quantum mechanics.

2.7 The Confusion About the Completeness of Classical Mechanics

Since Newton published "Mathematical Principles of Natural Philosophy" [4] , the development of classical mechanics has driven a great leap forward in human civilization, but classical mechanics only explained the law of interaction of gravity, yet

There is no explanation for the origin and mechanism of gravity, especially since astronomical observations show that the gravitational effects in galaxies are much stronger than what is predicted by the current mass-based gravity laws, leading people to doubt whether there might be deeper mechanisms and principles at work.

The problems mentioned above suggest that we need to think more deeply and explore the more fundamental principles and mechanisms hidden within the current physical theory framework, eliminating differences or contradictions between various theories and achieving a unified foundation for physical theory.

The universe is a unified whole, which is also why many scholars believe there must be a complete and unified fundamental law framework underlying it, and have been striving to uncover it.

2.8 The Incompleteness of the Standard Model and Cosmological Theories

The Standard Model is a successful theoretical framework in particle physics, but it still lacks a complete explanation for the origin and evolution mechanisms of particles, the internal structure of fundamental particles, and it contains errors in explaining the mechanisms of particle mass and spin. It also fails to provide reasonable explanations for the origin of gravity, the nature of charge, the essence of antimatter, and asymmetry.

The Big Bang theory of the origin of the universe cannot offer reasonable theoretical explanations in aspects such as the anomalous detection of the Hubble constant, singularities, the magnetic monopole problem, and symmetry breaking. These issues require further exploration and research.

The Basic Principles of Spatial Relativity

In current physical theories, the principle of the constancy of the speed of light and the relativity principle contain fundamental errors, mainly because they do not take into account the background reference frame of empty space. In reality, the speed of light, or the propagation and motion of electromagnetic waves and gravity, all occur through empty space, and their velocities are all relative to empty space.

Modern science has proven that empty space is an objective reality and exists in relative motion, so this inevitably affects the propagation of electromagnetic waves and gravity.

The special theory of relativity's principle of the constancy of the speed of light and the relativity principle did not consider this, which is why errors arose. In addition, modern physical theories also have errors in the logical determination of the relativity of simultaneity, incorrectly assuming that simultaneous events cannot be determined across different inertial frames. These defects and errors in verification must be corrected in the development of new theories.

Taking into account the above issues, we propose a new theory: the Theory of Spatial Relativity, also known as the New Ether Theory. The Theory of Spatial Relativity (New Ether Theory) is mainly based on the following several fundamental principles.

3.1 The Relationship Between Physical Laws and Observed Phenomena (Principle of Objectivity of Physical Laws)

(1) Objectivity of Physical Laws
Physical laws are objective, not dependent on human will, and not affected by observers. Regardless of who the observer is, or what motion the observer may have, the physical laws remain objective and unchanged.

Direct observation of physical events or phenomena is influenced by the observer. When different observers or observers in different states of motion observe the same physical event, their direct observations may differ.

For example, an object in empty space accelerates when subjected to a force, which is a physical law. This law does not change regardless of the observer or their state of motion. However, the direct observation of the phenomenon may vary depending on the observer's different states of motion or acceleration.

Another example is that the speed of light is constant in a uniform, flat empty space, which is a physical law. Observers in different states of motion may observe different speeds of light in a particular event, which is an observational phenomenon.

Differences in direct observations do not affect the objectivity and invariance of physical laws. Observers must adjust their observations based on their relative motion and other actual conditions.

In simple terms, phenomena can reveal the essence or distort it, requiring observers to use rational analysis to see through the phenomena and understand the essence.

This was a basic philosophical concept, but due to the principle of the constancy of the speed of light in special relativity, which confuses the distinction between physical laws and observed phenomena, it has caused significant confusion in understanding. Therefore, it is particularly necessary to clarify and explain. For detailed explanation of the principle of the constancy of light speed, see Principle 4: The Relativity of the Speed of Light (New Principle of the Constancy of Light Speed).

(2) Objectivity of Causality and the Time Arrow

According to the objectivity of physical laws, it necessarily leads to the objectivity and uniqueness of causality. That is, once an event occurs, its corresponding causality and causal chain are determined, objective, and unique, becoming established facts. They will not change due to differences in the observer's subjectivity, motion, or any changes in the motion of other substances, even if the observer or other substances move faster than light. Even if an observer first observes the effect and then the cause, it does not alter the objectivity, uniqueness, and determinism of causality, because what is directly observed is often just a surface phenomenon.

This is like specifying a unique direction for the universe that cannot be reversed. Even if time were reversed, for example, going back 30 years, the result would still be caused by physical processes under the current environment and current causal chain. It would be a further development according to physical laws under previous causal conditions, and would not change or reverse causality.

In this sense, the law of causality and causal chains are the strict time arrows that identify the process of the universe's operation.

As for the origin of causality, or the original causality of the universe, the first driving force of the cosmos, we currently have no correct understanding of it. It can be called "uncaused cause," meaning the initial, first cause that produces an effect without a prior cause. This "uncaused" is not truly without a cause, but rather refers to a mechanism and process that we are currently unable to observe or recognize. This uncaused cause is not only the initial driving force of the universe, but it also exists extensively throughout the universe's operation, meaning there are numerous uncaused effects in the cosmos, appearing suddenly from emptiness, similar to quantum fluctuations described in quantum theory. These uncaused effects continuously influence the operation of the real material world in the universe and are one of the fundamental driving forces behind the universe's operation and evolution. The mechanisms and principles behind the uncaused will be the key to determining whether the universe is deterministic or indeterministic.

In addition to the effect without cause, the causal chains in the physical world eventually lead to a cause without effect, meaning an event occurs but produces no result and has no impact on the physical world, as if the causal chain breaks here.

In fact, just like the causeless origin of an effect, this cause without effect is not truly without effect, but rather a special mechanism and function that our real world cannot detect.

All causal chains in the real world begin with an effect without cause and end with a cause without effect.

Let us denote the set of all observable matter and events in our real physical world as Φ , and the force field space as f, then the mathematical expression of the causal law in the real world is:

$$y = G_f(x), x, y \in \Phi$$

For the world to which cause and effect belong, which we cannot observe, we call it the virtual world. The collective space of all virtual matter and virtual events is denoted as $i\Psi$, and the virtual force field space is if. Then, the effect without cause can be expressed as:

$$\mathrm{y} = \mathrm{G}_{\mathrm{if}}\,(\mathrm{x}_{\mathrm{i}}), \mathrm{x}_{\mathrm{i}} \in \mathrm{i}\Psi, \mathrm{y} \in \Phi$$

The cause without effect can be expressed as:

$$y_i = G_f(x), x \in \Phi, y_i \in i\Psi$$

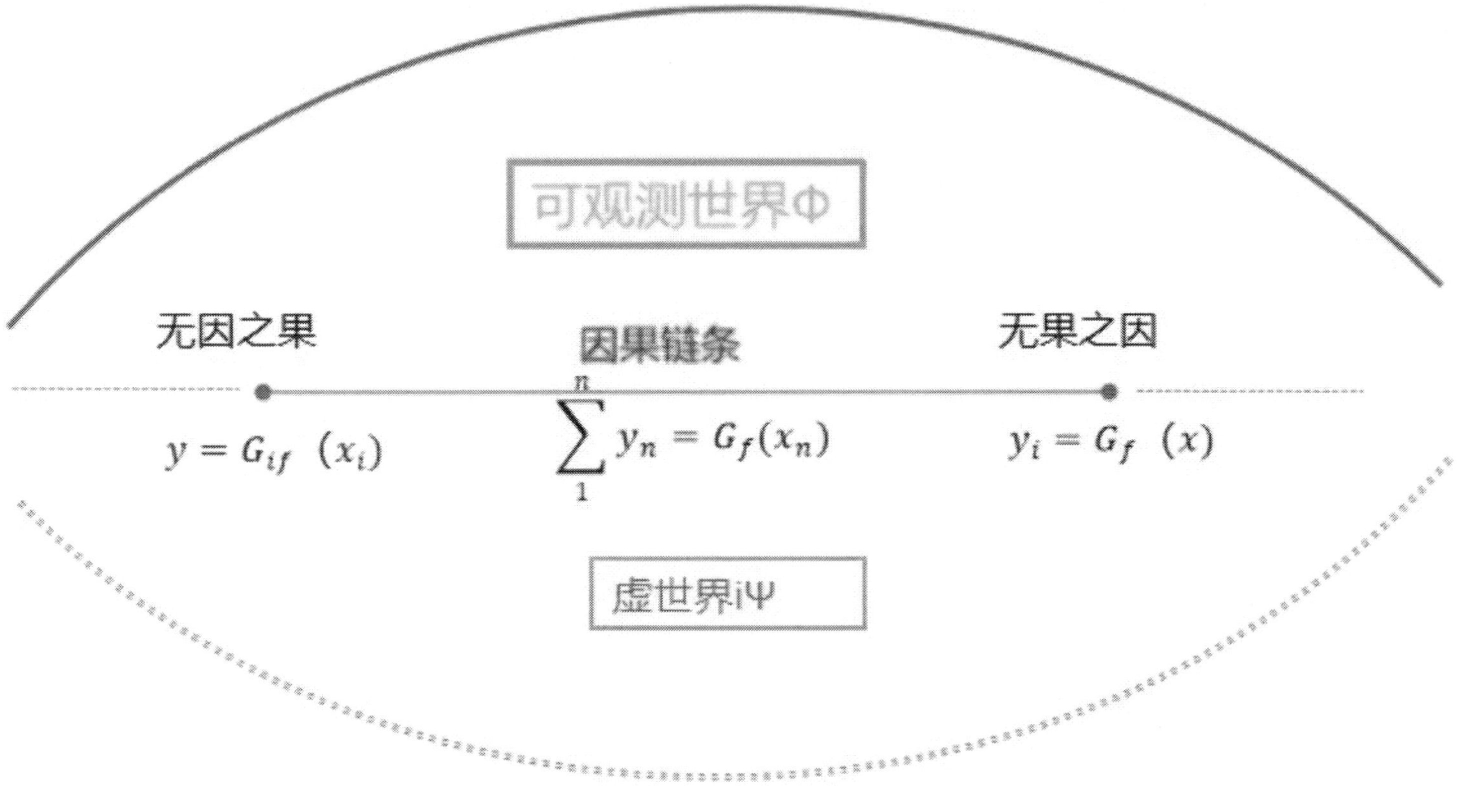

Figure 3-1 The causal chain of the observable world

Our observable real world is the material world composed of a vast and complex causal chain between the effect without cause G_{if} and the cause without effect G_f , as shown in Figure 3-1. The causal chains in the real world are all deterministic and predictable, so the determinism of our entire world depends on the form of causality G_{if} in the unobservable virtual world $i\Psi$:

$$y_i = G_{if}\,(x_i), x_i, y_i \in i\Psi$$

3.2 The objective reality of space (Principle of spatial objectivity)

(1) The vacuum space is material
Modern scientific theories suggest that electric fields, magnetic fields, and gravitational fields exist in vacuum space, and these fields are objective realities. Since these fields exist based on vacuum space, the vacuum space itself is also an objective reality. When light, particles, and other substances move within vacuum space, the vacuum space itself becomes a real reference frame, not merely a concept of emptiness.

From this perspective, it can also be said that vacuum space is material.
At the same time, since electromagnetic forces and gravity, among others, are transmitted and exert their effects through vacuum space, vacuum space itself should also possess mechanical properties and exert mechanical influences on all particles within it.

The void space discussed in this article refers to a superfluid substance with extremely low viscosity (nearly non-viscous), existing in the vacuum space (a space completely devoid of any matter), and possessing properties of motion and mechanics.
(2) Void space is consistent at both macro and micro levels
Whether it is the void space in the macroscopic universe or the void space within microscopic matter, around particles and atoms, it is the same objective existence.

Although differences in environment, density, and motion in macroscopic and microscopic scenarios may lead to slight variations in the properties of different void spaces, they do not obscure the essential unity of all void spaces. Regardless of being macroscopic or microscopic, void space serves as the medium for the excitation and propagation of fields such as magnetic fields, electric fields, electromagnetic waves, and gravity, and all follow the same propagation laws.

Macroscopically, light will bend around massive objects such as stars. Microscopically, light also bends around particles such as atoms. This is the reason for the existence of the light refraction phenomenon: the distortion effect of massive objects on the surrounding empty space, that is, the density of empty space has changed.
(3) Empty space is in motion
All matter is in motion.
The material nature of empty space determines that it must be in motion. In fact, it is precisely the motion of empty space itself that determines the motion characteristics of all matter within it.

Modern science has observed and confirmed that massive objects distort the surrounding empty space, proving that empty space is in motion. Its change from a flat and uniform state to a distorted state is the very essence of motion.

Quantum theory also suggests that there are endless quantum fluctuations in empty space, which is another indication that empty space exists in motion.

Proof of motion.
Modern astronomical observations prove that the motion of a mass body drags the empty space, which is known as the dragging effect of the inertial frame. This is also one of the proofs that empty space has motion.

At the same time, this dragging effect of mass bodies is not limited to celestial bodies in the universe, nor is it merely the dragging effect of the inertial frame. Rather, all material mass bodies, including objects in daily life and particles in the microscopic world, will cause distortion and dragging of the surrounding empty space, though the range and degree of the distorted and dragged empty space vary.

The famous Fizeau water flow experiment is an example of this dragging phenomenon.
The motion of the vacuum space includes microscopic fluctuations and vibrations, and macroscopic waves and flows. Among these, the flow type is similar to that of gases and other fluids, and can include irrotational flow, rotational flow, one-dimensional flow, two-dimensional flow, three-dimensional flow, incompressible flow, and compressible flow.

This motion of the vacuum space affects the matter within it, such as microscopic particles, and serves as an important source of their motion dynamics.

(4) The vacuum space is inhomogeneous and in relative motion

Because mass bodies distort the surrounding vacuum space, the vacuum space is inhomogeneous. Because mass bodies exist widely and universally in the universe, the inhomogeneity of vacuum space is also universal and widespread.

Or in other words, all "uniform flat" empty space is local and idealized.
Because the movement of mass bodies drags the empty space, the empty space in different regions is in relative motion. Due to the universality and absoluteness of mass body movement, the relative motion of empty space in different regions is universal.

Because mass bodies, or matter, in the universe are hierarchical, there are countless galaxy clusters under a large galaxy group, countless planets within galaxies, countless objects with different movements within planets, etc., which is common in the universe.

Existential. Therefore, the void space is also hierarchical, with each large area of void space containing a vast number of sub-void space regions that are uneven and in mutual motion, which is also universally present.

(5) The Nature of Vacuum Space

The material nature of vacuum space determines that it has a density. It can be distorted and dragged by mass bodies, leading to complex relative motion, so it should possess properties such as gradient, curl, velocity, and acceleration. Its movement also exhibits different forms, including waves, divergence, advection, and vortices, corresponding to different effects such as electric fields, gravitational fields, and magnetic fields.

The distortion of the vacuum space by mass bodies is essentially due to the gravitational nature of the mass bodies (for details on the gravitational essence, refer to the analysis in sections "5.3 Unified Origin of Forces: (4) Gravity" and "8.11 Principles of the Origin and Evolution of Forces"). The density of the surrounding vacuum space (aether) changes as a result, with the density decreasing as one gets closer to the mass body and increasing as one moves away from it, eventually returning to a uniform density state in the surrounding area.

The difference in the density of vacuum space affects its dielectric constant properties, thereby causing variations in the speed of light propagation within it. A vacuum space where the speed of light is the standard speed C (approximately 29.9792458 ten thousand kilometers per second) can be referred to as standard density vacuum space.

Vacuum space is dragged by mass bodies, resulting in motion. The closer to the mass body, the greater the degree of dragging, eventually reaching a state of synchronous motion, that is, stationary relative to the mass body's surface; the farther away from the mass body, the lesser the degree of dragging, eventually reaching a state of relative rest with the surroundings.

In summary, we can conclude that the void space has the following properties:
(1) Microstructure
The void space is an entity considered as a whole, with a microstructure inside it, meaning it is composed of a large number of microscopic particles or grains (for the sake of discussion, we can uniformly refer to them as wave particles), forming the overall superfluid characteristics of the void space.
(2) It has mass, volume, and density
Each basic element wave particle in the vacuum space has mass and size, so the vacuum space as a whole also possesses mass, volume, and density properties.
(3) It has force
The vacuum space exerts force on all particles within it, essentially being the interaction between microscopic wave

particles and microscopic particles.
(4) It has motion
All fundamental elements, wave particles, in the void space have motion, and the void space as a whole also has motion.
(5) Strong diffusivity
There is void space existing within the smallest microscopic particles of the universe and the starry sky. The void space has strong diffusivity and permeability.
(6) Void space and ether
In general, the void space can be considered or equated to "ether," but it differs from the old scientific concept of "ether" as uniformly distributed and completely stationary throughout the universe. Here, the void space is dynamic and non-uniform, similar to a special superfluid that exhibits various complex movements, changes, and physical properties. We can refer to this as the "New Ether Theory."

3.3 The Absolute Nature of Simultaneity (Principle of Temporal Objectivity)

(1) Absolute Simultaneity
Since each inertial frame can have multiple observers and each observer can also have multiple different measuring devices, all inertial frames can accurately determine the time at which an event occurs and establish the simultaneity of every event.

For example, if there were only one ground station on Earth, it might be difficult to accurately track the data of a satellite in space, but by using multiple ground stations to communicate and observe the same satellite simultaneously, more precise parameters such as distance and speed can be obtained.

Moreover, even if an observer perceives phenomena that differ from the actual situation, the true state can still be deduced through the application of regular patterns.

For example, when a person sees lightning and then hears thunder a few seconds later, it seems as though the two events are not simultaneous. However, based on known physical laws, he can deduce that the two events actually occurred at the same time, and even calculate the distance between the location of the lightning and his own position based on the time difference between the two.

Observations in fields such as the universe and astronomy also follow this principle, which is more in line with our normal way of thinking.

Therefore, the simultaneity of events is definite, unique, and absolute. Even if the observed results differ, it does not alter the determinacy of the time and sequence of events, nor does it affect the causality law.

On this basis, the properties of objects such as length and mass are definite, unique, and absolute for different inertial frames and observers, and do not change due to relative motion between the object, the inertial frame, or the observer.

Essentially, this is the distinction between physical laws and observed phenomena. Physical laws are fundamental and objective, and they do not change with different observers.

Observed phenomena are superficial; they can reflect the essence or distort it. They may manifest differently depending on the observer, but through reasoning and research, we can understand the underlying essence. Just as different people conducting the same experiment may have different errors, each person must carefully calculate and eliminate all errors to obtain accurate results and conclusions.

(2) Locality of time

Time is a measure of the continuity of material motion, and the differences in the material motion conditions in different regions of the universe determine the differences in time

The time in these regions will also be different, which is the locality of time.
Different motion situations that lead to variations in time measurement include:
(1) Differences in timing caused by objective environmental variations
As the Earth rotates once in approximately 23 hours 56 minutes and 4 seconds, and Mars rotates once in about 24 hours 37 minutes and 22 seconds, using a day as a unit of time, a day on Mars would be about 1.029 Earth days. Therefore, the flow rate of time differs between the two planets.
(2) Differences in timekeeping due to variations in timekeeping instruments
More accurate timekeeping relies on specialized instruments, but differences in these instruments can lead to variations in time records.

For example, with a water clock that uses dripping water, the speed of the water droplets falling is not only affected by the water level and the size of the opening, but also by the gravitational conditions of the planet they are on. Differences in these conditions can result in variations in the actual outcome of dripping timekeeping, leading to different time records.

More precise timing devices, such as atomic clocks, have their timing principles affected by the motion of microscopic particles like atoms in the microcosm. When changes occur in the temperature, voltage, or electromagnetic interference of the timing device, the accuracy of the atomic clock will be impacted, leading to timing errors, and so on.
(3) Different time flow rates caused by differences in space itself
The density and motion state of the empty space itself differ, which will have a collective effect on all substances within the space, leading to an overall change in the motion of all substances within the space.

If the fluctuation frequency f of the empty space is higher, the motion of all particles will also intensify, increasing the temperature T of the matter. The changes in macroscopic phenomena caused by the motion of various substances will also accelerate, resulting in an overall faster time flow rate t, that is:

$\Delta t \propto f(3.5)$

Or, in other words, for two regions that are otherwise identical, the time at which the same event occurs is counted

Time relationship is:

$$\frac{\Delta t_A}{\Delta t_B} \propto \frac{f_A}{f_B}$$

That is $\Delta t_A = \mathrm{K} \Delta t_B$
A proportionality coefficient for the time flow rate between two regions. The time differences caused by the other two reasons are also consistent with this proportionality.

The density of the void space decreases as the fluctuation frequency and temperature increase, that is:

$$\rho \propto \frac{1}{f}$$

Or:

$$\rho \propto \frac{1}{T}$$

Therefore, there is also:

$$\frac{\Delta t_A}{\Delta t_B} \propto \frac{\rho_B}{\rho_A}$$

In addition, the advective motion of the void space will also affect the motion of material particles. The greater the velocity of the spatial advection V, the shorter the interaction time between particles, and the faster the interaction manifests, giving the illusion that time is speeding up. Let the time flow rate be Δt_0 when the advective speed of the void space relative to this region of material particles is 0, and Δt_v when the advective speed is V. Then:

$$\Delta t_v = \frac{\Delta t_0}{\sqrt{1 - \frac{V^2}{C^2}}}$$

Equation 3.7 is the equation of time locality. From the equation, we can derive $\Delta t_v > \Delta t_0$. The time flow rate conforms to the Lorentz transformation (the specific calculation is similar to the "Misuse of the Principle of the Constancy of the Speed of Light" in section 2.1, where the speed of light was observed in the experiment of a spaceship moving relative to Earth).

However, this is under ideal conditions. In reality, changes in the flow velocity of the void space can also cause changes in the motion velocity of microscopic particles and the fluctuation velocity of space, all of which will affect the speed of time flow.

For humans, the main reason for perceiving a significant change in the speed of time is the third type of influence, which is the change in the speed of time caused by the relative motion of the void space itself. This is because it can result in some areas experiencing accelerated aging while others experience time slowing down, which is the most direct perception of time for humans. The relative motion between humans and the void space can be either the active flow change of the void space or the active movement of humans relative to the void space, such as accelerating or moving at high speeds in a spaceship.

But regardless of which situation it is, it does not affect the absoluteness of simultaneity, because according to the relationship of time flow speed in different regions:

$$\Delta t_A = \mathrm{K} \Delta t_B$$

Equation 3.8 represents the time synchronization relationship between different time regions (where K is a proportionality coefficient). It can be seen that when two events occur simultaneously, $\Delta t_B = 0$, at this moment, the Δt_A in other different regions will also be zero, so they are simultaneous as well.

3.4 The Relativity of the Speed of Light (New Constancy Principle of Light Speed)

(1) Medium for Light Wave Propagation
Light waves require a medium for propagation, and that medium is the vacuum space. The speed of light is constant in uniform, straight vacuum space. That is, the speed of light varies in different density levels of vacuum space. In the standard density vacuum space, which is spacious and uniform in interstellar regions, the speed of light is constant at C, approximately 299,792.458 kilometers per second.

Because the speed of light is related to the permittivity and permeability of vacuum space, the formula is:

$$C = \frac{1}{\sqrt{\mu_0 \varepsilon_0}}$$

Therefore, the permittivity of vacuum space is related to the density of that region of vacuum space, and can be considered as the influence and manifestation of vacuum space density on its electromagnetic properties. In different density levels of vacuum space, the permittivity will vary, and its value is inversely proportional to the space density.

The density of the vacuum space in which light propagates at speed V is ρ_v , and the permittivity is ε_v . Then:

$$\varepsilon_v \propto \frac{1}{\rho_v}$$

Therefore, according to the light speed formula, when only considering the density of the vacuum space (without considering the effect of changes in vacuum space density on other properties that may influence the change in light speed), the speed of light is proportional to the square root of the density of the vacuum space it propagates through, that is:

$$C \propto \sqrt{\rho}$$

Where ρ is the density of the vacuum space in which light propagates.
Therefore, the speed of light increases with the increase of the vacuum space mass density ρ .
More analysis can be found in the section "8.10 Principles of Electromagnetic Wave Generation and Propagation: (4) Principle of Electromagnetic Wave Speed".
(2) The essential nature of the wave-particle duality of light [5]
Photons, along with all other microscopic material particles, are particle-like, and their motion is also that of particle motion.

However, as analyzed in Principle Two (Principle of Spatial Objectivity) earlier, the vacuum space is an objective reality, similar to a superfluid like the open sea. It inherently possesses complex waves and movements, and all particles located within the vacuum space are influenced by the movement of the vacuum space itself.

Therefore, photons and other microscopic particles do not maintain strict linear motion when moving through such a complexly moving vacuum space. At the same time, due to the wave-like motion generated in sync with the particles in the vacuum space, all microscopic particles exhibit noticeable wave-like properties.

Therefore, the linear motion of photons and all other microscopic particles, combined with the wave-like motion of the vacuum space, ultimately forms the apparent nature of both particle and wave characteristics, which is the wave-particle duality of particle motion.

At the same time, due to the complex motion influences in the void space, the wave nature of photons and other particles propagating within it will not be a strict single-direction linear vibration; they will have both vertical and horizontal vibrations. Depending on the difference in the amplitudes of vertical and horizontal vibrations, the overall manifestation of light and other particle waves will show different vibration modes such as circular polarization, elliptical polarization, and linear polarization.

The speed of light we refer to is the linear speed along the direction of propagation.
Because of the limitations of the influence of void space fluctuations, the smaller the mass of a particle, the more pronounced the effect on it, thus exhibiting more distinct wave properties. Conversely, the larger the mass of a particle, the smaller the effect, and its wave nature diminishes accordingly.

The wavelength of a particle conforms to the de Broglie formula:

$$\lambda = \frac{h}{p} = \frac{h}{mv}$$

Wherein, λ: particle wavelength, h: Planck constant; p: particle momentum, m: particle mass; v: particle linear velocity.

The interaction of space waves with particles also creates a resistive effect, so as the running distance increases, the speed and kinetic energy of the particles will decrease, the wavelength will increase, and for light, this results in redshift.

The mass and momentum of the basic elements of space waves (which can be called waveons) can be inversely calculated based on their effects on the wavelengths and frequencies of various particles, and parameters such as the Planck constant

can be determined using the energy formula $E = hf$.
(3) The relationship between the speed of light and the motion of the light source
The relationship between the speed of light and the light source should be considered in two different scenarios.
One scenario is that a photon is a rigid fundamental particle emitted by a source, so like all other massive objects, it would possess a velocity component V in the direction of motion of the emitter at the moment of emission. At the same time, because the speed of a photon in empty space is constant at C (unless otherwise specified), it would have a velocity of C in the vacuum space

Vacuum space (standard density vacuum space) causes the direction of light propagation to change, as shown in Figure 3-2.

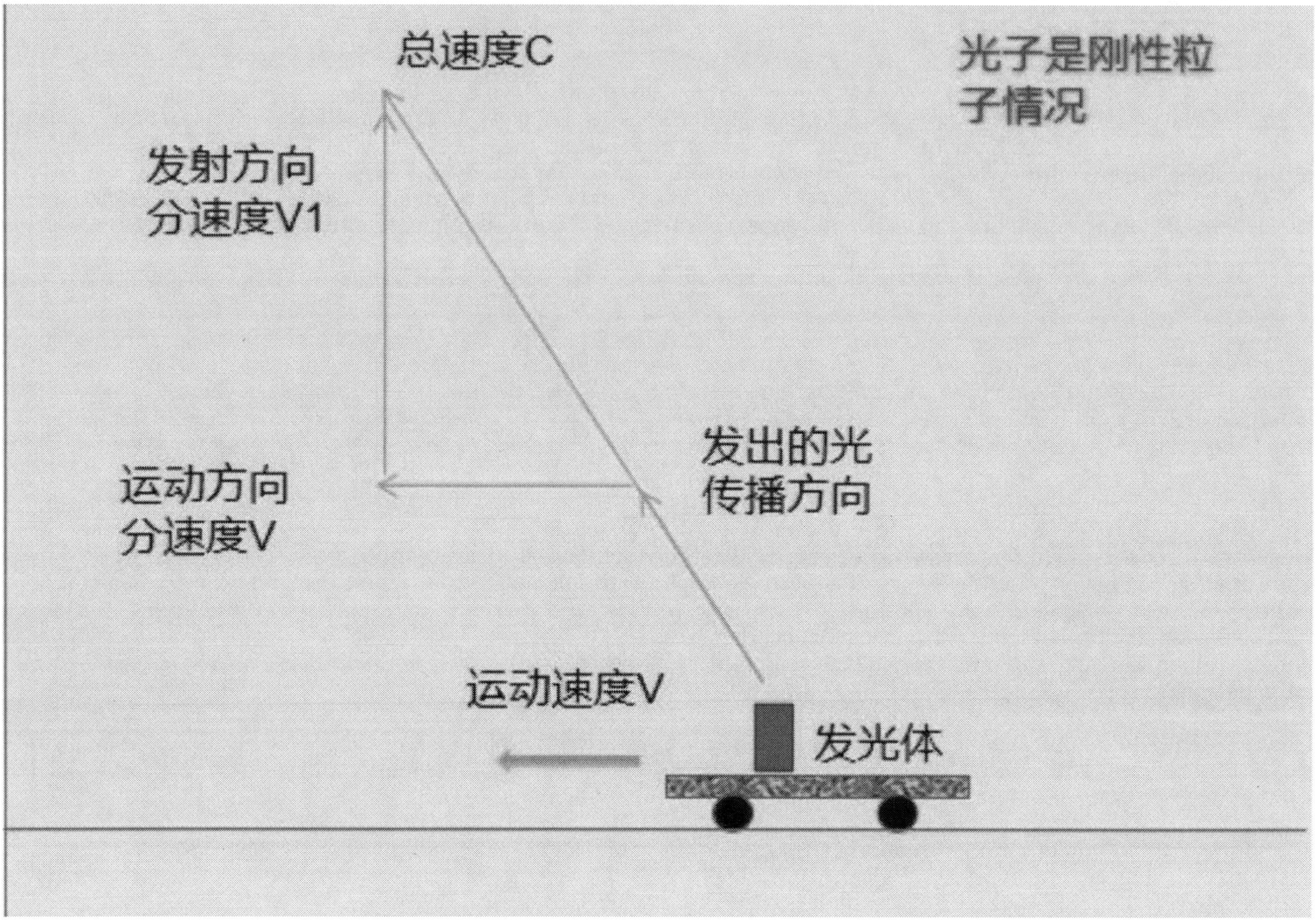

Figure 3-2 The relationship between the speed of light and the speed of the light source in the case of rigid particles

The light source moves horizontally at speed V , and the light it emits vertically will have the same horizontal speed V as its initial velocity. Since its total speed C remains constant, the vertical velocity V1 will decrease, specifically:

$$V_1 = \sqrt{C^2 - V^2}$$

The second case is that a photon is a soliton, or solitary wave, with a special stable coherent structure excited in the vacuum space. In this case, since its propagation speed in the vacuum space is constant at C, the magnitude and direction of its speed are not affected by the motion of the excitation source, as shown in Figure 3-3.

In this scenario, the emitting body moves horizontally at the speed of V , and the light emitted vertically will not have the same horizontal initial velocity. That is, the horizontal speed of the light is zero, because its total speed remains constant at C, so its vertical speed is C.

However, considering the distortion and dragging effect of the emitting body on the vacuum superfluid, its motion speed will also influence the direction of the excited light wave, similar to the first case mentioned above. This effect mainly exists within the range of vacuum space distorted and dragged by the emitting body, and the influence will quickly disappear once it is outside this range.

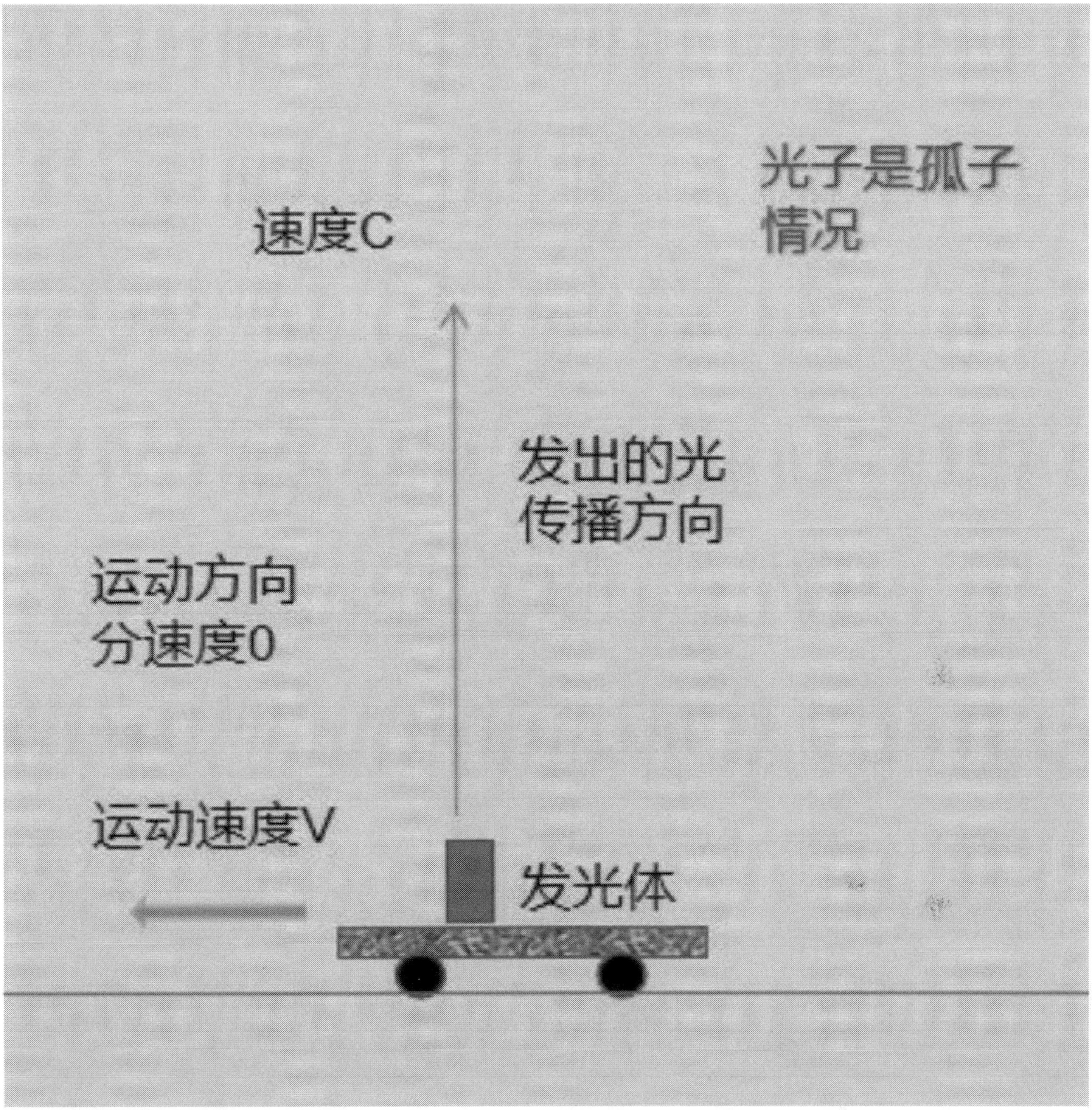

Figure 3-3 The relationship between light as a soliton and the velocity of the emitting body

Photons are either rigid particles or coherent structures similar to solitons, which require further experimental verification and confirmation. Considering the microscopic particle nature of photons, we can also regard solitons as rigid bodies. Therefore, in the following discussion, we will consider photons as solitons excited in the vacuum space, possessing a stable coherent structure. The distinction between these two cases only affects the specific formula calculations and does not impact the fundamental principles and analysis of spatial relativity that follow.

Regardless of the case, the speed of light is relative to the vacuum space it is in, and depends only on the spatial density or vacuum permittivity of that space, not on the motion of the light source. This can be referred to as the principle of invariant speed of light.

Just as the propagation speed of water waves and sound waves remains approximately constant in water and air respectively, according to their respective formulas, their speeds depend only on certain parameters of the medium they are in. Their speed is also relative to the medium in which they propagate, and is independent of the source.

(4) The relationship between the speed of light and the observer (the principle of relativity of the speed of light)

The motion of light is subject to the principle of relativity, just like the motion of all objects, meaning that the speed of light is also relative; it propagates at a constant speed relative to the uniform and flat empty space it is in;

However, empty space is local and has relative motion, so the speed of light can vary in different empty space regions or relative to different areas of empty space, and is no longer constant.

For example, suppose the uniform flat empty space region A and region B, A are moving relative to B at a speed of V. Assume a light beam traveling in the same direction as V within region A, its speed remains constant at C within A. However, relative to empty space region B, this light beam will show a change in speed of V, meaning that an observer at rest in region B will see the speed of this light relative to themselves as

$C_1 = C \pm V$ (同向"+", 反向"-")

That is, the observed speed of light conforms to the Galilean transformation, as shown in Figure 3-4.

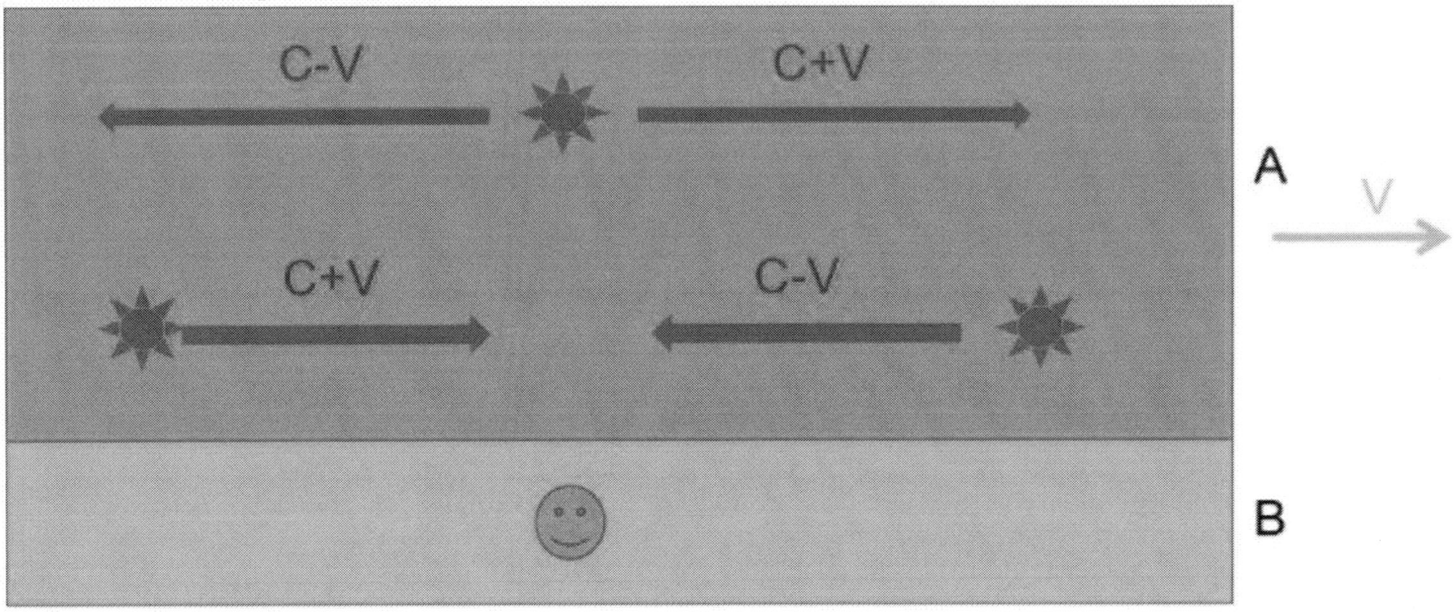

Figure 3-4 The Relativity of the Speed of Light and Motion in Void Space

Even within the same void space, the observed speed of light can appear to change due to the motion of the observer. The actual observed speed of light and the observer's speed also conform to the Galilean transformation. For instance, in the B region of void space, if the observer's speed relative to the void space is V, then the speed of light propagating in the same B region and in the same direction as the observer would be observed as:

$C_1 = C \pm V$ (同向"-", 反向"+")

That is, the observed speed of light also follows the Galilean transformation, as shown in Figure 3-5.

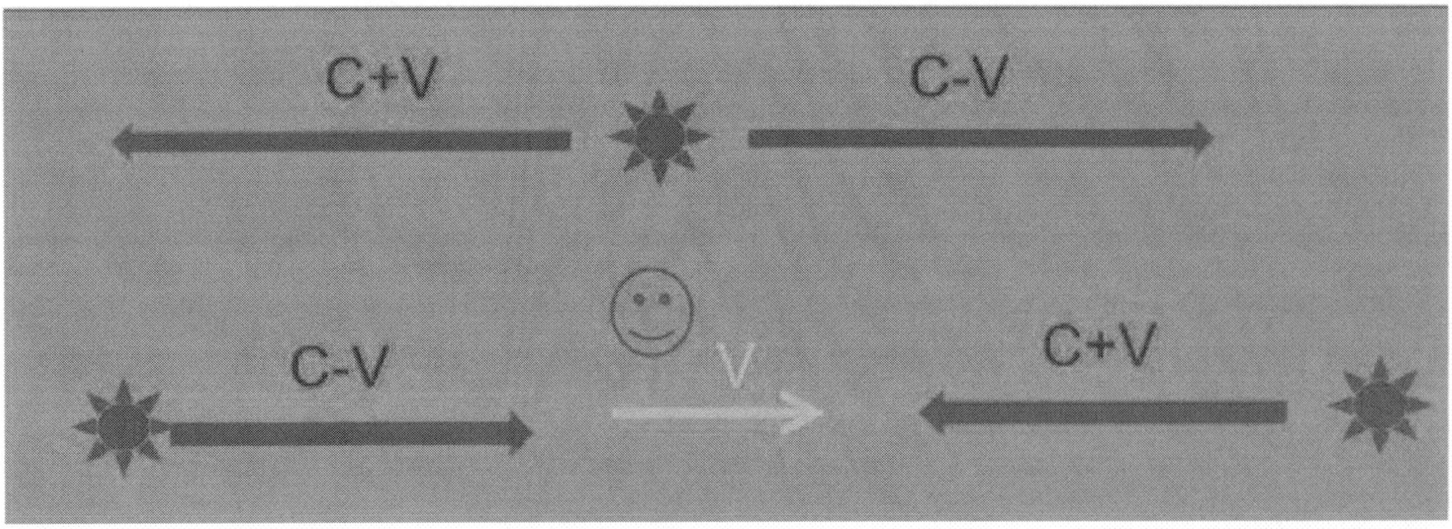

Figure 3-5 The Relativity of the Speed of Light and Observer Motion

Not only light, but all electromagnetic waves, including the propagation of electricity, magnetism, and all other forces, follow the same laws.

(5) Variation of light speed and the principle of refraction

Since the permittivity of the vacuum space is inversely proportional to the density of the vacuum space (Equation 3.9), it can be concluded that the speed of light is proportional to the density of the vacuum space it is passing through. Therefore, when light propagates into vacuum spaces of different densities, its speed changes.

At the same time, changes in the density of the vacuum space will cause the energy density of the light wave per unit volume to change, which will affect the amplitude of the light wave, causing it to change.

When entering a high-density vacuum space from a low-density vacuum space, the unit volume energy density of light increases, and its propagation speed increases; conversely, when entering a low-density vacuum space from a high-density one, the unit volume energy density of light decreases, and its propagation speed decreases. The speed of light is proportional to the square root of the density of the vacuum space in the region.

Refraction occurs when the direction of light propagation forms an angle less than 90 degrees with the boundary between two vacuum spaces of different densities. The principle of refraction can be explained by Huygens' principle, but Huygens' principle cannot account for the phenomenon where light waves of different frequencies have different refraction angles. This requires analysis based on the wave-particle duality of light. See section "8.10 Principles of Electromagnetic Wave Generation and Propagation": (3) "Principles of Refraction of Electromagnetic Waves".

Light rays passing near massive objects such as stars in the cosmos bend. In daily life, light entering different media like glass and water bends. Similarly, in the microscopic world, light bends when passing around particles such as atoms. Essentially, all these phenomena are manifestations of the same principle of refraction.

The speed of light in standard density vacuum space is the standard speed of light, denoted as C. Let the vacuum density of standard density vacuum space be ρ_c , and let the speed of light in a certain medium be v, with the average vacuum density in that medium being ρ_v . Then the refractive index of the medium is:

$$n = \frac{c}{v} = \sqrt{\frac{\rho_c}{\rho_v}} = \sqrt{\frac{\varepsilon_v}{\varepsilon_c}} \quad (3.10)$$

The above-mentioned regions of empty space are all ideal uniform, flat, and relatively stationary empty spaces. For the propagation of light within a medium, the influence of the medium's molecules must also be considered, that is, the local changes in the density and motion of empty space around the molecules, and their impact on the average density and motion of the entire medium's empty space. For gases and liquids, since the distance between molecules is sufficiently large, these effects are generally not considered. However, for solid media, because the distance between molecules is small, the changes in the density of empty space around the molecules have a greater impact on the propagation of light, especially when there are special molecular structures (such as molecular crystal structures) within the solid, which can lead to more complex effects on light propagation.

3.5 The Absolute Nature of Motion (New Principle of Relativity)

(1) The Absolute Nature of Motion

In mathematical calculations, motion between objects can be considered relative, but when it comes to the propagation of light, electromagnetic waves, and factors such as electricity, magnetism, and gravity, it is necessary to consider their motion relative to the empty space in which they exist.

Because the propagation and motion of light, electromagnetic waves, and gravity are based on the void space, which is a real and objective reference frame, it is necessary to consider the actual motion of each object relative to this void space reference frame in order to accurately determine their relative motion with respect to light, electromagnetic waves, and gravity.

The motion speed of each object relative to the void space they are in is unique and cannot be arbitrarily assumed; it can be considered as their "absolute" motion speed within this void space.
The motion speed of each object relative to the void space they are in is unique and cannot be arbitrarily assumed; it can be considered as their "absolute" motion speed within this void space.

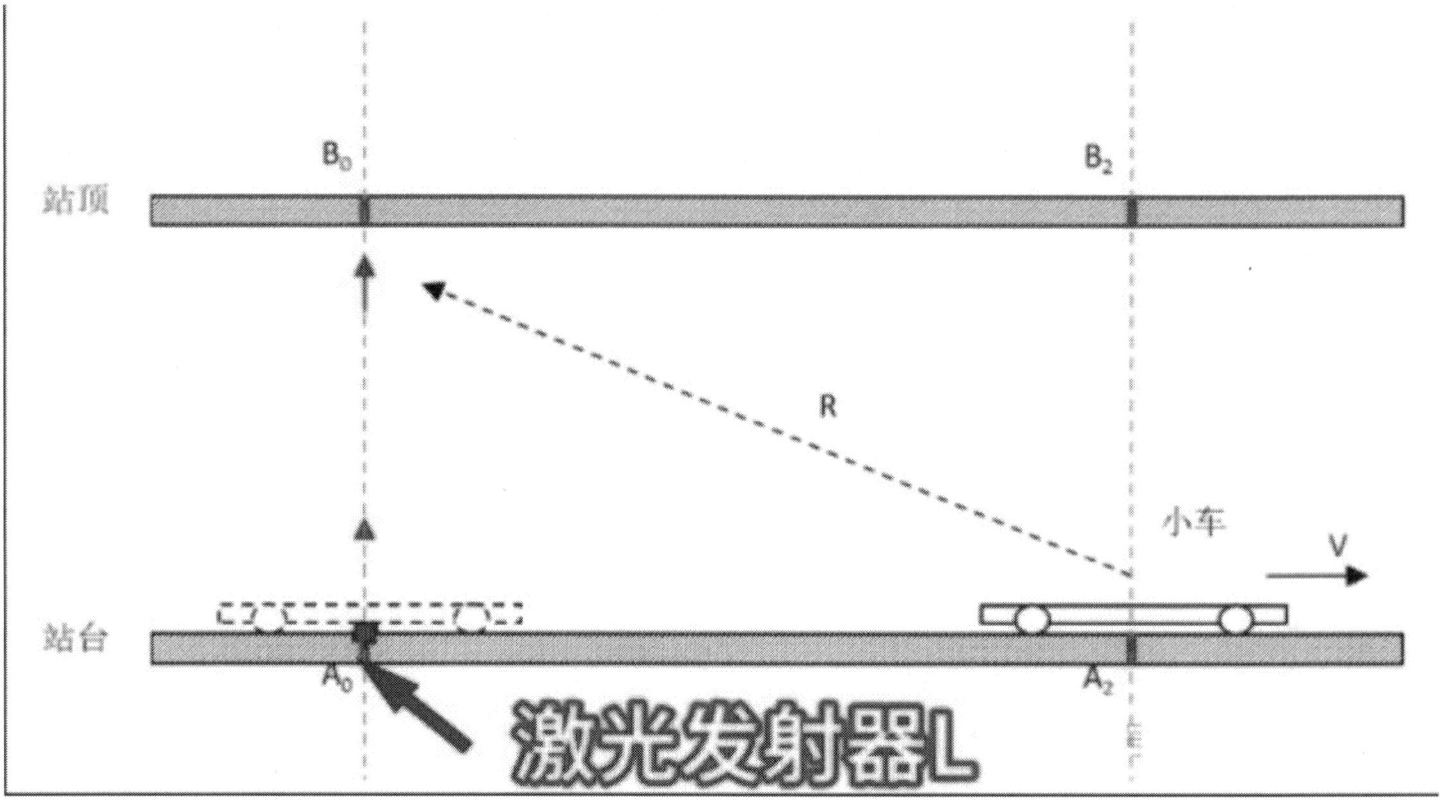

Figure 3-6 Station Laser Emission Hypothetical Experiment

We assume there is a station and a small car in the empty space. The car moves to the right relative to the station at a speed of V . There is a laser emitter at the " A_0 " point of the station. When the car passes the " A_0 " point, it triggers the laser, which shoots a laser beam L vertically downward onto the platform. As shown in Figure 3-6, where will this laser beam finally hit on the top of the station?

The relative motion speed between the car and the station is V. We assume both are in the same uniform empty space. If the station is stationary relative to the empty space and the car moves at speed V relative to the empty space, then both the observer on the car and the observer on the station will see the laser beam L hitting the " B_0 " point on the top of the station vertically.

However, if the car is stationary relative to the empty space, and the station is moving to the left at speed V relative to the empty space,

In this case, when the " A_0 " point of the station passes the car, the laser emitter is triggered. At this moment, the car and the station will observe the laser beam L hitting the " A_0 " point on the top of the station vertically.

What would an observer on the platform see regarding the motion of the laser L emitted by the laser?
We assume the platform is transparent in front and back, meaning it does not drag or affect the surrounding empty space. Therefore, since the speed of light is constant at C relative to the empty space and independent of the speed of the source, when the platform moves to the left at a speed of V, the horizontal speed of the laser L remains zero and is unaffected; it still propagates vertically upward. As the platform moves to the left, the laser L will eventually hit the B_2 point on the platform's ceiling, as shown in Figure 3-7.

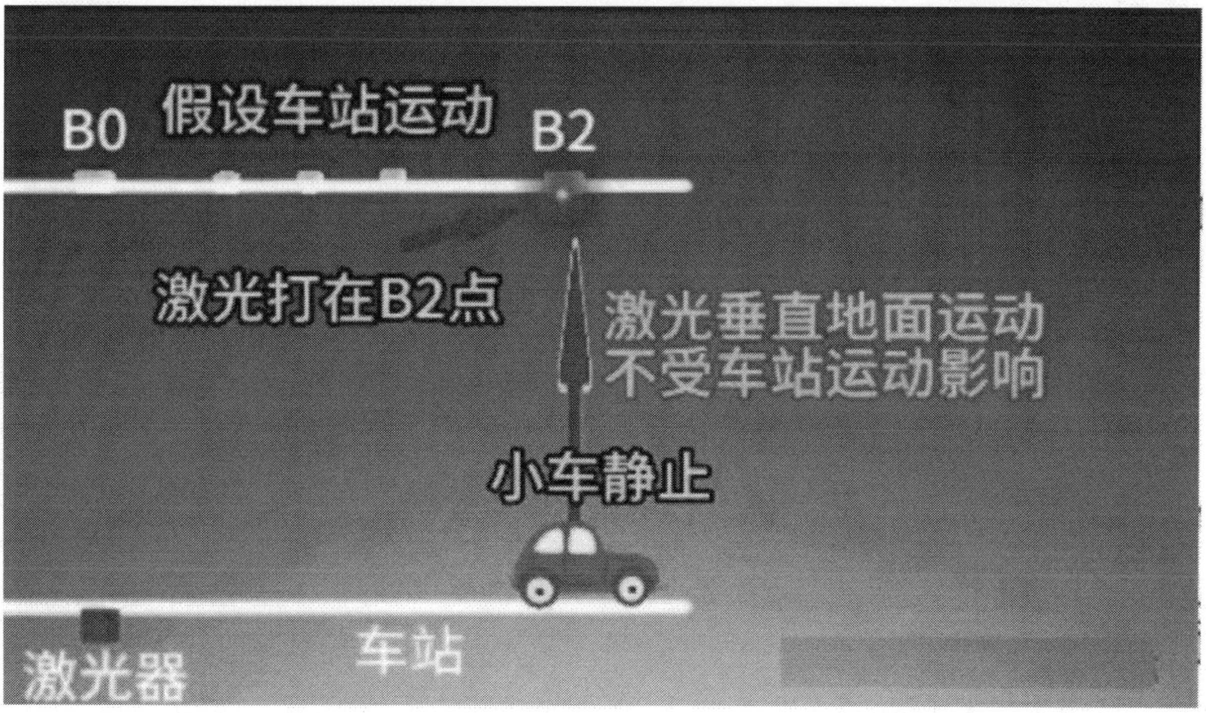

Figure 5: Figure 3-7 The laser's vertical propagation direction is independent of the horizontal movement of the station

Therefore, the point on the station's roof where the laser L will finally hit depends only on the relative motion speed between the station and the void space, and is unrelated to the relative motion speed between the station and the train.

Assuming the station's motion speed relative to this void space is V_1 , and the station's height is H, then when the laser hits the top of the station, the distance the station has traveled is $L_1 = V_1 * \mathrm{H/C}$. The laser L will hit a position that is L_1 away from the point " B_0 " on the roof in the opposite direction of the motion speed " V_1 ".

Therefore, as long as the station's height and its relative motion speed to the void space remain constant, the point where the laser hits the station's roof will be fixed. Regardless of the train's speed relative to the station, it will not affect the point where the laser hits the roof. However, if the station's motion speed relative to the void space changes, the point where the laser hits the roof will also change.

Of course, this is under the assumption that the station is open and does not drag the surrounding empty space. If the station is enclosed, its movement would drag the internal empty space, which would also affect the propagation of the laser. Assuming the dragging coefficient of the internal empty space by the station is K, the velocity of the internal empty space being dragged would be $\mathrm{K}V_1$, thereby giving the laser propagating within it a horizontal velocity of KV_1 . Thus, when the laser hits the top of the station, the distance the station has moved becomes:

$$L_1 = V_1(1-K)*H/C$$

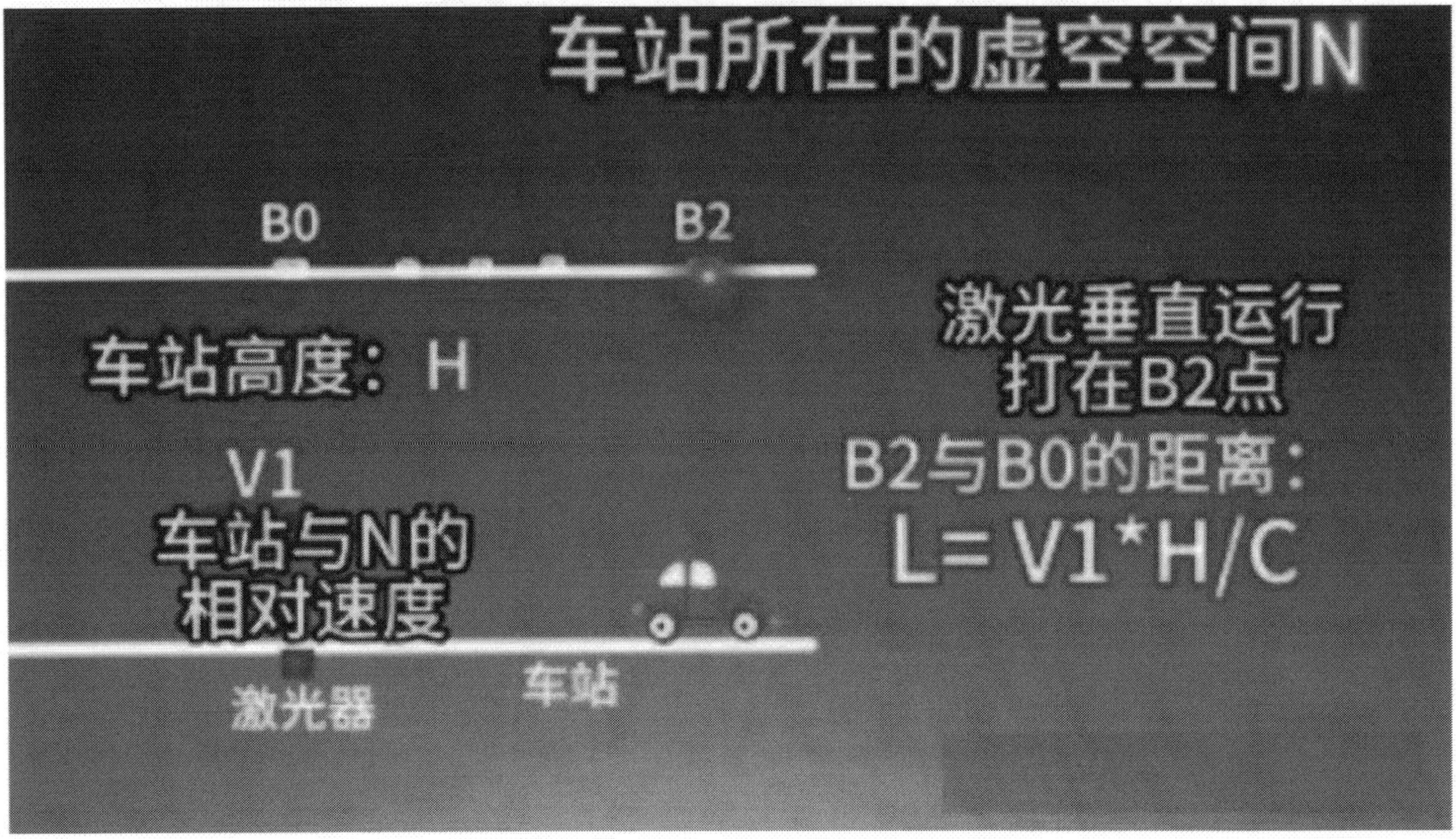

Figure 3-8: Relationship between the laser's impact point and the station's motion speed

It is visible that if the enclosed station drags the internal empty space with a coefficient of $K=1$, meaning complete dragging where the internal empty space moves in sync with the station, then no matter what speed V_1 the station moves relative to the external empty space, the position where the laser hits the top of the station will remain unchanged, always being directly above the laser source, i.e., point B_0 . However, if the dragging coefficient is not 1, then as the station moves at different speeds relative to the external empty space, the position where the laser hits the top of the station will also change, as shown in Figure 3-8.

The above analysis is based on the assumption that photons are excited solitons. If photons are emitted particles, their initial direction is affected by the motion of the emitting body, and a similar analysis can be applied. If the station is fully transparent front and back, the laser emitted will always hit the point B_0 at the top regardless of the station's motion relative to the vacuum space. However, if the station is enclosed and has a drag coefficient K for the internal vacuum space, the laser will hit the point B_0 in the opposite direction of the station's motion.

Above, the distance is:

$$\mathrm{L}_1 = \mathrm{V}_1(1-\mathrm{K})\frac{\mathrm{H}}{\sqrt{\mathrm{C}^2-\mathrm{V}^2}}$$

(2) Relativity of Motion

As analyzed above, when involving factors such as light or electromagnetic waves, it is necessary to consider the object's relative motion speed with respect to this void space, and we cannot arbitrarily assume their speed. This relative motion speed with respect to the void space can be referred to as the object's "absolute speed" in that region.

Of course, due to the locality of void space and the relative motion between different regions, as well as the fact that each void space is part of a larger void space region, the object's motion speed relative to various different void space regions will also vary.

Without considering the relative motion with respect to void space, an object's motion speed depends on the selected reference object or reference frame, meaning its motion speed is relative to the reference object or frame. The motion speed of an object will also differ relative to different reference objects and frames. In fact, void space itself is also a reference object or frame for an object's motion, but it is considered as a special reference frame due to its significant influence on the propagation of light, gravity, etc. (we can call it a medium reference frame).

In all reference frames, the form and content of physical laws are definite and do not change, but because different reference frames and observers have different motion conditions, the observed surface phenomena will vary. It is necessary to consider the influence of one's own motion to deduce the true situation behind the surface phenomena.

The velocity transformation between different inertial frames applies the Galilean transformation.

3.6 The Relativity of Fields and Empty Space (Principle of Spatial Relativity)

All fields, whether electric, magnetic, or gravitational, are modes of motion in empty space and are excited and propagated through empty space.
(1) Threshold of Empty Space Excitation

The material nature of the vacuum space determines a threshold for its excitation into motion; not all charge movements can generate a magnetic field or electromagnetic wave.

(2) Magnetic Field Excitation Threshold

Let the threshold velocity for vacuum space to generate a magnetic field under standard space density ρ_c be V_B, the threshold distance be L_B, and the threshold time be T_B. A magnetic field is only generated when the relative velocity $V > V_B$ of the motion exceeds the threshold, and the motion distance $L > L_B$ or motion time $T > T_B$ meets the threshold conditions.

The magnitude of the magnetic field is given by the Biot-Savart law:

$$B = \frac{\mu_0}{4\pi}\frac{q\vec{v} \times \vec{r}}{r^3}$$

Where v is the relative motion velocity of charge with respect to the vacuum space.

(3) Electromagnetic wave excitation threshold

The threshold acceleration for electromagnetic wave motion in the vacuum space under standard spatial density ρ_c is a_Q, the threshold motion distance is L_Q, and the threshold motion time is T_Q. Electromagnetic waves will be excited only when the relative motion acceleration $a > a_Q$ between it and the charge, and the motion distance $L > L_Q$ or motion time $T > T_Q$ meets these conditions.

The transverse electric field strength of the electromagnetic wave at a position making an angle of θ with the direction of motion of the charge q and at a distance r is:

$$E_\theta = \frac{qa \sin\theta}{4\pi\varepsilon_0 c^2 r}$$

Transverse magnetic field strength is:

$$B_\psi = \frac{qa \sin\theta}{4\pi\varepsilon_0 c^3 r}$$

where acceleration a is the acceleration of the charge relative to the surrounding vacuum space.

(4) Electric field

A charge in vacuum space will generate an electric field. The electric field strength under standard space density ρ_c is:

$$\mathrm{E} = \frac{\mathrm{KQ}}{\mathrm{r}^2}$$

(5) Relativity of Motion

The motion of the charge relative to the vacuum space is relative. If the charge is stationary, but the vacuum space it is in is moving or accelerating, it will also generate a magnetic field and electromagnetic waves.
(6) Relative Motion of Vacuum Space
Because of the locality and relative motion of vacuum space, when the electric field, magnetic field, and electromagnetic waves generated by the charge propagate from one vacuum space to another that is in relative motion, their strength and direction of propagation will change. The extent of the change depends on factors such as the density and relative velocity of the two vacuum spaces.
(7) Relativity of Maxwell's Equations
Maxwell's equations are equations that describe the motion of electromagnetic waves, so they also conform to the above description of "relative empty space." However, due to the widespread misunderstanding of this equation set and the motion of electromagnetic waves it describes, an additional explanation is provided here.

Maxwell's equations describe the motion of light or electromagnetic waves, which moves relative to empty space. Therefore, Maxwell's equations describe the situation of electromagnetic waves moving in a uniform flat empty space, or in other words, their motion relative to a uniform flat empty space, rather than the motion of electromagnetic waves relative to an observer or any other object.

The vacuum space itself is an objective real reference frame, so Maxwell's equations are only valid for vacuum space, specifically for the vacuum space where light propagates. Observers or reference frames in other vacuum spaces that are in relative motion with respect to this one will observe electromagnetic waves, electric fields, and magnetic fields that differ from the predictions of Maxwell's equations. The deviation depends on the relative motion between the two vacuum spaces and their density differences. These differences can be transformed through velocity and density changes to make the equations consistent with the observed results. The velocity transformation follows the Galilean transformation.

Currently, people believe that Maxwell's equations are universal and consistent for all reference frames. This

view is correct in that Maxwell's equations describe the physical laws of electromagnetic waves relative to the reference frame of empty space, which is universal and unchanging; however, it is wrong in confusing the distinction between physical laws and observed phenomena, assuming that all directly observed results, such as the speed of light, electricity, and magnetism, seen in different inertial reference frames, completely conform to Maxwell's equations. In reality, directly observed surface phenomena may reflect the true nature of things or distort it, requiring us to analyze and discern them dialectically.

The main cause of this error is the famous Michelson-Morley experiment's results, but historically, the analysis and understanding of these results were also incorrect.

3.7 The Relationship Between Mass-Energy and Spacetime (Principle of Conservation of Mass-Energy)

(1) There are two incorrect understandings and expressions of the relationship between mass and energy in the current theory

The first is the misunderstanding of the mass-energy equation $\mathrm{E} = \mathrm{MC}^2$, which is considered as a conversion equation between mass and energy. For example, in nuclear fission, the mass of matter appears to be lost, and the lost part of mass is transformed into energy. Therefore, mass is not conserved.

But according to this statement, mass is missing, so mass is not conserved. If energy is also increased, can we also say that energy is not conserved? Of course, this understanding is inappropriate. In reality, energy has not increased; it was previously stored within the material as intrinsic potential energy and was released during the fission process, converting into photon kinetic energy and potential energy, which is a transformation of energy. Therefore, energy is conserved in this process.

The total amount of mass and energy in the universe remains constant. From this perspective, if energy is conserved, mass should also be conserved.

Therefore, the mass lost in nuclear fission has not disappeared, but has been converted into the mass of the emitted photons.

The current theory claims that photons have no mass, which is inappropriate. Photons do have radiation pressure, momentum, and energy.

Quantity, therefore, must have mass. All particles that have momentum, pressure, are affected by gravity, and can convert kinetic energy with matter must certainly have mass.

Photons have no mass. This is an incorrect assumption caused by the current erroneous spacetime theories and understanding that "particles moving at the speed of light must have zero mass."

On the other hand, according to the relativistic mass formula, as the speed of matter increases, its mass also increases. This would result in mass appearing out of nowhere and no longer being conserved, and further, it would lead to energy appearing out of nowhere and no longer being conserved, because an increase in mass necessarily brings an increase in kinetic energy. Especially when matter approaches the speed of light, mass and energy tend toward infinity, which clearly violates the conservation of mass and energy.

According to our previous principles of spatial objectivity and temporal objectivity, the total energy of the universe should remain constant and should not result in non-conservation of mass or energy.

(2) Correct Understanding of the Mass-Energy Equation

The true meaning of the mass-energy equation $E = MC^2$ is that the energy of a photon with mass M is MC^2 , i.e., the energy of a photon is expressed as the product of its mass and the square of the speed.

Then, why is the energy of a photon $\mathrm{E} = \mathrm{MC}^2$, while the kinetic energy of a common macroscopic object is $\mathrm{E} = 1/2\mathrm{MV}^2$?

Kinetic energy of matter: $E = \frac{1}{2}mv^2$ Photon energy: $E = mc^2$

Figure 3-9: Photon Energy and Kinetic Energy of Matter

In fact, by looking at the names of these two energies, we can see the difference. As shown in Figure 3-9, $\mathrm{E} = 1/2\ \mathrm{m}\ \mathrm{V}^2$ refers to the kinetic energy of matter, while $\mathrm{E} = \mathrm{mC}^2$ refers to the energy of a photon. Energy is not only kinetic energy.

We know that light exhibits wave-particle duality. Assuming the photon has mass m, its particle-like motion at the speed of light gives it kinetic energy $\mathrm{E} = 1/2\mathrm{mC}^2$. At the same time, the photon is also a wave, moving in a wave form, so it also possesses potential energy. Since light waves are transverse waves, their kinetic and potential energies are equal. Therefore, the photon has $\mathrm{E} = 1/2\mathrm{mC}^2$ potential energy. Thus, we obtain the photon's energy as the sum of its potential and kinetic energy, that is, $\mathrm{E} = \mathrm{mC}^2$. As shown in Figure 3-10.

The wave-particle duality of light photons are particles

Therefore, photons have kinetic energy: $E_1 = \frac{1}{2}mc^2$
Light travels as a wave. It is a transverse wave
Therefore, photons have potential energy:
(Transverse wave kinetic energy equals potential energy)
Therefore, the total energy of a photon is:

$$E_2 = \frac{1}{2}mc^2$$
$$E = mc^2$$

Figure 3-10 Total Energy of a Photon

It can be said that if stones and other objects in our macroscopic world also follow the wave-particle duality principle and exhibit wave-like motion, they would also have equal kinetic and potential energy, that is, total energy of $E = mV^2$. However, if we only consider their kinetic energy, it would be $\mathrm{E} = 1/2\mathrm{mV}^2$.

So, this is the true meaning of the mass-energy equation $E = \mathrm{mC}^2$. This formula merely indicates that the energy of a photon is the sum of its kinetic and potential energy, arising from its dual nature as a wave and a particle.

This equation is not about the conversion of mass into energy, but rather about the transformation of nuclear material with mass m into a large number of photons with total mass m, converting the potential energy contained within the m mass material into the energy of these m mass photons.

(3) The Principle of Mass-Energy Conservation

Without involving the causal interactions related to the virtual world mentioned in the first fundamental principle (the Principle of Objective Physical Laws)

In a closed system of the real world, mass is conserved, energy is conserved, and the total mass-energy is conserved.

Photons have mass. Let the photon mass be m, and when photons propagate freely in the void space, their velocity is C, so their kinetic energy is $\mathrm{E}_1 = 1/2\mathrm{mC}^2$. Since light waves are transverse waves, they also possess the same potential energy, that is, potential energy $\mathrm{E}_2 = 1/2\mathrm{mC}^2$. Therefore, their total energy is $\mathrm{E} = \mathrm{E}_1 + \mathrm{E}_2 = \mathrm{mC}^2$.

When photons are accumulated inside particles or within the void space around particles and become part of the particle's mass, their energy is also accumulated internally, becoming the intrinsic potential energy of the particle. The mass of the photon also becomes part of the particle's mass.

When a particle undergoes a reaction, fission, etc., and releases these photons, the intrinsic potential energy of these photons is also released, allowing the m mass photons to carry mC^2 energy back into the external free void space.

Throughout the process, particles may lose mass and energy, but the overall system maintains mass conservation, energy conservation, and the conservation of mass-energy.

(4) Objectivity of Spacetime Parameters

According to the previous principles of spatial objectivity, temporal objectivity, and the conservation of mass-energy, we can consider that in the same constant vacuum space environment, an object's mass, size, and time are objective and do not change on their own due to variations in velocity or other conditions.

In other words, within the same constant vacuum space environment, when an object's velocity increases or decreases, its mass, size, and time remain unchanged.

The current spacetime theory and the concept of mass-energy equivalence have been discarded, the length-velocity equation is wrong (even according to special relativity, it should be length expansion rather than contraction, this error caused the incompatibility between special relativity and the gravitational formula, forcing the development of general relativity, leading further and further away. See Chapter 2.1: Misuse of the principle of the constancy of light speed. The time-velocity equation's Lorentz transformation (including the Lorentz transformation of mass and length) is just a mathematical calculation and correction method, which is a calculation and correction under ideal conditions.

4 Introduction to Spatial Electrodynamics [6]

4.1 The Relativity of Charge, Electromagnetic Field and Void Space [7]

(1) A century-old physics puzzle

In the physics community, there has been a puzzle that has confused all scientists for over a century: why do electrons orbiting the nucleus not emit electromagnetic waves?

According to classical theory, accelerated charges produce electromagnetic waves. Since electrons are constantly accelerating as they orbit the nucleus, they should continuously emit electromagnetic waves. However, this phenomenon has not been observed in experiments. Moreover, if electrons continuously emit electromagnetic waves, they would lose energy and quickly fall into the nucleus. Such a model of matter would not be stable, which clearly contradicts our physical reality.

To resolve this issue, quantum theory revised the motion of electrons around the nucleus. They are no longer considered to move in circular orbits, but instead exist as a probability cloud around the nucleus. This means that electrons may be located anywhere within a certain region around the nucleus, with their exact position being probabilistic rather than certain. We cannot determine where they are, but they randomly jump from one point to another.

This theory avoids the paradoxes associated with electrons orbiting the nucleus, but it uses the concept of probability clouds to sidestep the fundamental nature of electron motion. It does not explain how electrons actually move around the nucleus. It is merely a mathematical description, not an explanation of the true nature of electron motion.

Then why do we observe in our daily life that electrons undergoing accelerated motion radiate electromagnetic waves, while electrons orbiting the nucleus also undergo accelerated motion but do not radiate electromagnetic waves?

(2) Conditions for electron motion to radiate electromagnetic waves

Here, we need to analyze clearly under what circumstances the accelerated motion of an electron will result in the radiation of electromagnetic waves.

In fact, accelerated motion of an electron does not necessarily lead to the radiation of electromagnetic waves. For example, an electron that is stationary on the ground does not emit electromagnetic waves around it. However, at this moment, this electron is also undergoing accelerated motion by following the Earth in its orbit around the Sun. So, can we say that the electron is also in accelerated motion at this moment? Then why does it not radiate electromagnetic waves?

Some people say that according to the principle of relativity, electromagnetic waves can only be observed when there is relative acceleration between an electron and the observer. However, just like this electron that is stationary on the ground, no matter how the observer accelerates, they will still not observe it radiating electromagnetic waves. Only when this electron accelerates relative to the ground will it radiate electromagnetic waves. At this time, regardless of whether the observer is stationary or moving relative to the electron, the electromagnetic waves radiated by the electron will not disappear.

Therefore, the phenomenon of an electron radiating electromagnetic waves is unrelated to the observer's motion, and also unrelated to the relative motion between the observer and the electron.

Then, why does an electron radiate electromagnetic waves when it accelerates relative to the ground, but does not radiate when it is stationary relative to the ground and follows the Earth's acceleration relative to the Sun?

The reason lies in our basic principles of space relativity: the principle of space objectivity and the principle of space relativity. The vacuum space is an objective reality, so it must be local and have motion. A charge will only radiate electromagnetic waves when it accelerates relative to the vacuum space it resides in, and when the speed, acceleration, amplitude, etc., of this motion reach the threshold of the vacuum space.

For example, when a charge is stationary on the Earth's surface, according to the basic principle of spatial relativity 2: the principle of spatial objectivity, the Earth exerts a dragging effect on the surrounding empty space, causing the empty space to have relative motion. The closer the empty space is to the Earth, the smaller its relative velocity, and the empty space closest to the ground can

It is considered to be completely dragged, so this void space is stationary relative to the Earth's surface.
Therefore, when charges near the ground are stationary relative to the ground, they are also stationary relative to this void space and will not trigger the emission of electromagnetic waves from the void space. In this case, even if the charges follow

the Earth in its accelerated motion around the Sun, it does not change the fact that they remain stationary relative to the void space they are in, so they still do not emit electromagnetic waves.

In this case, even if the observer accelerates relative to the charge, it does not change the fact that the charge is stationary relative to this empty space, and the electromagnetic wave is still not observable.

Only when a charge accelerates relative to this surrounding vacuum space will it emit electromagnetic waves. At this time, regardless of whether the observer is stationary or moving relative to the charge, it does not change the charge's motion status, so electromagnetic waves can be observed by all observers, unless the observer's relative speed in this vacuum space exceeds the speed of light.

(3) Characteristics of Microscopic Electron Motion

Why don't electrons orbiting the atomic nucleus emit electromagnetic waves?

This requires considering the basic principle of space relativity 6: the motion trigger threshold of the vacuum space mentioned in the principle of space relativity is related to the density of the vacuum space. Only when the relative velocity, acceleration, amplitude, etc., of the charge in this vacuum space reach this threshold will it trigger the corresponding motion of this vacuum space, resulting in electric fields, magnetic fields, or electromagnetic waves.

For example, in a still pond, if a pebble is stationary, when the pebble starts moving, we will observe waves forming in the water. However, if the movement of the pebble is very small, even smaller than the distance between it and the surrounding water molecules, it will not trigger the movement of the surrounding water molecules, and thus no water waves will be produced in the pond.

That is, the movement of pebbles in a pond only triggers the formation of water waves when it reaches a certain threshold, such as when it exceeds the distance to surrounding water molecules.

Similarly, the relative motion of charges and empty space also requires such thresholds. Different threshold conditions exist to excite different motion patterns in empty space.

Our macroscopic observation of charge movement produces electric fields and electromagnetic waves, because the threshold for triggering motion in empty space is very small, and the amplitude of charge movement we observe macroscopically is usually much greater than this threshold.

However, at the microscopic atomic level, electrons orbit the nucleus, and their acceleration and movement amplitude are generally not enough to reach the threshold for exciting electromagnetic waves in empty space, so they do not produce electromagnetic waves.

The electromagnetic waves we refer to here are continuous electromagnetic waves radiated in space, not individual photons resulting from changes in particle structure.

(4) Experimental Verification Plan

According to the previous analysis, the vacuum space is an objective reality that can be "distorted" by mass bodies to produce density gradient changes, can be dragged by mass bodies to produce motion, and even may exist in more complex relative motion.

The propagation of electromagnetic waves has a constant speed of C relative to the vacuum space they are in, not relative to any other inertial frame or observer.

Charge produces a magnetic field or radiates electromagnetic waves only when it has relative motion with respect to the surrounding vacuum space. If the charge remains stationary relative to the vacuum space, even if it is moving at high speed or accelerating relative to other inertial frames or observers, it will not produce a magnetic field or radiate electromagnetic waves.

If the observer is stationary relative to the charge, but the charge is moving or accelerating relative to the surrounding vacuum space, then a magnetic field or electromagnetic waves will still be produced in the space.

We can design a determining experiment to verify this.

(1) Magnetic Field Observation Experiment

Since matter can twist and drag the empty space, we can design a closed box with a charge inside, and have both move together on an orbit, examining the magnetic field generation under different motion patterns.

Step 1:

A closed box with charges moving at a constant speed along an orbit.

At this moment, due to the dragging effect of the box on the internal empty space, the charges remain relatively stationary with respect to this surrounding empty space, so no magnetic field is produced around the charges.

At this moment, if there is an observer, whether moving relative to the charges or stationary, they will not observe a magnetic field.

Of course, this is under the assumption that the material of the box can ensure complete dragging of the internal vacuum space. If it cannot fully drag, but only partially drag, then the charge will still have relative motion with respect to the vacuum space inside, thus generating a magnetic field, as shown in Figure 4-1.

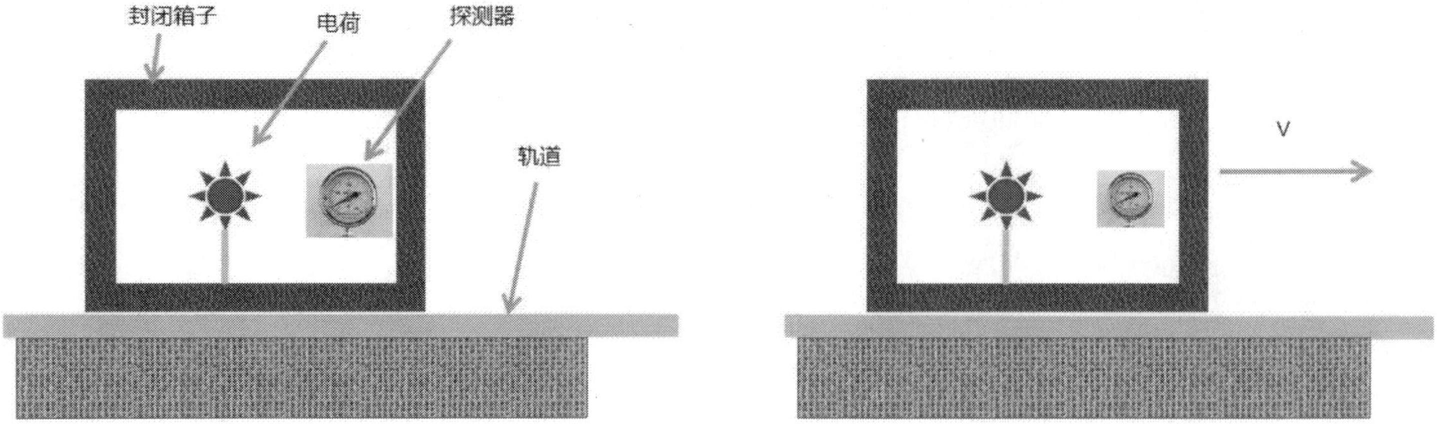

a）实验装置示意图

b）木箱运动拖曳虚空空间超流体与电荷产生相对运动，检测磁场和电磁波情况

Figure 4-1: Schematic of the space dragging experiment

Let the dragging coefficient of the box material with respect to the vacuum space be K, and its common velocity be V. Then the relative actual velocity of the charge with respect to the vacuum space inside the box is $(1-K)V$.

Let the charge have a charge quantity q, and its position inside the box be point O. Take any point P in the space inside the box, and let OP be $= \mathrm{r}$. Then the magnetic field at point P is:

$$B=\frac{\mu_0}{4\pi}\frac{q(1-k)\vec{v}\times\vec{r}}{r^3}$$

Let the unit vector in the direction of $\overrightarrow{\mathrm{r}}$ be $\overrightarrow{\mathrm{e_r}}$, the magnetic field at point P can also be written as:

$$\mathrm{B}=\frac{\mu_0}{4\pi}\frac{\mathrm{q}(1-\mathrm{k})\overrightarrow{\mathrm{v}}\times\overrightarrow{\mathrm{e_r}}}{\mathrm{r}^2}$$

The magnitude of the magnetic field can be written as:

$$B=\frac{\mu_0}{4\pi}\frac{q(1-k)v\sin\theta}{r^2}$$

The angle between $\vec{v}$ and $\vec{r}$.
It is assumed here that the motion of the empty space inside the box has no effect on its magnetic permeability μ_0. If there is an effect, it needs to be corrected synchronously.

Step 2:

Remove the front and back ends of the box, making it transparent front and back, and move it at the same constant speed as the charge.
At this moment, the box is still moving at a constant speed, but because it is now transparent front and back, it no longer exerts a dragging effect on the internal vacuum space. The charge inside is therefore moving relative to the surrounding vacuum space, and a magnetic field will be generated around the charge at this moment.

The specific size of the magnetic field can also be calculated as in Step 1, just with the dragging coefficient $K=0$, returning to the normal magnetic field formula:

$$\mathrm{B}=\frac{\mu_0}{4\pi}\frac{\mathrm{qv}\times\overrightarrow{\mathrm{r}}}{\mathrm{r}^3}$$

At this moment, the observer can detect a magnetic field whether the charge is moving or stationary relative to them. The presence or absence of the box's movement does not affect the generation of the magnetic field.

(2) Electromagnetic Field Observation Experiment

The experiment is designed similarly to Experiment One. A closed box is created with a charge inside, and both move together on a track to examine the generation of electromagnetic fields under different motion patterns.

Step One:

A closed box and a charge move along the orbit with the same acceleration.

At this moment, due to the box's dragging effect on the internal empty space, the charge remains relatively stationary with respect to this surrounding empty space, so no magnetic field is produced around the charge.

At this moment, if there is an observer, whether moving or stationary relative to the charge, they will not observe an electromagnetic field.
Of course, this is under the assumption that the material of the box can fully drag the internal empty space. If it cannot fully drag but only partially, then the charge will still have relative motion with respect to the empty space, thus producing an electromagnetic field.

Let the drag coefficient of the box material on the vacuum space be K, and let the common acceleration of the box be a. Then the relative actual motion acceleration of the charge with respect to the vacuum space inside the box is $(1-K)a$, so the magnitude of the resulting electromagnetic field is similar to Experiment 1, and the electromagnetic field formula should be corrected based on the actual motion acceleration.

If the motion of the empty space inside the box affects its permittivity ε_0, then a synchronous correction is needed.
Step two:
Remove the front and back ends of the box, making it transparent at both ends, and move it with the same acceleration as the charge.
At this moment, the box is still moving at a constant speed, but because it is now transparent at both ends, it no longer exerts a dragging effect on the internal empty space. The charges inside are therefore moving relative to the surrounding empty space, and an electromagnetic field will be generated around the charges at this point.

The specific electromagnetic field magnitude can also be calculated following step one, with the drag coefficient $K=0$, returning to the normal electromagnetic field formula.

At this moment, the observer can detect the electromagnetic field whether the charge is moving or stationary relative to them. Including whether the box is moving or not will not affect the generation of the electromagnetic field.
(3) Experiment on the Relativity of Motion between Charge and Empty Space
The above two experiments tested that the relative motion of a charge with respect to empty space can produce a magnetic field, and relative acceleration can produce an electromagnetic field. However, here the question arises: is the relative motion between the two entities equal and mutual, or is it only the active motion of the charge relative to empty space that produces a magnetic field and an electromagnetic field?

Based on our daily experience, the motion of the two should be relative and equal, but it still needs experimental verification.

We can also base on the above experimental conditions, the box is completely sealed, the charge is located in the internal space of the box, and both are stationary on the track.

Step one:

Let the charge move at a constant speed V along the track inside the box, and a magnetic field can be generated in the box; let the charge accelerate at an acceleration a, and an electromagnetic field can be generated in the box. This is correct.

Step two:

Let the charge be stationary, and the box begin uniform motion along the track with velocity V. Let the drag coefficient of the box on the internal empty space be K, then the relative velocity of the internal empty space with respect to the charge is $(1-K)V$. At this time, observe whether a magnetic field is produced.

Next, let the box begin accelerated motion along the track with acceleration a. Then the relative acceleration of the internal empty space with respect to the charge is $(1-K)a$. At this time, observe whether an electromagnetic field is produced.

If both a magnetic field and an electromagnetic field are observed to be produced under the respective conditions, it indicates that the motion between the charge and the empty space is completely relative. Whether it is the charge or the empty space that is moving, it can generate magnetic and electromagnetic fields.

If no magnetic and electromagnetic fields are observed, it indicates that the motion between charge and empty space is not relative. Only when the charge moves relatively to empty space will magnetic and electromagnetic fields be generated.

If, in two steps, the charge and empty space move at the same relative velocity, but the observed strengths of the magnetic and electromagnetic fields are inconsistent, then it is necessary to determine whether there is an error in the box's dragging motion relative to empty space, or whether the motion of empty space has affected its permittivity and permeability parameters.

(4) Improved Michelson-Morley Experiment

In 1887, American physicists Albert Michelson and Edward Morley conducted multiple experiments consecutively, hoping to measure the Earth's velocity relative to the ether, but all results were negative.

The famous Michelson-Morley experiment

In fact, whether this experiment proved that the ether does not exist as an objective reality or merely showed that the Earth's surface moves almost synchronously with the ether has been a subject of significant academic debate.

Here we can design an experiment to determine what exactly caused the results of this experiment.
The experimental setup is similar to the above experiment, except that the charge on the track is replaced with the detection system of the Michelson-Morley experiment, which detects the change in interference fringes when moving along the track.

Step One:

The experimental system is stationary in orbit, as it is in relative static equilibrium with the vacuum space, so it does not interfere with the fringes.

Step Two:

The experimental system is moving at a constant speed in orbit, at which moment it also produces a constant relative motion with the vacuum space, so interference fringes can be observed. The number of fringes corresponds to the speed of the experimental system.

Step 3:

Place the experimental system inside a closed box, and move them together at the same constant velocity v on the orbit. At this moment, because the box drags the internal empty space, the experimental system remains stationary relative to the empty space, so no interference fringes will be produced.

If the box cannot completely drag the internal empty space, then according to its drag coefficient K, the relative velocity between the internal empty space and the experimental system is $(1-K)_v$, and the number of interference fringes observed by the experimental system will also be less than in Step 2.

Through this experiment, we can draw the following conclusion: when the Michelson-Morley experiment is stationary relative to the ground, no interference fringes are observed because the experimental system is stationary relative to the ground's empty space, or in other words, the ground is stationary relative to the surrounding empty space. This is due to the Earth's dragging effect on the surrounding empty space. When the experimental system starts moving relative to the ground, it also begins to move relative to the surrounding empty space.

Motion occurs, so interference fringes will be produced.
This result also shows that the historical use of this experiment to deny the existence of "ether" was incorrect.
(5) Improved Trouton-Noble Experiment
The Trouton-Noble experiment, like the Michelson-Morley experiment, was used to detect the Earth's motion relative to the ether, but both yielded negative results.

In fact, the academic community has always been in controversy over whether these two experiments proved the non-existence of ether as an objective reality, or merely showed that the relative motion between the Earth and ether is almost synchronized, making it undetectable.

Here we can design an experiment to test what causes the results of this experiment.
The experimental setup is similar to the previous one, except that the charges on the orbit are replaced with the experimental system consisting of two charges connected by an insulating rod, as in the Trouton-Noble experiment, to detect the rotation of the two charges when moving along the orbit with acceleration.

Step One:

Two charges are stationary on the orbit. At this moment, since the two charges are in relative static equilibrium with the empty space, they do not produce a magnetic field or additional magnetic force, and thus do not rotate.

At this moment, regardless of whether the observer is in relative motion or at rest with respect to the charge system, the observer can see the charge system as stationary, with no deflection and no magnetic field produced.

Step Two:

An experimental system composed of two charges moves at a constant speed along an orbit, at which moment it also produces a constant relative motion with the vacuum space, so a magnetic field will be generated around it, causing the charge system to experience a torque and thus rotate.

At this moment, regardless of whether the observer is in relative motion or at rest with respect to the charge system, the observer can detect this magnetic field and the rotation of the charge system.

The same effect occurs when the experimental system accelerates.

Step three:

The experimental system is placed in a closed box and moves at the same constant speed v in orbit. At this moment, because the box drags the internal empty space, the charge system remains stationary relative to the empty space, so no magnetic field is produced, and the two charges do not rotate.

If the box cannot completely drag the internal empty space, then according to its drag coefficient K, the relative velocity between the internal empty space and the charge system is $(1-K)v$, and the magnetic field produced by the two charges and the torque between them will be less than in step two.

At this moment, regardless of whether the observer is in relative motion or at rest with respect to the charge system, the observer can see this magnetic field and the rotation of the charge system.

Letting the box and the experimental system accelerate together also produces the same effect.
Through this experiment, we can conclude that in the Trouton-Noble experiment, no magnetic field or rotation was observed when the system was stationary relative to the ground, because the experimental system was stationary relative to the surrounding empty space, or equivalently, the ground was stationary relative to the surrounding empty space, which was due to the Earth's dragging effect on the surrounding empty space. When the experimental system starts moving relative to the ground, it also moves relative to the surrounding empty space, thus generating a magnetic field and causing rotation.

This result also indicates that the historical use of this experiment to deny the existence of the "aether" was incorrect.
It is important to note that the above experiments should take into account the effect of the box's material on its dragging ability of the internal empty space, as well as the changes in the corresponding dielectric constant and magnetic permeability parameters when the empty space is dragged.

The dragging ability of the box on the internal empty space is influenced by its material properties such as mass, density, refractive index, and transparency. The greater the mass, density, and refractive index, and the smaller the transparency, the stronger the dragging effect on the internal empty space. Conversely, the weaker the dragging effect. The dragging coefficient K is between $0-1$. When the dragging ability is stronger, the K value is larger. When complete dragging occurs, K equals 1, and when there is no dragging at all, K equals 0.

If the drag coefficient is insufficient due to the material of the box, the box can be filled with water or other media to enhance the drag capability on the empty space.

To avoid the influence of air flow on the experimental results, the entire system, including the box and charges, can be placed in a vacuum.

If the experimental precision is sufficient, the above experiment can be used to detect parameters such as the electromagnetic wave excitation threshold in empty space (see basic principles of spatial relativity 6, principle of spatial relativity).
(6) Experiments in Extreme Conditions
Considering the extremely low detectability of the void space and the dragging capability of mass bodies, when the dragging capability of a mass body is too weak to form a sufficient effect on the void space, the experimental results cannot be obtained. In other words, in the above experiments, if the dragging coefficient K of the closed box for the void space is approximately equal to 0, the dragging effect cannot be presented, and the desired experimental results cannot be achieved.

For example, assume an ideal spherical and uniform mass body, whose complete dragging distance for the void space is $1/1000$, and the limiting distance for partial dragging is 1% , that is, within a range of one-thousandth of the radius of the mass body from its surface, the void space can be completely dragged; within a range of one-hundredth of the radius of the mass body from its surface, the void space can be partially dragged; and beyond the range of one-hundredth of the radius, the dragging effect is almost negligible. (Taking the Earth's radius of 6300 kilometers as an example, the complete dragging distance of the Earth for the surrounding void space is 6.3 kilometers, and the limiting distance for partial dragging is 63 kilometers. This is just an assumption for illustration and does not consider the density effect. The actual dragging distance still needs to be verified through experiments.)

In this case, if the distance between molecules and atoms of the experimental container material is too large, exceeding the diameter of atoms or molecules, the container material's overall dragging effect on the internal empty space will be extremely weak. Additionally, since the velocity of the empty space is not less than the speed of light, once it leaves the surface dragging effect of the container material, the effect will quickly decay, making it almost undetectable and not affecting the experimental results.

Under this extreme condition, we can still obtain the experimental detection results through the first two steps of experiments four and five,

and make a judgment on the objective reality and relative motion of the empty space.
We can also make a determination through a simple control experiment.
Let electric charges and a magnetic needle that detects a magnetic field move in sync at a constant speed on an orbit, and see if the magnetic needle can detect a magnetic field. Then let both move in sync with acceleration, and see if an electromagnetic wave can be detected. If both a magnetic field and an electromagnetic wave can be detected, it indicates that both are related to the relative motion between electric charges and the vacuum space, and are independent of the observer's motion state.

Or see the following two experiments.
(7) Fluid Drag Experiment
Due to the extremely high permeability of the vacuum space superfluid, ordinary material containers may not be able to create sufficient flow of the vacuum superfluid inside, thus failing to produce relative motion between the vacuum superfluid and electric charges, which would prevent the observation of the experimental phenomenon.

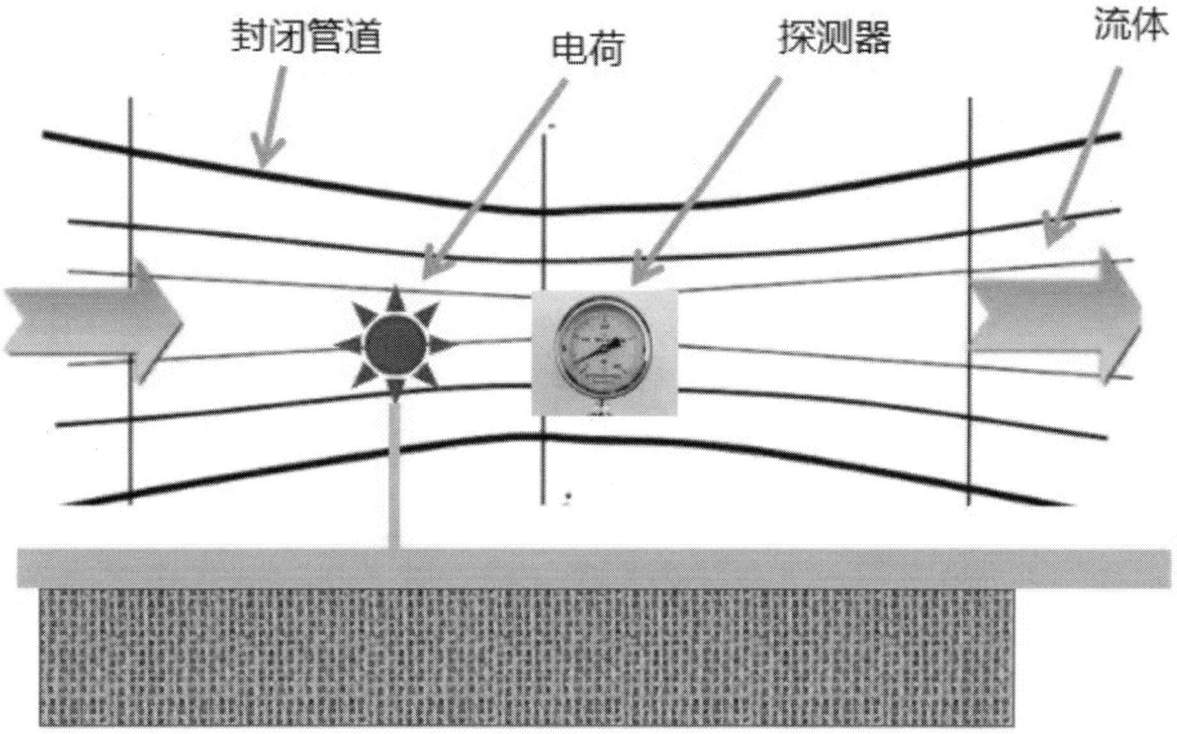

Figure 6: Figure 4-2 Fluid Drag Experiment Setup

In this case, we can utilize the drag effect of a fluid medium on the superfluid vacuum space to replace the box in the above experiment with a fluid, using the flow of the fluid to substitute the motion of the enclosed box dragging the vacuum space, and perform the experiment to detect the relative uniform motion or acceleration of charge with respect to the vacuum space

The corresponding magnetic field or electromagnetic wave effect, as shown in Figure 4-2.
The Fizeau fluid flow experiment has confirmed that flowing fluids can drag the surrounding empty space, so we can place a charge in a fluid, keeping the charge stationary while the fluid flows. Because the fluid drags the empty space, the empty space also appears to move relative to the charge, thereby exhibiting magnetic or electromagnetic wave effects.

The actual motion speed of the superfluid vacuum space dragged by a fluid, needs to be calculated according to the drag coefficient of the fluid. If the fluid drag coefficient is K, then the relationship between the speed of the vacuum space being dragged and the fluid speed is:

$$V1 = V * K$$

Where, V is the fluid flow speed, and $V1$ is the motion speed of the vacuum space generated by the fluid drag.
The vacuum space being dragged by the fluid moves relatively to stationary charges, thus being able to generate a magnetic field or electromagnetic waves.

The selection of liquid media can include water, oil, or other more suitable liquids according to experimental needs, which can form a better drag effect and avoid other interfering factors.
(8) Charge Oscillation and Rotation Experiment
In order to verify that electromagnetic waves are generated by the relative motion between charges and empty space, the following two experiments can be used for verification.

The charge oscillation producing electromagnetic waves experiment is shown in Figure 4-3(a). Detector A is fixed together with the charge on the oscillation table, while Detector B is fixed on the workbench. When the oscillation table moves back and forth on the rail, the charge will generate an electromagnetic wave. At this time, observe the electromagnetic wave detection data from Detector A and Detector B.

According to the theory of spatial relativity, if detector B can detect electromagnetic waves, then detector A can also detect electromagnetic waves. That is, whether a detector can detect electromagnetic waves is unrelated to whether it is stationary relative to the oscillating charge, but instead depends on whether there are electromagnetic waves in the space. When the charge oscillates and produces electromagnetic waves,

When a charge is at rest or in motion relative to the detector, the electromagnetic wave can be detected.

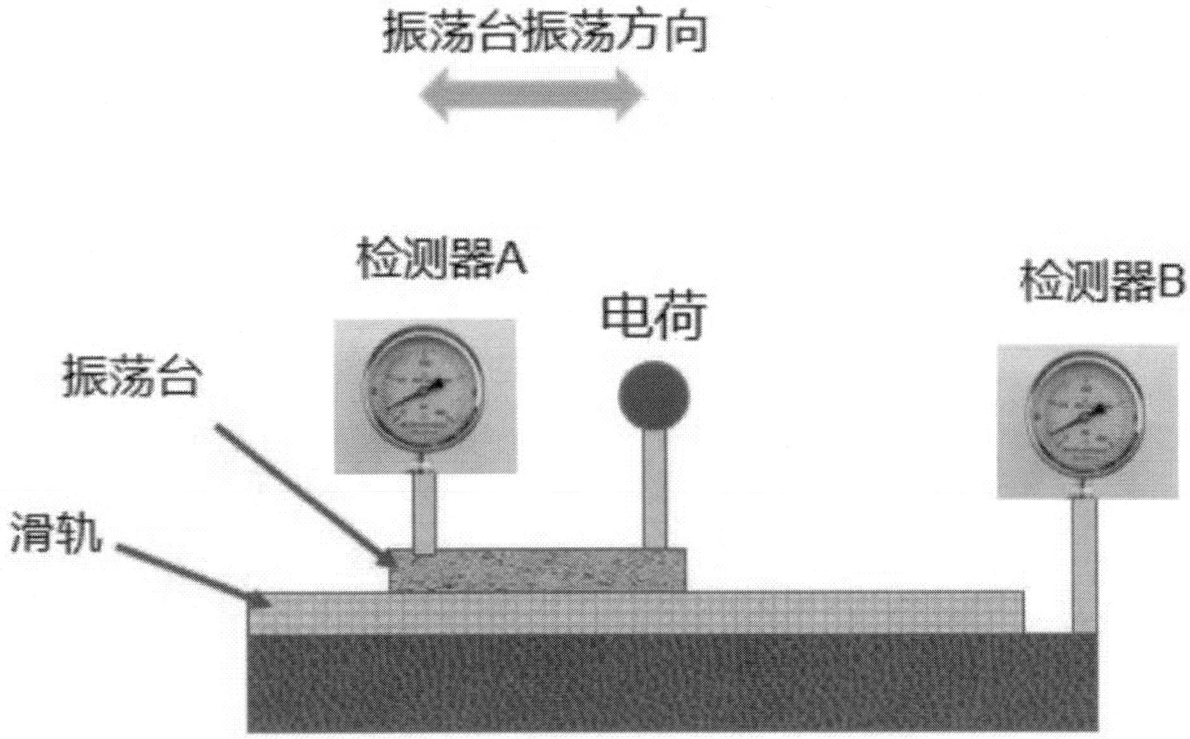

Figure 7: a) Charge oscillation generates electromagnetic waves for detection

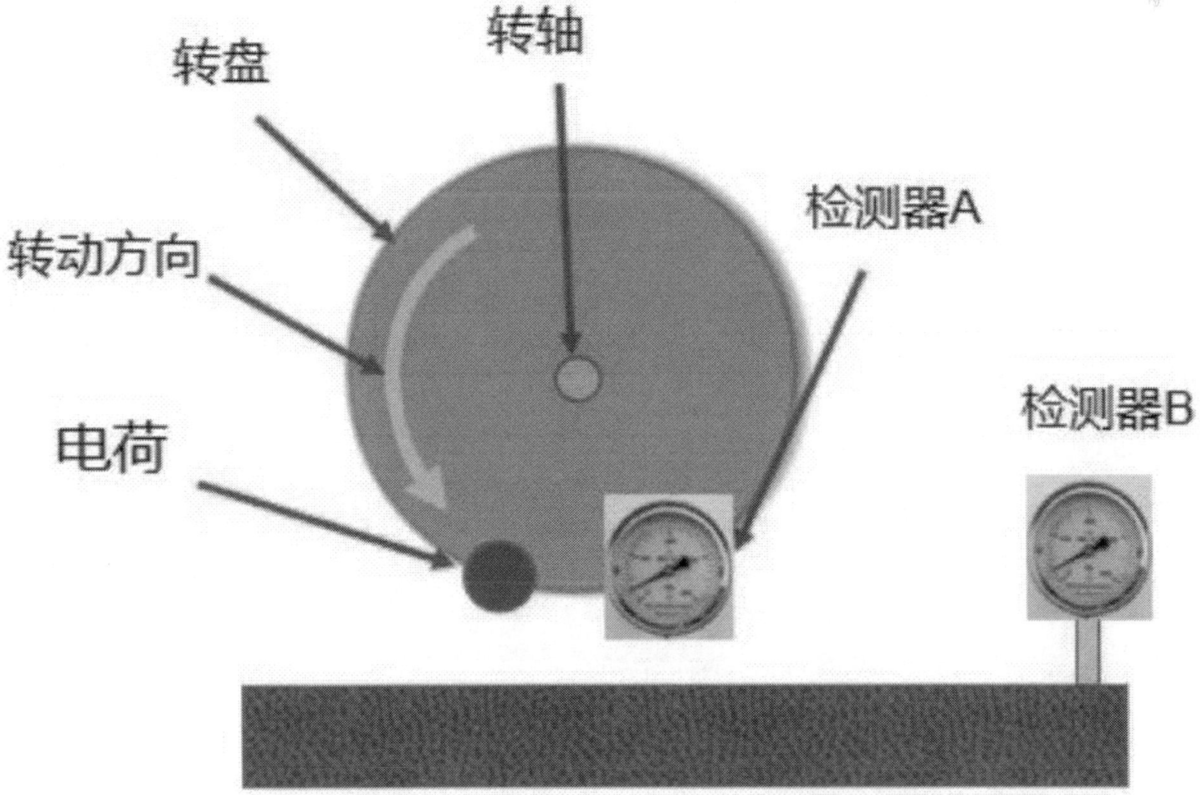

Figure 8: b) Charge rotation produces electromagnetic wave detection

Figure 4-3 Observation experiment of electromagnetic wave generation by relative motion of charges

The experiment showing that a rotating charge produces electromagnetic waves is illustrated in Figure 4-3(b). Detector A is fixed together with the charge at the edge of the turntable and rotates with it, while Detector B is fixed on the workbench. When the turntable rotates at high speed, the charge undergoes accelerated motion and thus emits electromagnetic waves. The direction of these waves will be within a slight angle along the tangent of the turntable edge where the charge is located. Therefore, by precisely adjusting the positions of the two detectors, one can detect this electromagnetic wave. That is, when stationary detector B can detect the electromagnetic wave emitted by the rotating charge, the detector A, which rotates together with the charge and is relatively stationary with respect to it, should also be able to detect the electromagnetic wave.

It is the accelerated motion of the charge relative to the surrounding space that generates the electromagnetic wave, not relative to the observer. When the charge accelerates relative to the surrounding space and emits electromagnetic waves, any detector or observer, whether stationary or moving relative to the charge, can observe these waves, provided their relative velocity with respect to the charge does not exceed the speed of light, i.e., the speed of electromagnetic wave propagation.

The experimental setup shown in Figure 4-3(a) can also be adjusted so that the charge (oscillating platform) moves at a constant speed, and Detector A is aligned parallel to the charge. This setup is used to detect the magnetic field produced by the motion of the charge, verifying the view of space relativity: that the magnetic field generated by a moving charge is due to its motion relative to the surrounding space, not relative to the observer. When the charge moves at a constant speed relative to the surrounding space, even if the observer is stationary relative to the charge, the magnetic field can still be observed.

Relative motion to the observer does not produce the magnetic field; it is the motion of the charge relative to the surrounding space that does.

The above two experiments can be conducted in a vacuum environment, reducing the influence of air movement.

4.2 Realizing the Principle of Superluminal Speed

There are two ways to accelerate an object: one is external force, and the other is internal force.
External force is applied by other substances outside the object through pursuit or collision, thereby achieving acceleration, such as in a particle accelerator.

Intrinsic force is a part of the object's internal matter that exerts a forward accelerating force by moving backward in reverse motion, such as in a rocket or space ship powered by a fuel engine.

(1) Extrinsic force acceleration

Taking a particle accelerator as an example, the acceleration effect of extrinsic force is limited by the speed limit of the accelerating force itself. As the speed of the accelerated particles approaches the upper limit of the accelerating force, the effective accelerating force becomes smaller. As the particles approach the speed of the accelerating force infinitely, the effective accelerating force will also become infinitely small.

This will result in a speed limit for the accelerated particles, which is the speed limit of the accelerating force (in the case where the accelerating force carries particles, such as photons, with a mass less than the mass of the accelerated particles).

We can illustrate this with a hypothetical experiment.

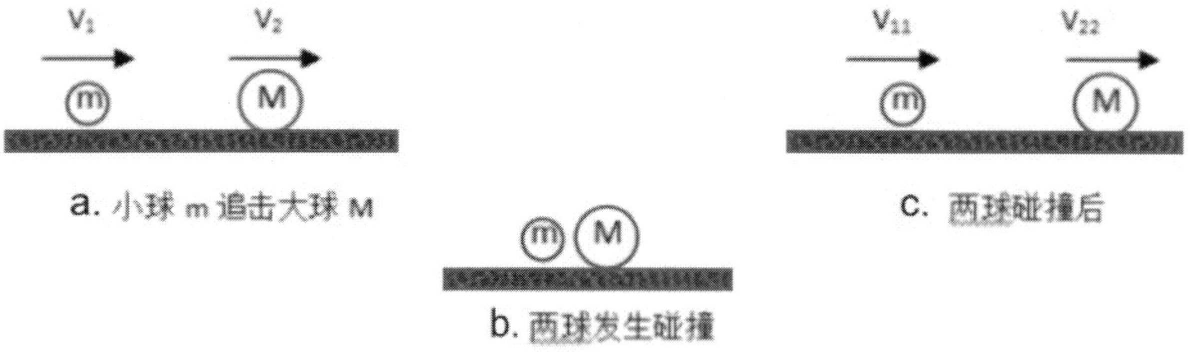

Figure 9: Figure 4-4 Two-ball collision acceleration experiment

We consider an infinite number of small balls accelerating the large ball by colliding with it sequentially.
Let the mass of the small ball be m, and the mass of the large ball be M. The mass of the small ball is much smaller than that of the large ball. The small ball and the large ball move in the same direction with velocities V_1 and V_2, respectively, and the velocity of the small ball is greater than that of the large ball, i.e., $V_1 > V_2$. In this case, the small ball m will catch up and collide with the large ball M.

By the principle of action and reaction, the large ball M can be accelerated, as shown in Figure 4 - 4.
We can simply calculate it. Assuming an ideal situation where the collision follows both the conservation of momentum and energy principles, and after the collision the small ball m and the large ball M move with velocities V_{11} and V_{22}, respectively, the increase in the velocity of the large ball after the collision $\triangle V2$ is:

$$\Delta V_2 = V_{22} - V_2 = \frac{2mV_1 - mV_2 + MV_2}{m + M} = \frac{2\,m\,(V_1 - V_2)}{m + M}$$

The masses of the large ball and the small ball M, m are constant, so the increase in the velocity of the large ball after the collision depends only on the relative velocity between the two balls $(V_1 - V_2)$.

When we use multiple identical small balls to accelerate a large ball through successive collisions, the relative velocity between the large ball and the small balls (V_1 - V_2) decreases with each collision, resulting in smaller increments of velocity gained by the large ball each time.

In other words, in this classical collision acceleration model, the maximum velocity limit of the large ball is the velocity of the small ball used to accelerate it, V_1.

Over time, it will approach the velocity of the small ball infinitely closely but will never reach this limit, as shown in Figure 4-5.

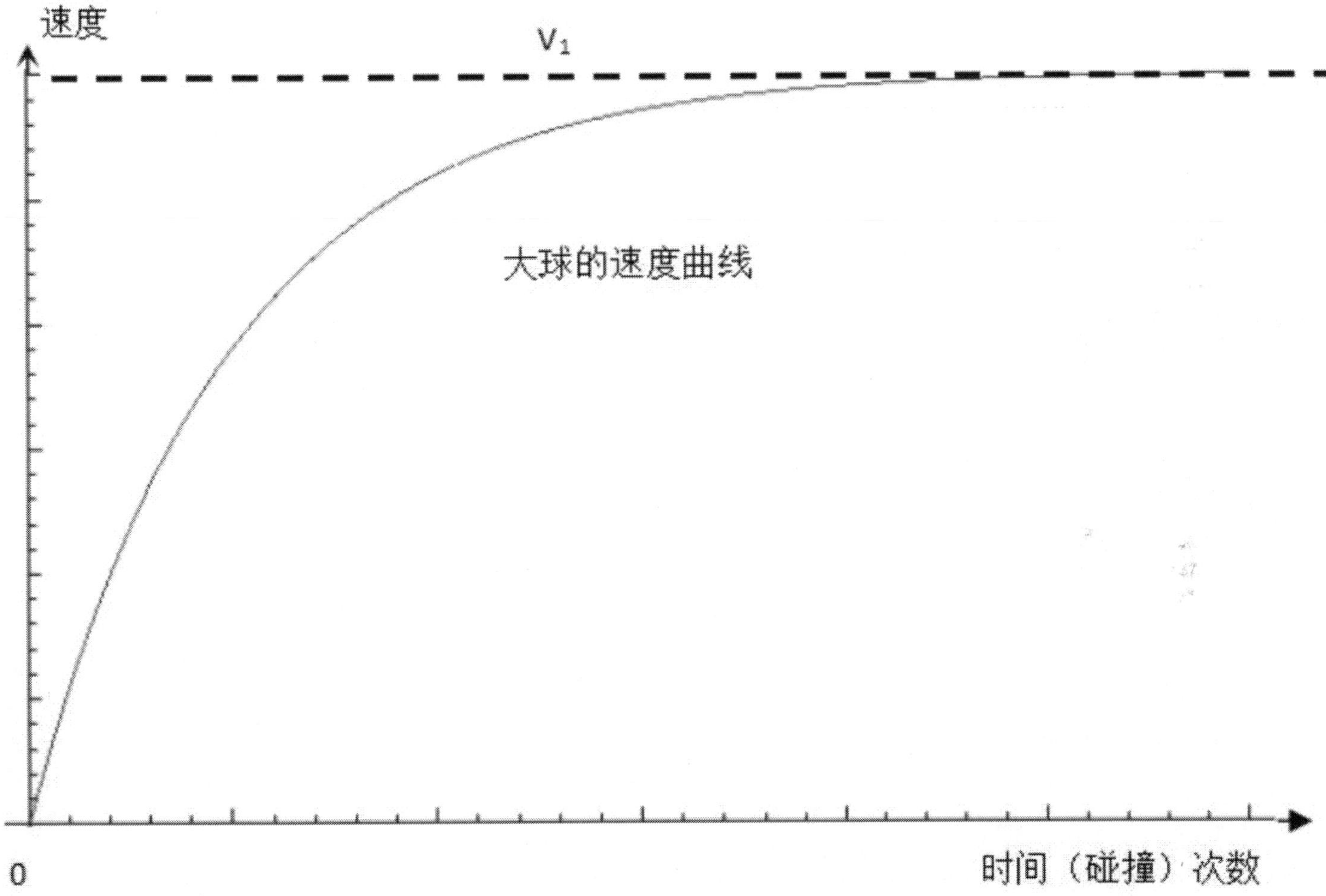

Figure 4-5 Velocity curve of the large ball under continuous collision acceleration by multiple small balls

This situation can also be interpreted as follows: as the large ball M approaches the small ball, the acceleration force F applied by the small ball through collision becomes smaller, resulting in the acceleration of the large ball approaching zero infinitely, as shown in Figure 4-6.

$$F = Ma = M\frac{\triangle V_2}{\triangle t} = \frac{2mM\,(V_1 - V_2)}{(m + M)\triangle t}$$

Because the small ball is increasingly unable to catch up with the large ball, it becomes more difficult to apply an accelerating force to the large ball through collisions.

Therefore, the ultimate speed limit of the large ball accelerated by the continuous impact of countless small balls is the speed of the small balls, which cannot be exceeded.

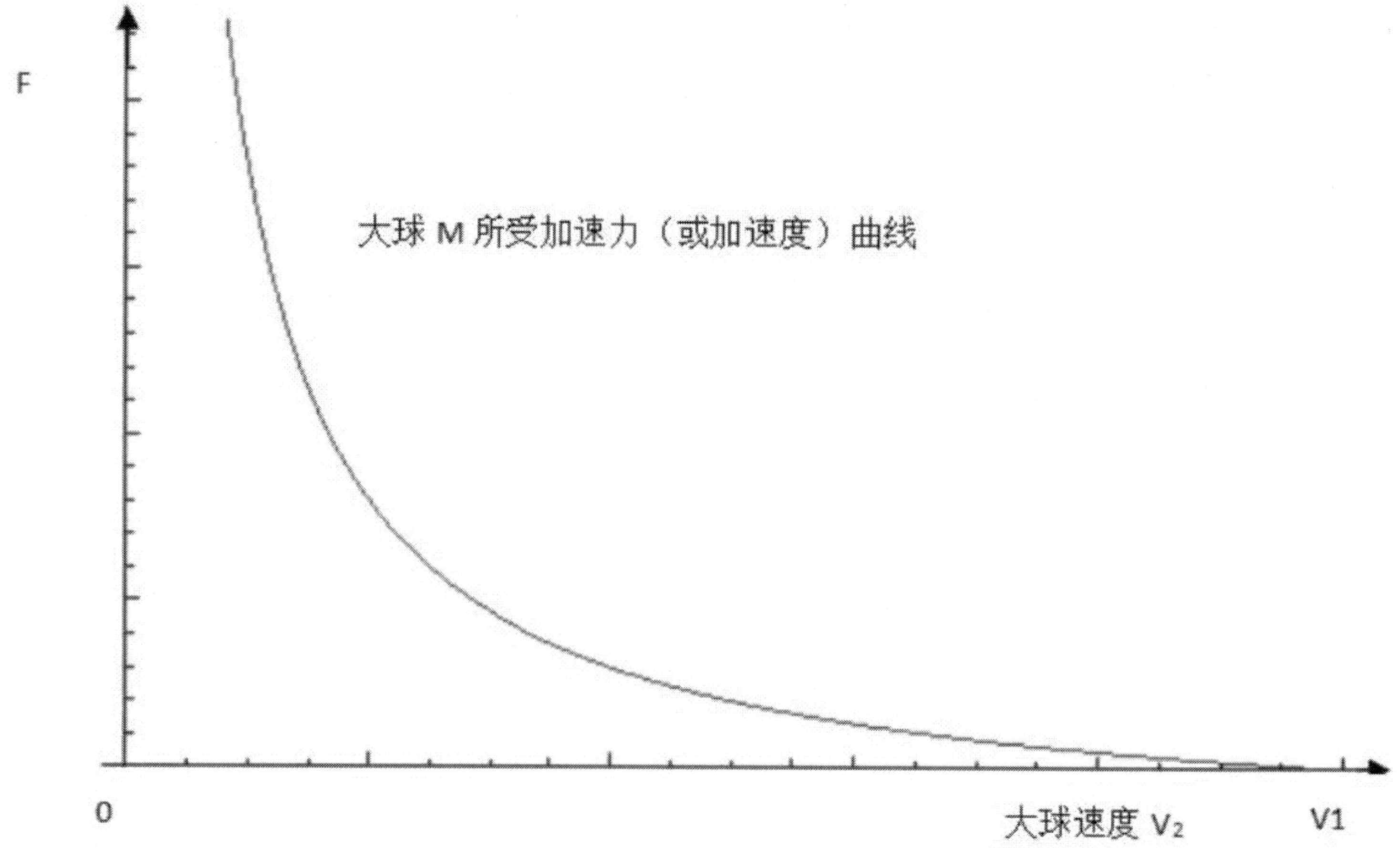

Figure 4-6: Curve showing the change in acceleration force experienced by the small ball as the large ball's speed increases

When we use a super accelerator to speed up particles, the principle is fundamentally similar. The electromagnetic force used to accelerate particles has a speed limit of light, so as the speed of the accelerated particles approaches the speed of light, it gradually approaches the speed limit of the electromagnetic force, resulting in smaller and smaller acceleration, approaching zero.

Therefore, in this mode, the speed of the accelerated particles can only approach the speed of light infinitely, but cannot reach or exceed it.

There are two approaches to accelerate particles to exceed the speed of light using an external force mode.
One method is to change the mass ratio between the accelerating particle and the accelerated particle, making the mass of the accelerating particle much greater than that of the accelerated particle. In this case, the accelerating particle collides with the accelerated particle at a speed close to the speed of light. Through the law of conservation of momentum, the accelerated particle can be instantly accelerated to exceed the speed of light.

However, due to the light burst effect when exceeding the speed of light, the duration of superluminal motion may be very short, but it can still be observed through phenomena such as superluminal glow.

The second method is through the relative motion of the vacuum space. In the A region of vacuum space, particles are accelerated to near-light speed, and if this A region of vacuum space has a relative motion speed with respect to the B region of vacuum space, then in the B region's vacuum space, it may be observed that the particles' speed has exceeded the speed of light.

In this case, there would be no observation of a similar luminous phenomenon like light burst, because the accelerated particles, relative to the A region of vacuum space they are in, have not exceeded the speed of light.

The main difficulty of this mode lies in how to form a region of relative moving vacuum space through effects such as the dragging of mass bodies.

In addition, the speed limit of acceleration force only affects one aspect of the problem of objects exceeding the speed of light; there are other influencing factors that we will discuss in more detail in the overview of spatial wave dynamics.

(2) Intrinsic Acceleration

Take a spaceship accelerated by a fuel engine as an example.
A spaceship accelerated by a fuel engine, in this case, although the fuel engine accelerates the spaceship essentially through electromagnetic forces, since the fuel is integrated with the spaceship and moves in sync, the fuel remains stationary relative to the spaceship even as the spaceship's speed increases to near the speed of light.

Assuming the spaceship is in a completely dragged state internally, the acceleration force from burning fuel backward, which is the electromagnetic force, remains at the speed of light relative to the spaceship. Therefore, the acceleration force does not weaken as the spaceship's speed increases, allowing for continuous effective acceleration until it exceeds the speed of light.

The challenge with this method lies in the efficiency of the engine and the amount of fuel it can carry. It also requires considering the effects of gamma radiation when approaching or exceeding the speed of light, and whether this radiation weakens the spaceship's dragging effect on the internal vacuum space, thereby affecting the acceleration performance.

(3) Experimental Verification

Now that particle accelerators cannot accelerate particles to the speed of light, it is due to the relativistic mass increase effect when approaching the speed of light, or because the speed limit of the accelerating force, that is, electromagnetic force, is the speed of light. The two would result in significantly different growth curves for the particle's speed, which can be detected through experiments.

4.3 Interpretation of the Michelson-Morley Experiment

(1) Explanation of the Michelson-Morley Experiment Results

Historically, the Michelson-Morley experiment was designed to detect the existence of the ether and to measure the Earth's motion

The velocity of motion relative to the ether. However, after numerous experiments, they never detected any difference in the speed of light in different directions.

This experimental result once caused great confusion, and its explanation requires considering the nature of light and empty space, which can be explained by the basic principle 2 (principle of spatial objectivity) and basic principle 4 (new principle of light speed invariance) of spatial relativity.

The speed of light is constant in a uniform flat vacuum space, and vacuum space can be dragged by the motion of mass objects. That is, the vacuum space around the Earth's surface is dragged and moves together with the Earth due to its motion, thus achieving a relative state of rest with the Earth's surface.

In this case, the speed of light moving in this empty space would be the same in all directions, with no differences. This is also the reason why the Michelson-Morley experiment did not detect the anisotropy of light speed.

Historically, Stokes' complete drag theory, developed on the basis of Fresnel's partial drag theory, is actually similar to this, and could explain the experimental results, as shown in Figure 4-7. However, in the end, Einstein's special relativity became the mainstream explanation, negating the ether and drag theories.

However, in the development of general relativity, Einstein once again restored and acknowledged the ether, and general relativity had to admit the dragging effect of space, which is essentially the Stokes drag theory explanation. This is the same as our fundamental principle 2 (Principle of Space Objectivity) and fundamental principle 4 (New Constancy of Light Speed Principle), which involve the relativity of light speed, the motion of empty space, and the dragging effect.

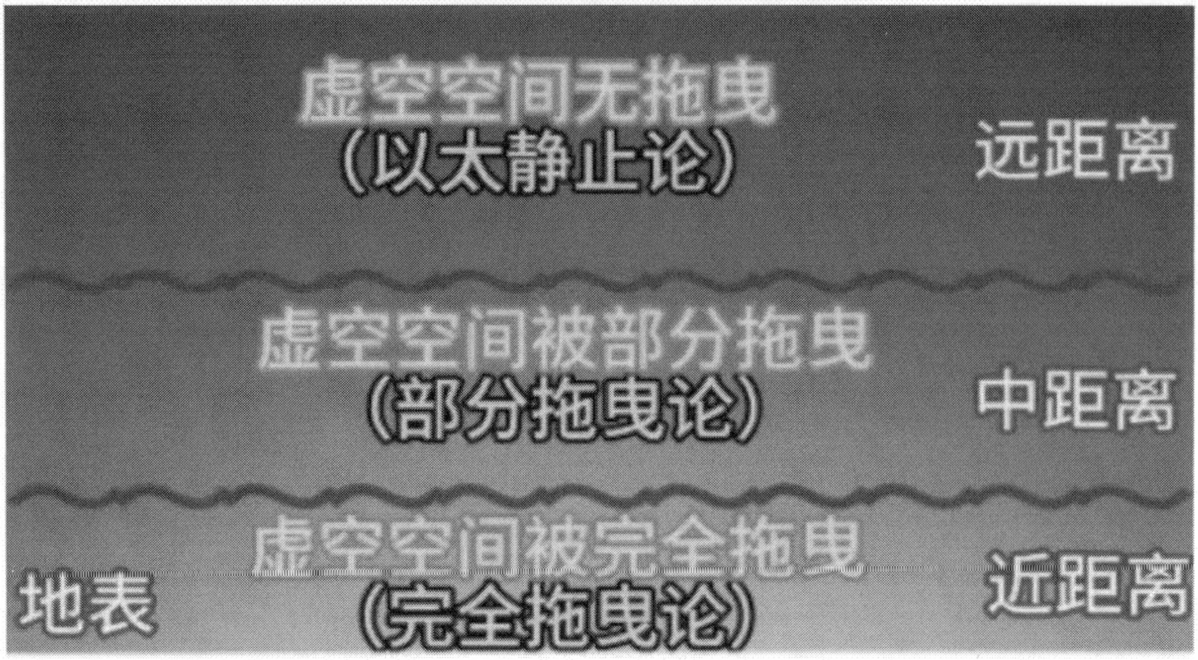

Figure 10: Figure 4-7 Stokes' Complete Drag Theory

(2) Improvements to the Michelson-Morley Experiment

Due to the dragging effect of Earth, the empty space at the surface is relatively stationary, so the Michelson-Morley experiment could not detect the anisotropy of the speed of light. For this, there are two possible improvements:

The first is to make the experimental system move itself, thereby creating a relative motion with respect to the empty space. At this time, observing the anisotropy of the speed of light should theoretically be easier, thus verifying the objective reality of empty space and confirming the relative motion between the experimental system and empty space, which is known as the "aether wind."

For example, placing the experimental system on a car and moving it together, to see if interference fringes can be detected, confirming the car's motion relative to empty space, as shown in Figure 4-8.

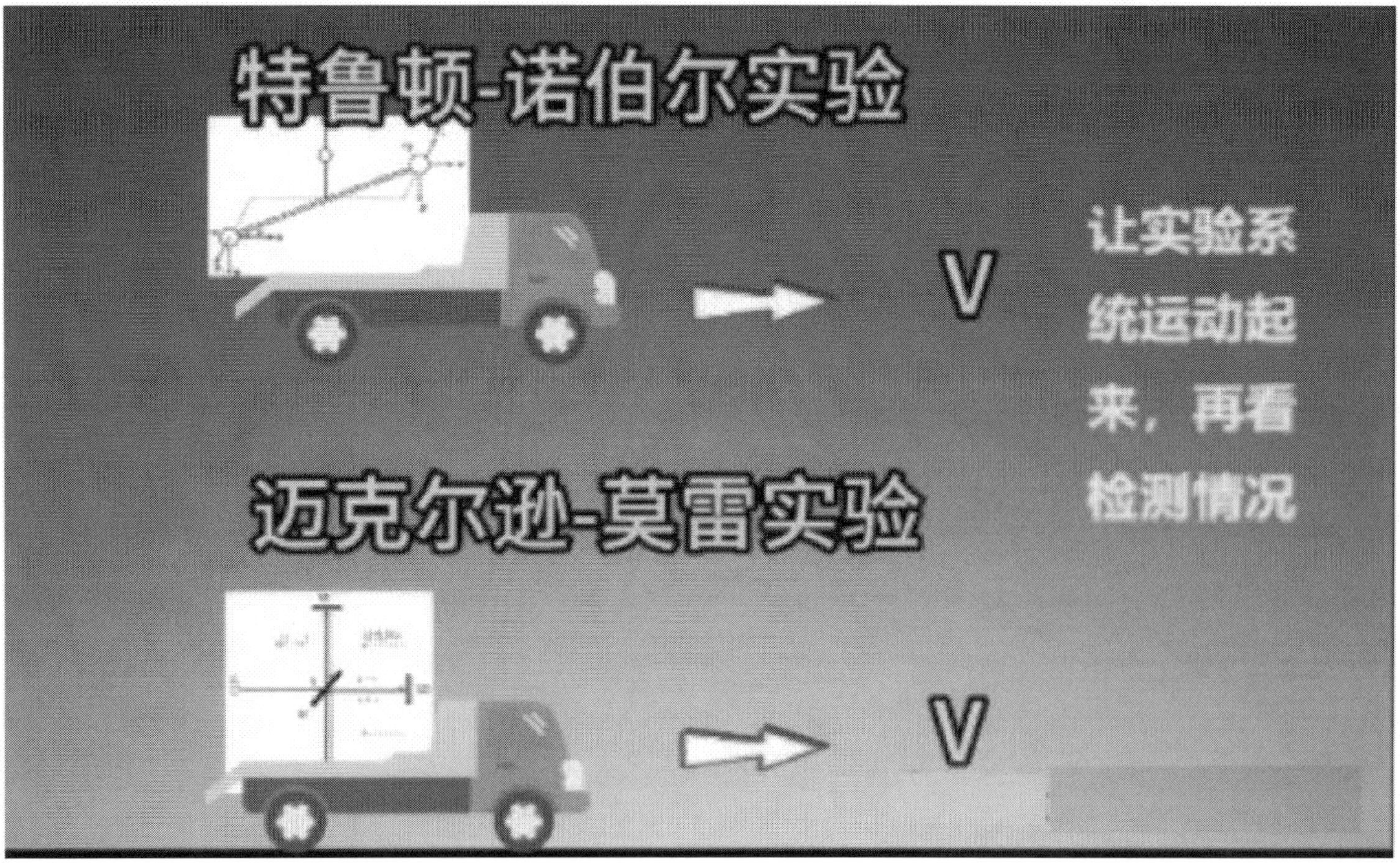

Figure 4-8 Improved Truitt-Noble Experiment and Michelson-Morley Experiment

The second approach is to place the experimental system far away from the Earth's surface, such as on a satellite or space station in space hundreds of kilometers above the Earth. In this region, the distance from the Earth's surface is sufficient to greatly reduce the dragging effect. As the experimental system moves with the satellite or space station, it will inevitably have relative motion with the surrounding empty space. At this time, the interference fringes of the experimental system can be observed to confirm the relative motion of the satellite or space station with respect to the empty space.

It is important to note that during the experiment, in addition to considering the dragging effect of the Earth on the vacuum space, the dragging and influence of the experimental system, equipment, and all surrounding objects on the vacuum space in the test area should also be taken into account, to ensure experimental accuracy. For example, regardless of whether the experimental system is placed on a vehicle on the ground or on a satellite in space,

On the space station, it is necessary to ensure that the exposed areas in outer space are protected from the drag effect caused by cars, satellites, and space stations on the internal vacuum space, to avoid affecting the experimental results.

Other experiments, such as the Trouton-Noble experiment, can also be explained by the same principle. The experimental system can be improved to allow it to be both at rest and in motion relative to the vacuum space, thereby verifying the effect of "ether wind" caused by movement through the vacuum space.

Of course, all of this assumes that when the experimental system moves relative to the vacuum space (or the so-called ether), there would be observable results. Otherwise, if the system does not produce the expected results when moving relative to the vacuum space, it would be inappropriate to expect these experiments to detect the Earth's motion relative to the vacuum space.

4.4 Explanation of Faraday's Paradox

In 1831, the physicist Michael Faraday deduced Faraday's Law of Induction, but he encountered many difficulties when applying this law to explain Faraday's Paradox, as shown in Figure 4-9.

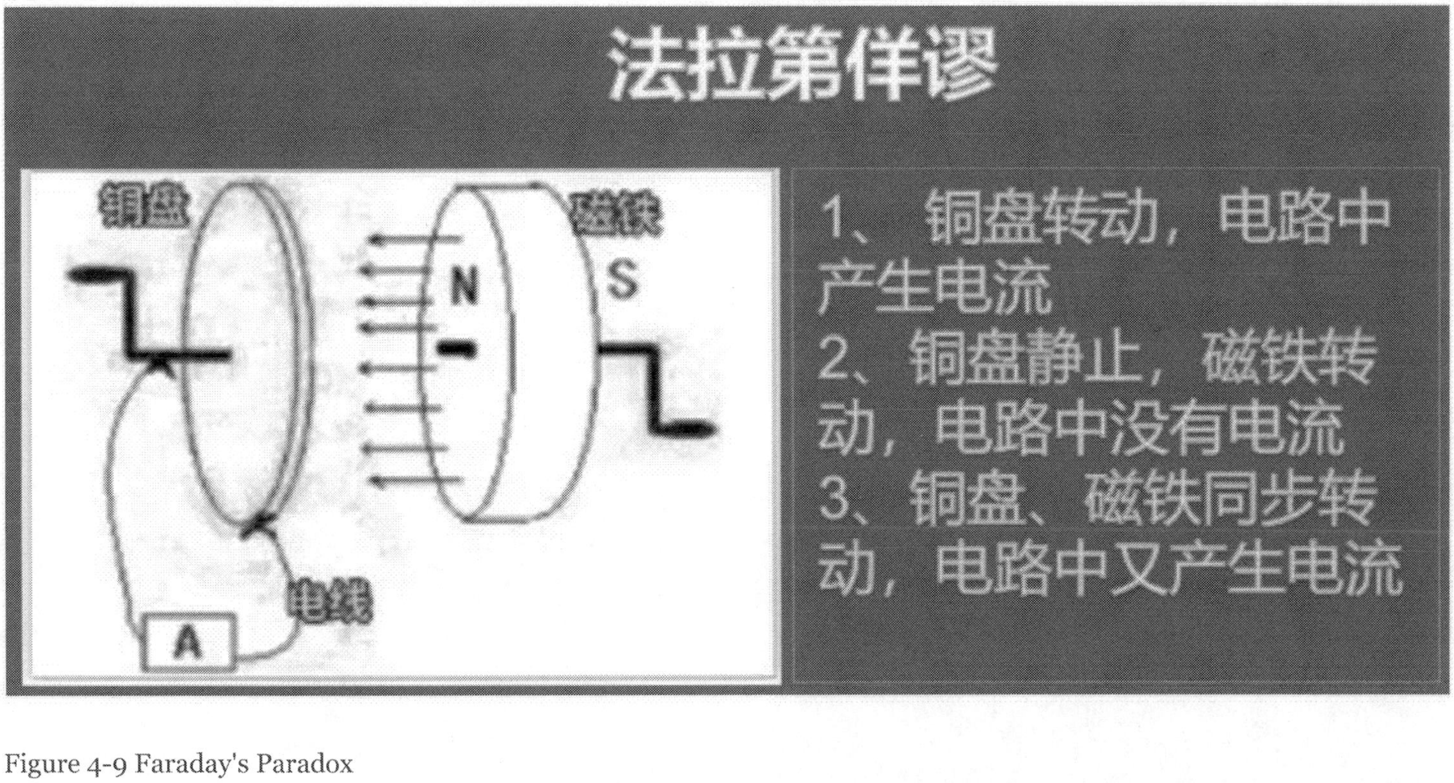

Figure 4-9 Faraday's Paradox

We know that Faraday's law of electromagnetic induction states that a conductor placed in a changing magnetic flux will generate an electromotive force, that is, an induced electromotive force. The magnitude of the induced electromotive force is proportional to the rate of change of the magnetic flux passing through this circuit.

We assume there is a copper disc connected by wires from its edge to its central axis, forming a closed circuit, and a magnet is placed opposite to the disc.

When the copper disc rotates, the magnetic flux within its area does not change, yet an electromotive force is generated and current flows in the circuit. However, when the copper disc is stationary and the magnet rotates around the axis of symmetry, no current is generated in the circuit.

The magnet rotates and the copper disk rotates, yet the results are different.
In fact, this question can be understood according to the basic principle 5 (New Relativity Principle) and basic principle 6 (Space Relativity Principle) of spatial relativity.

The transmission of electric and magnetic fields, and the motion of electromagnetic waves, are relative to the vacuum space, not to any other observer or moving object. The motion of electromagnetic waves is also independent of the movement of the source.

Therefore, in this experiment, the magnetic field emitted by the magnet moves relative to this vacuum space and is independent of the magnet itself. Thus, the rotation of the magnet does not affect the already emitted magnetic field and will not change the magnetic field around the copper disc. Of course, considering the dragging effect of mass on vacuum space, the emitted magnetic field will move in sync with the dragged vacuum space within that range.

However, in this experiment, the magnet's mass is very small, its dragging ability is very limited, and its effect on the movement of the magnetic field is also very subtle and can be neglected. Therefore, the rotation of the magnet does not cause a noticeable change in the already emitted magnetic field. At the same time, due to the symmetry of the magnet's own magnetic field, the newly emitted magnetic field is almost identical to the previous one, with an almost zero time interval between them, making it appear continuous and consistent. Thus, the rotation of the magnet does not result in the rotation of the magnetic field, and consequently, there is no relative motion between the magnetic field and the copper disc, so the copper disc does not cut through the magnetic field lines, and therefore, no current is generated in the copper disc.

For example, consider an electric fan swaying while blowing air. This is not the fan carrying the air with it as it rotates, but rather the old air disappearing and new air being blown out during the swaying motion.

The rotation of the copper disc will cut through the magnetic field, thus generating an electric current. There is no contradiction between the two experimental results.

This is consistent with the N theory in the modern academic interpretation that "the magnetic field does not rotate with the magnet," although the N theory cannot explain the reason for this phenomenon. The actual reason can be understood according to the basic principle 5 (new relativity principle) and basic principle 6 (space relativity principle) of spatial relativity.

4.5 The Limitations of Lorentz Transformation and Its Complete Form

The origin and significance of the Lorentz transformation have been analyzed in the previous chapters. We can illustrate this with a hypothetical experiment.

For example, suppose we have a spaceship moving at a uniform speed V relative to Earth. We assume that the spaceship completely drags its internal empty space, meaning the empty space inside the spaceship is stationary relative to the spaceship. In this case, the speed of light observed by the observer inside the spaceship is C, and light is also vertically reflected between two mirrors, as shown in Figure 4-10.

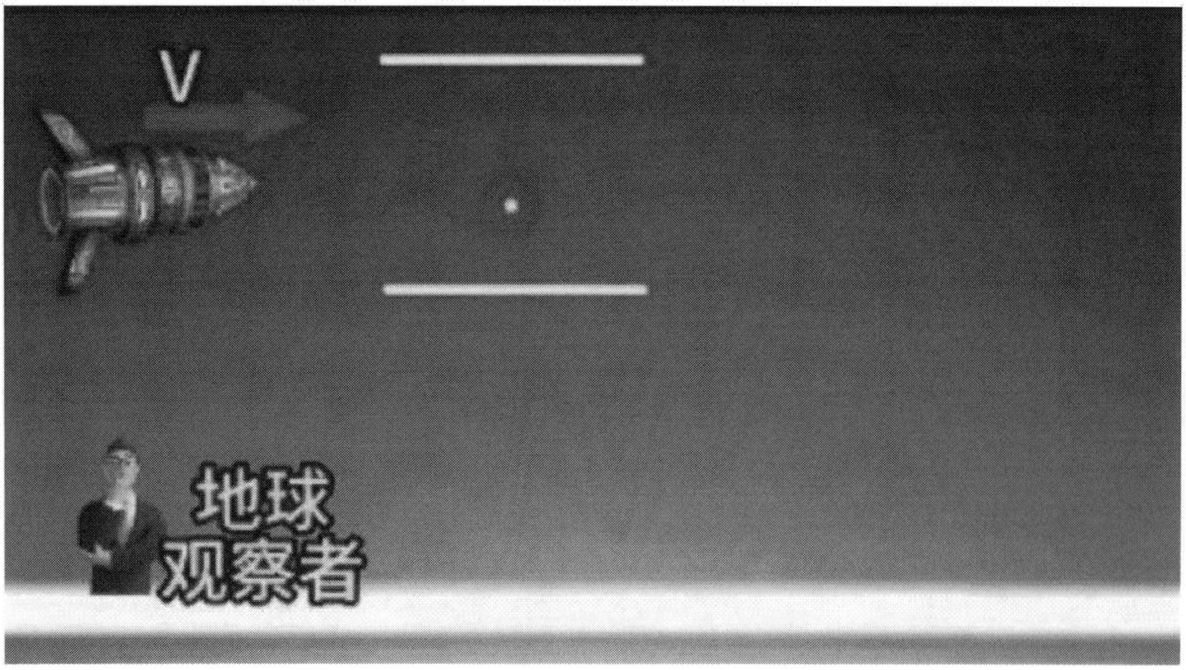

Figure 11: Figure 4 - 10 Light pulse vertical propagation inside the spaceship

However, to a stationary observer on Earth, the path of light appears as a slanted line, no longer a vertical motion, as shown in Figure 4-11.

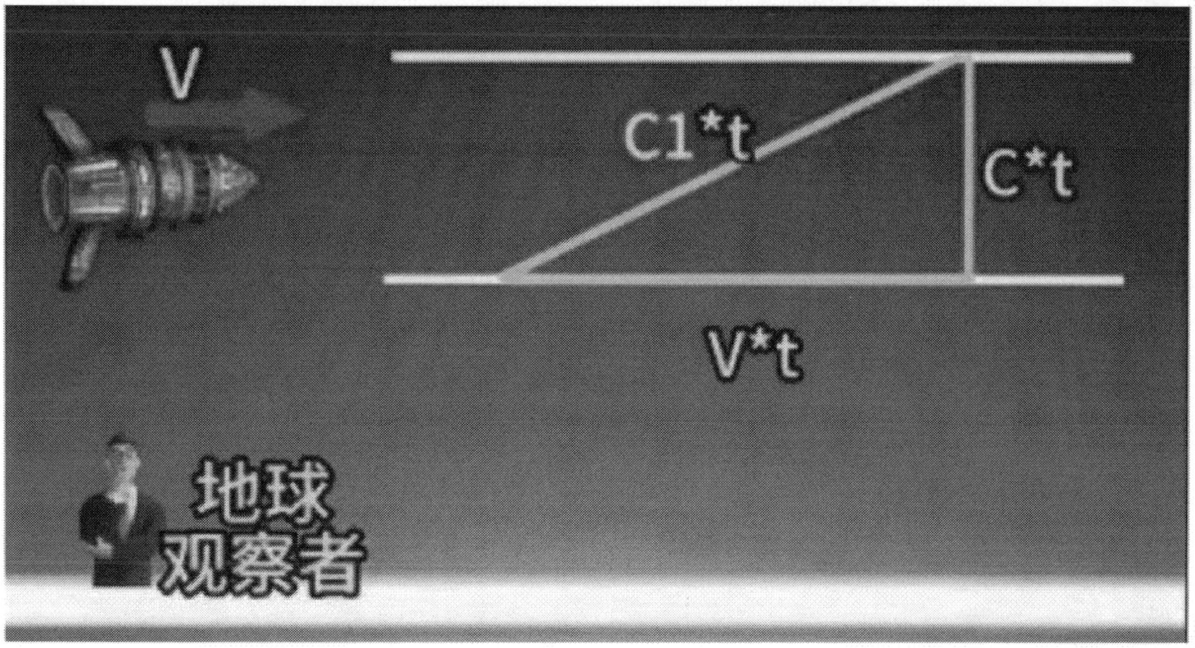

Figure 12: Light pulse propagation direction observed by a stationary observer as shown in Figure 4-11

For an external observer, the true speed of light, which is the actual speed of light, can be obtained through the synthesis of velocity V and C using our daily Galilean transformation

$C_1 = \sqrt{C^2 + V^2}$

But if we assume the speed of light is constant here, that is, force C 1 to be equal to C, then this mathematical calculation system will immediately lead to a contradiction, making the Pythagorean theorem invalid.

To solve this problem, the other two variables, length and time, must change. For example, time along the direction of the spaceship's motion, t_1 , can change, that is, $(c^*t)^2 + (v^*t_1)^2 = (c^*t_1)^2$,

The calculation can be obtained: $t_1 = t/\sqrt{1-(v/c)^2}$

This is the time transformation in special relativity, where $1/\sqrt{1-(v/c)^2}$ is the Lorentz factor, and this transformation is known as the Lorentz transformation, as shown in Figure 4-12.

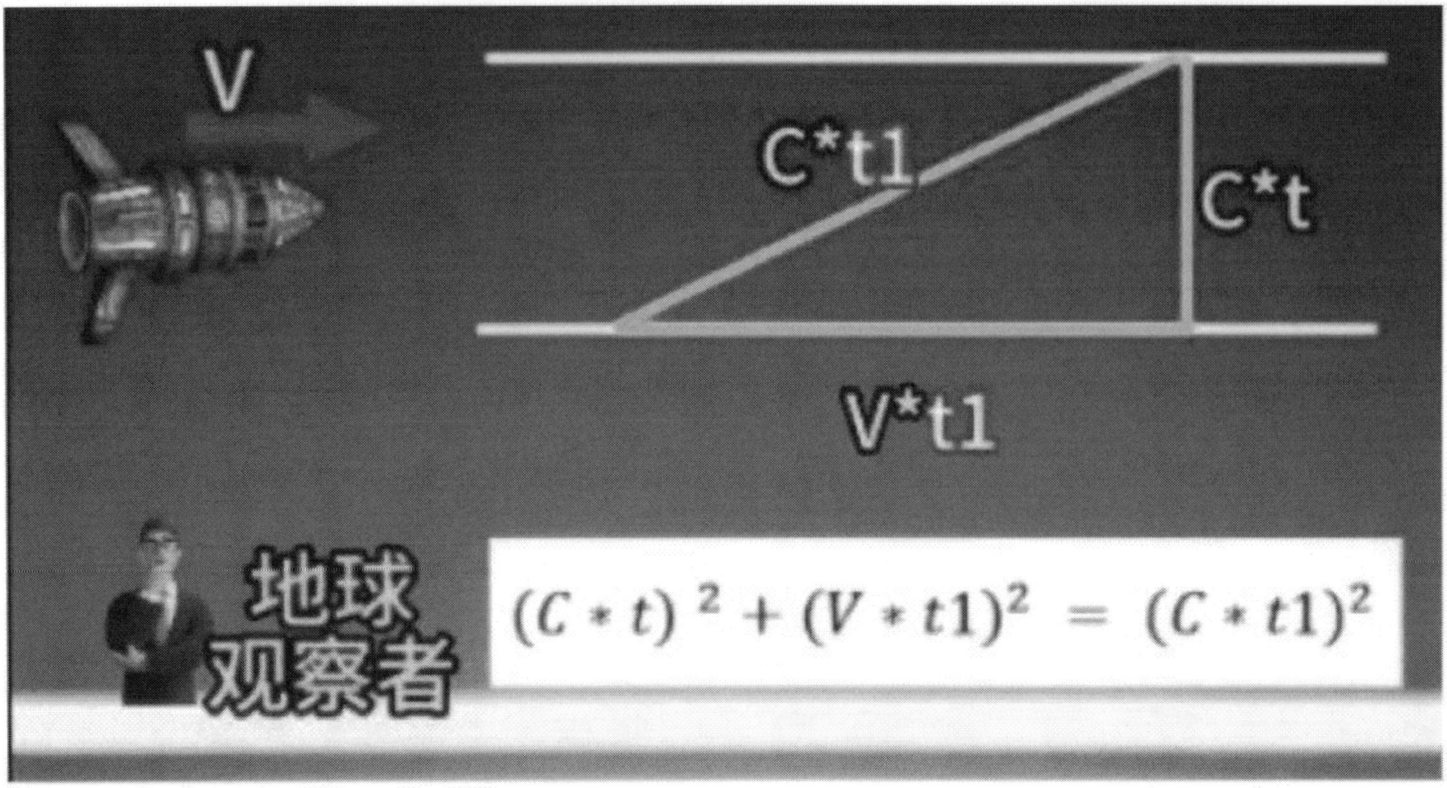

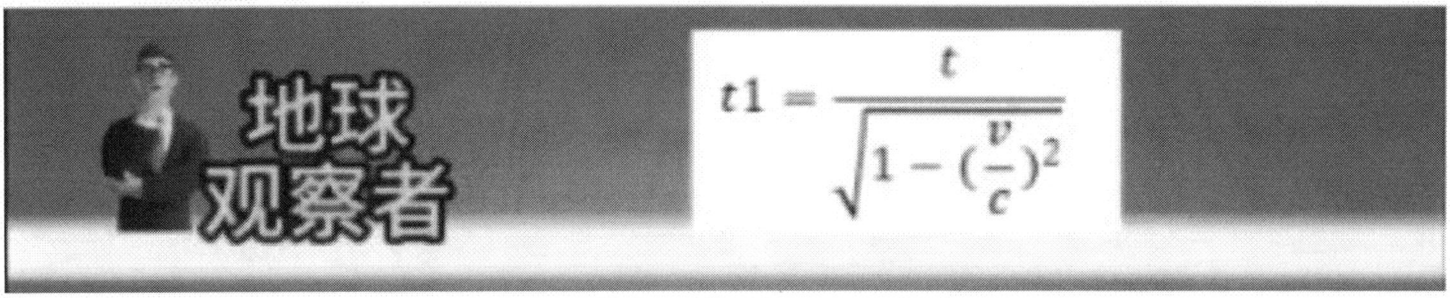

Figure 4-12 Time observed by a stationary observer under Lorentz transformation

In this transformation, t_1 is greater than t, the time experienced by the ground observer is longer than the time experienced by the observer on the spaceship, making it seem as if the spaceship's observer's time is slowing down. This is the phenomenon of time dilation in motion.

Based on this, we can further deduce the effects of length contraction, mass increase, etc., for moving objects (the length contraction effect in relativity is incorrect, which leads to incompatibility with gravitational formulas and eventually forces the development of general relativity, moving further away from the original theory; see section 2.1 "Misuse of the Principle of the Constancy of the Speed of Light" for details).

Therefore, this is the reason for the appearance of the Lorentz transformation in special relativity: in order to keep the speed of light C constant, time t must change, and consequently, length 1 and mass m must also change, so that this mathematical framework remains consistent.

The issue here is that, as we analyzed earlier, the speed of light is constant at C relative to the vacuum space it is in, not relative to all observers. This is a misunderstanding and misapplication of the principle of the constancy of the speed of light in special relativity.

The fundamental reason lies in that when dealing with motion involving light, electricity, magnetism, and even gravity, one must consider the background vacuum space reference frame, determine the true "absolute" motion velocity of different objects relative to the vacuum space, and clarify whether one is moving, the other is stationary, or both are moving to some extent, rather than arbitrarily assuming the motion status of both.

Motion is absolutely relative and can be arbitrarily assumed, which is a misunderstanding of the principle of relativity and also the root cause of many paradoxes in special relativity.

It can be said that the principle of the constancy of the speed of light and Lorentz transformation in special relativity is merely a mathematical calculation system, which distorts the physical essence and is not the actual physical reality. However, as a self-consistent mathematical framework, it can also be an effective tool in our hypothetical experiments. When we do not consider or fail to properly understand the role and impact of vacuum space in relative motion, or are unable to handle the errors it causes, using this mathematical system can appropriately address the problem.

This is also the reason why special relativity can effectively function in fields such as GPS navigation accuracy correction and quantum field theory.

However, we should also note the conditions under which the relativistic effects take effect, as we have imagined

This situation is an experiment conducted in a specific, ideal, uniform, and flat empty space, and the mass body completely drags the surrounding empty space.
(1) Empty space is uneven and different regions have relative motion; our above hypothetical experiment assumes that both parties in relative motion are in the same region of empty space, and this empty space is uniform and flat.
(2) The empty space can be distorted and dragged by massive objects. In our hypothetical experiment, the movement of the spaceship is set to completely drag the empty space inside it. Since light propagates at a constant speed relative to the empty space it is in, it is only in this scenario that light can be observed to bounce vertically off a mirror inside the spaceship.
(3) The direction of light propagation inside a moving spaceship is perpendicular to the direction of relative motion, and in this case, the Lorentz transformation can be conveniently derived using the Pythagorean theorem. If it is not perpendicular, then the cosine theorem must be used, and the angle between the light and the spaceship's motion will be introduced, resulting in a transformation that includes angular data, no longer being the standard Lorentz transformation.

The above three conditions can be considered as the conditions under which the relativistic corrections perform their best. If any of these conditions is not met, it will lead to a deviation in the correction effect of relativistic effects, even a significant error.

It can be said that the Lorentz transformation or the effects of special relativity are an approximate solution that satisfies the above conditions when considering the relationship between electromagnetic laws and relative motion with respect to empty space.

It is effective in cases such as GPS correction and quantum field theory because these scenarios can relatively well satisfy the above conditions.

This is also why many people recognize that relativity contains many errors and problems, yet strangely remains useful. The mass-energy equation and Lorentz transformation were already derived by people based on classical mechanics before the emergence of special relativity. Einstein merely provided them with an incorrect spacetime interpretation, but this incorrect interpretation does not affect the usefulness of the two formulas in their respective contexts.

In fact, it is difficult to fully satisfy the above three conditions. For example, different satellite materials result in varying degrees of drag on the internal empty space, leading to incomplete dragging. Additionally, the degree of drag exerted by the Earth on the empty space in the satellite receiving station area can also vary, and the uncertainty of these factors will weaken the effect of relativistic effect corrections, as shown in Figure 4-13.

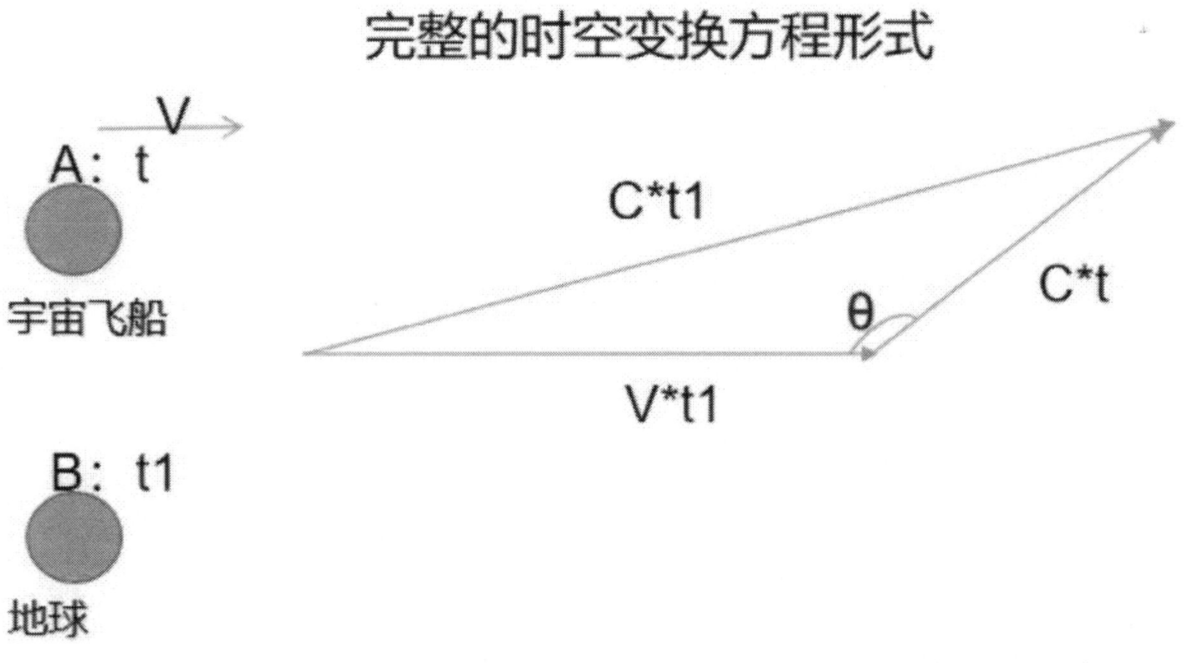

Figure 13: The complete form of the spacetime transformation equations, Figure 4-13

(1) Considering the angle between the spaceship's velocity and the direction of internal light illumination as θ , the Lorentz transformation formula should be adjusted accordingly:

$$(ct_1)^2 = (ct)^2 + (vt_1)^2 - 2ctvt_1 \operatorname{Cos} \theta$$

The calculated spacetime transformation equations in this case:

$$t_1 = \frac{\sqrt{(2ctv \, \mathrm{Cos}\, \theta)^2 + 4\,(c^2 - v^2)c^2t^2} - 2ctv \, \mathrm{Cos}\, \theta}{2\,(c^2 - v^2)}$$

When $\theta = 90$ degrees, the above equation simplifies to the standard Lorentz transformation:

$$t_1 = \frac{t}{\sqrt{1 - \left(\frac{v}{c}\right)^2}}$$

(2) If we consider the drag coefficient of the spaceship A in the vacuum space to be $K(0 \sim 1, 0$ (0 for completely non-dragging, 1 for completely dragging), then V in the above formula should be adjusted to $V * K$.
(3) Consider A and B existing in the same uniform empty space, but each moving at different velocities relative to the empty space

The degree is v_A and v_B . If the spaceship A moves relative to Earth at a speed of v_A , although the Earth observer B is stationary relative to Earth, the Earth's surface drags the vacuum space with a drag coefficient less than 1, i.e., not completely dragging, then the stationary observer B also has a relative velocity to the vacuum space, which is v_B . Therefore, the velocity V in the above formula needs to be further adjusted to $(v_A \mathrm{K} - v_B)$. Let $\mathrm{V1} = (v_A \mathrm{K} - v_B)$, then:

$$t_1 = \frac{\sqrt{(2ctV_1 \, \mathrm{Cos}\, \theta)^2 + 4\left(c^2 - V_1^2\right)c^2t^2} - 2ctV_1 \, \mathrm{Cos}\, \theta}{2\left(c^2 - V_1^2\right)}$$

When the spatial curvature degree $\theta = 90$ and the spaceship drag coefficient $K = 1$、$v_B = 0$ are considered, the above formula simplifies to the standard Lorentz transformation:

$$t_1 = \frac{t}{\sqrt{1 - \left(\frac{v}{c}\right)^2}}$$

(4) The situation becomes more complex when the vacuum spaces in which A and B are located are in relative motion or when the vacuum space is distorted (i.e., the space density is uneven), as these are among the factors that can cause deviations in related observational and experimental data.

The above adjustments are merely further approximations, corrections, and refinements based on the Lorentz transformation model. The correct treatment should be carried out according to the relative motion speeds of both parties with respect to the vacuum space, the relativity of light's motion relative to the vacuum space, as well as the relative motion and density of the vacuum space itself, following methods such as the Galilean transformation for processing and correction.

The current quantum mechanics, quantum field theory, and Maxwell's equations for electromagnetic waves are all based on special cases within a single stationary and homogeneous empty space. If any of the above conditions change, they need to be corrected and adjusted.

Non-Relativistic Spatial Relativity Calculation and Correction of Satellite Navigation Errors

The sources of errors in satellite navigation systems such as Beidou and GPS include errors caused by the satellites themselves (such as

Satellite clock error, satellite ephemeris error, and relativistic effect error, as well as errors caused by the signal propagation environment and errors introduced by the receiver itself, among many other factors. Relativistic effects occur due to the difference in gravitational environments between two objects and their different velocities, leading to a relative time deviation between them.

In the process of navigation positioning, since the positioning satellites are located in the space segment and the user receiver is on the ground, the two are in different gravitational environments. Moreover, the operating speed and trajectory of the positioning satellites are completely different from those of the user receiver, so positioning errors will inevitably occur due to these corresponding relativistic effects. Generally, a relativistic effect correction model is used to correct relativistic errors. However, since we understand the essence of relativistic effects as the relative motion of objects in the void space and the relative motion relationship between light and void space, we can make more precise considerations and corrections based on this principle.

For example, the Earth, as a massive body, will distort and drag the surrounding void space. As the distance increases, this distortion and dragging effect weakens. In terms of manifestation, this means that as the distance from the ground increases, the distortion coefficient Z of the Earth on the void space decreases, leading to an increase in the density of the void space ρ , until the distortion coefficient Z decreases to 0 and the density of the void space gradually increases to the standard void space density ρ_c . At the same time, the dragging coefficient K of the Earth on the void space also decreases, causing the relative motion speed between the void space and the ground to increase. Eventually, the dragging coefficient K decreases to 0, and the relative motion speed of the void space with respect to the Earth increases to the Earth's orbital speed (approximately).

Therefore, in this case, even if the satellite is stationary relative to the ground in space, it is still in relative motion with respect to the surrounding void space. Moreover, even if the signal receiving station is on the ground, the dragging coefficient at its altitude may be less than 1, meaning that the Earth does not fully drag the void space in this region.

The specific situation is shown in Figures 4-14 and 4-15:

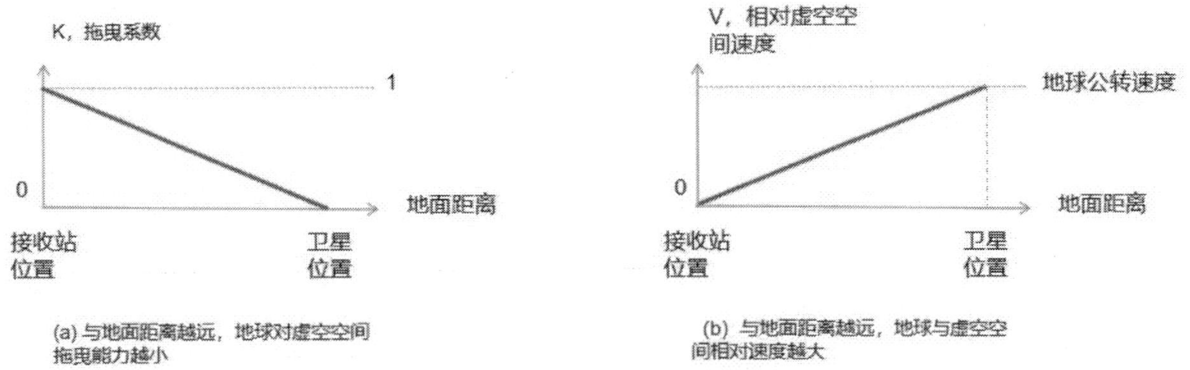

Figure 14: Figure 4-14 shows the change in the empty space dragged by the Earth as the distance from the ground increases

Figure 15: Figure 4-14 shows the change in the dragging of empty space by Earth as the distance from the ground increases

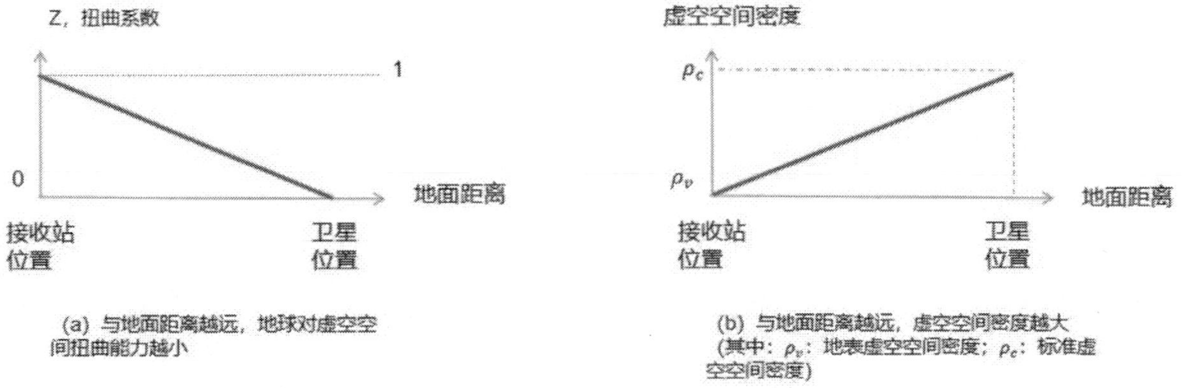

Figure 4-15 shows the variation of vacuum space density with increasing distance from the ground

Therefore, we can consider the impact of electromagnetic wave signal transmission based on the changes in this drag coefficient and vacuum space density coefficient, as well as the effects of Earth's motion velocity and gravity, to correct satellite communication errors, rather than assuming the speed of light is constant and handling it according to the effects of special relativity.

Of course, this assumes that the satellite is stationary relative to the ground. If the satellite itself is moving relative to the ground, the diagram would need to be adjusted accordingly. Additionally, this also assumes that the drag coefficient and distortion coefficient change linearly with distance from the Earth's surface. The specific change curves, as well as the actual drag and distortion coefficients for the Earth's influence on the satellite and the ground station, would still need to be tested and confirmed through real experiments.

4.7 Space Lenses and Refraction Effects

mass distorts the surrounding vacuum space, essentially due to the thermal motion of the mass body and the vacuum space's

Interaction causes the motion effect of the surrounding empty space, which is manifested as a decrease in the density of the empty space. Therefore, there is a density gradient in the empty space around a mass body, with the density of the empty space decreasing as it gets closer to the mass body and increasing as it moves farther away, eventually approaching the standard density of empty space.

This density gradient change will create a lens effect in the surrounding empty space. Since all electromagnetic waves, gravity, etc., are waves propagating in empty space, they will all experience refraction when passing through the vicinity of a mass. Essentially, it is the refraction of space waves.

Take light as an example.
Light undergoes refraction effects when passing around massive objects such as macroscopic planets, objects, and microscopic particles, due to the space lens phenomenon. The size of the refraction angle is related to the degree of distortion of the empty space by the mass object, which in turn depends on the mass and density of the object. The greater the mass and density of the mass object, the larger the space density gradient, and the greater the refraction angle when light passes through the space lens.

(1) Deflection Effect of a Single Mass

The density of the empty space around a single mass body changes in a spherical and ellipsoidal manner, similar to a single convex lens. Therefore, light passing through a single mass body will be deflected twice toward the center of the lens, once when entering the lens and once when exiting, resulting in the overall deflection of light, as shown in Figure 4-16.

Figure 16: Spatial wave passing through parallel interface mass body

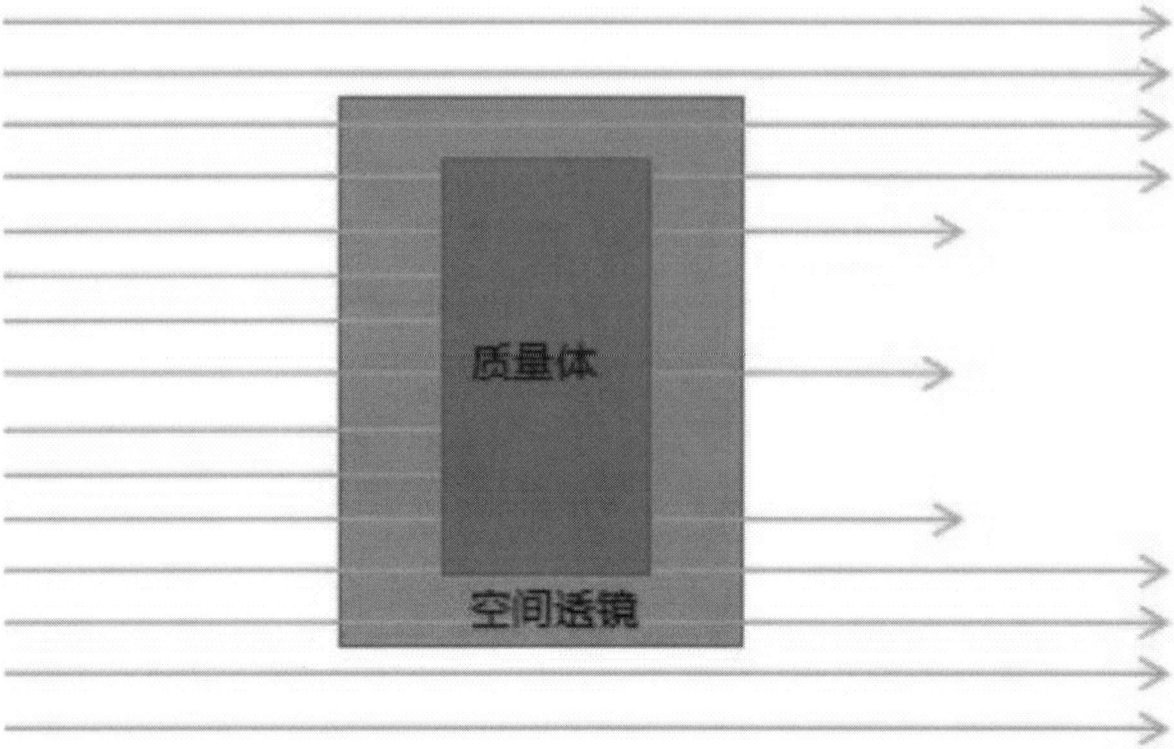

Figure 17: Spatial wave passing through a mass body with non-parallel interfaces

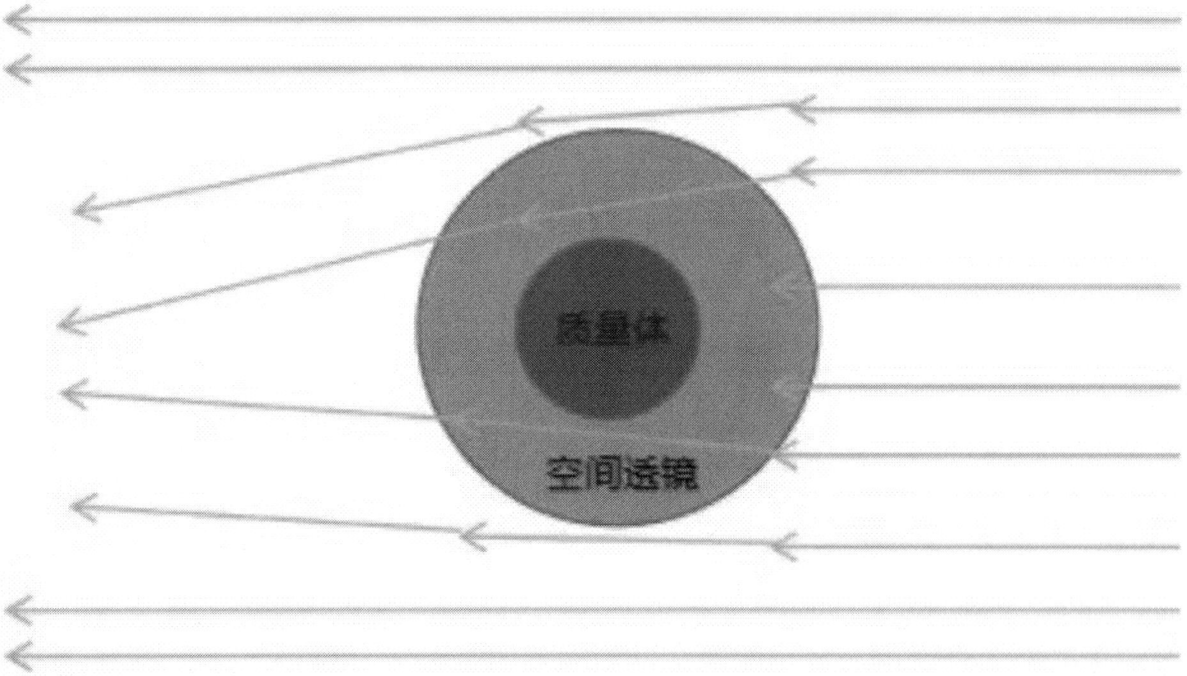

Figure 4-16 Refraction of light (space wave) passing through bodies of different masses

When light passes by a single planet, object, particle, etc., it will bend, which is this phenomenon

Example.
In 1919, astronomer Arthur Eddington led a team to measure the deflection of starlight by the Sun during a total solar eclipse, and they observed a slight deviation in the positions of stars around the Sun compared to their standard positions in the star catalog, which was another instance of experimental verification.

(2) Deflection Effect of Mass Groups

A group of multiple mass bodies gathered together, in addition to the density gradient formed by each individual mass body's distortion of the empty space, creates an overall distortion of the empty space surrounding the mass group as a whole. This results in an overall density gradient in the empty space around the mass group, forming a mass group space lens, as described by the Gladstone-Date law. In this case, when light passes through the mass group, it also experiences deflection, which is influenced not only by the individual space lenses of the mass bodies it passes through, but also by the overall space lens effect of the mass group.

For example, when starlight passes through surrounding galaxies, it is deflected due to the gravitational lensing effect of mass clusters. When light passes through water, glass, etc., which are molecular mass clusters, it also gets deflected. When light passes through planar media such as flat glass, it is deflected twice and then returns to its original parallel direction. Other mass clusters with non-parallel interfaces cause a change in the vacuum density, resulting in a lensing effect, so the direction of the outgoing light differs from that of the incoming light, leading to focusing or diverging effects, as shown in the Lorentz - Lorenz formula:

$$n^2 = \varepsilon - \frac{2\rho}{\varepsilon} - \rho$$

Where $ n $ is the refractive index in air, $ \varepsilon_0 $ is the refractive index in the medium, and $ \varepsilon_1 $ is the dielectric strength in the context.
The Lorentz-Lorenz formula can accurately calculate the refractive index of a substance within a geometric body.
For solid mass groups composed of molecules, if there is a special crystalline structure between the molecules, it will cause more complex spatial density variations inside, and also affect the deflection and path of light, leading to inconsistencies

Different deflection effects.
The greater the overall mass and density of a mass group, the larger the density gradient around its space, and the greater the refraction angle of light passing through the spatial lens.

4.8 Drag Effect and Spatial Motion

The void space is in motion. The distortion (density gradient change), waves, and other forms mentioned earlier are all manifestations of this motion. Another characteristic of void space is that it can be dragged by mass bodies to form motion. In addition to the drag effect caused by a single mass body, a group of masses can also produce an overall drag effect on the void space.
(1) Drag Effect of a Single Mass Body
Taking a single mass body, such as a planet in the universe, as an example, a planet will distort the surrounding void space and drag the surrounding void space along with it. The degree of drag exerted by the planet on the surrounding void space is related to the degree of distortion it causes in that void space. In other words, the greater the distortion of the void space by the planet, the greater the drag effect.

Assume the surrounding space is stationary around a planet (i.e., no other relative motion except that dragged by the planet), let the drag coefficient K be $0-1$, the planet's velocity be V_0, and the velocity of the dragged space be V_1, then:

$$V_1 = KV_0$$

The relative motion speed between the void space and the planet V_2 is:

$$V_2 = (1-K)V_0$$

When the degree of dragging is maximum, the dragging coefficient K is 1, meaning the dragging of the vacuum space by the planet is complete. The surrounding vacuum space moves in sync with the planet, at the same speed, and is relatively stationary with respect to the planet's surface, i.e., the relative motion speed is 0.

When the degree of dragging is minimal, the dragging coefficient K is 0, meaning the dragging effect of the planet on this empty space is zero

Frictionless, the surrounding vacuum space moves relative to the planet at a speed equal to the planet's speed.
At the same time, the degree of drag or dragging ability of the planet on the vacuum space is related to the degree of distortion the planet causes to the vacuum space. The closer the vacuum space is to the planet's surface, the greater the distortion, meaning the smaller the spatial density of the vacuum space. At this point, the speed of light in the vacuum space is set as V, and the density of the vacuum space at this moment is denoted as ρ_V. The farther away from the planet, the lower the distortion of the vacuum space, meaning the higher the spatial density of the vacuum space, eventually reaching the standard density ρ_C of the vacuum space where the speed of light is C.

Let the mass of a planet be M, and the density of the empty space at a distance R from the planet's surface be ρ_V. Then:

$$\rho_v \propto \frac{\mathrm{R}^2}{\mathrm{M}}$$

The dragging ability of a planet on the vacuum space, that is, the dragging coefficient $ K $, is inversely proportional to the vacuum density at that location:

$$K \propto \frac{1}{\rho_v} \propto \frac{M}{R^2}$$

So the dragging ability of a planet or a single mass on the surrounding empty space depends on the planet's mass and the distance between the region of empty space and the planet. Let the dragging coefficient K be related to the ratio of the planet's mass and the distance by the coefficient G, then the relative motion velocity between the empty space and the planet V_2 is:

$$V_2 = \left(1 - G\frac{M}{R^2}\right)V_0$$

The results of the Michelson-Morley experiment and the Trouton-Noble experiment discussed in our previous chapters are examples of this dragging phenomenon. As the distance from the Earth increases, the Earth's ability to drag the empty space weakens, and corresponding results from the two experiments would be observable.
(2) Tidal Drag Effect of Mass Clusters
Similar to the space distortion and space lens effect, mass clusters will form an overall drag on the surrounding empty space. The drag effect depends on the mass size of the cluster, its mass distribution, density, and other related factors.

For example, in the case of a uniformly refracting medium such as water, these mass group parameters can be approximately represented by the refractive index n.

In 1851, French scientist Armand Fizeau conducted an experiment to test whether the speed of light would be affected in moving media such as flowing water, known as the Fizeau's flowing water experiment, as shown in Figure 4 - 17.

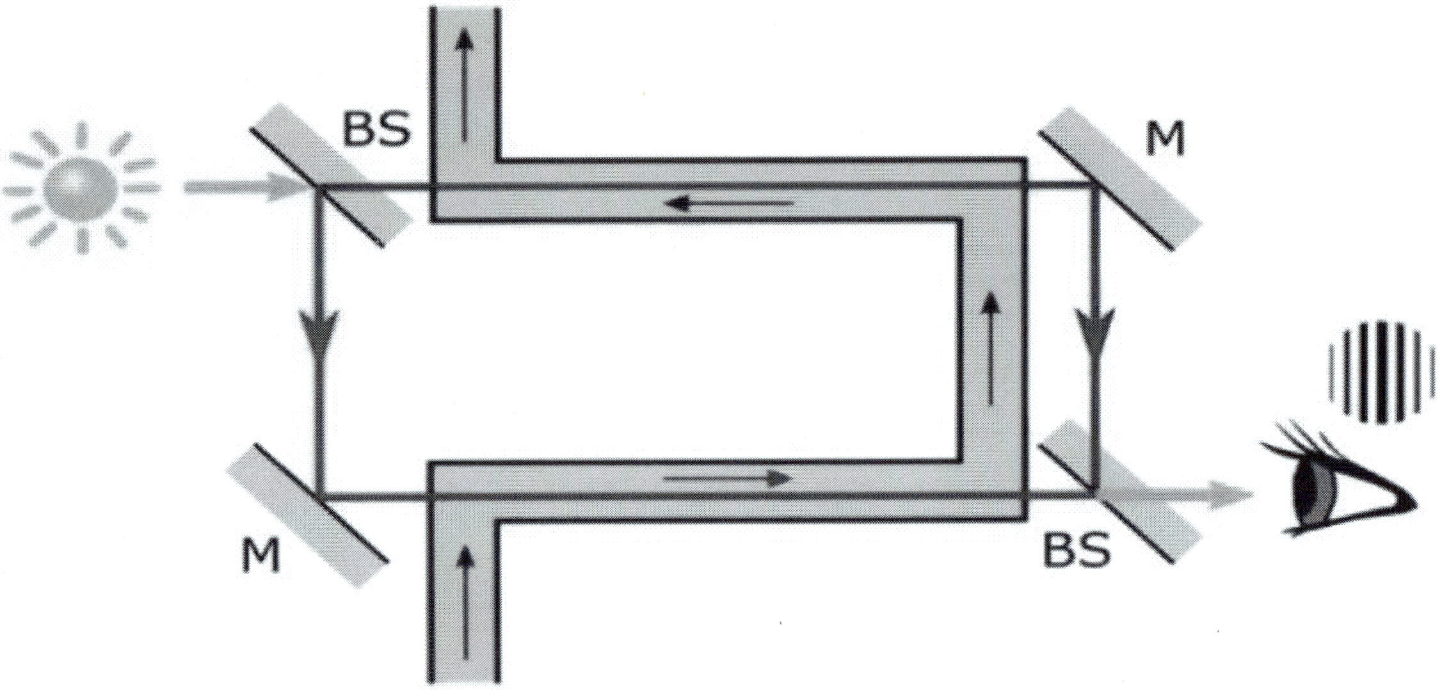

Figure 4-17 Fizeau's River Experiment

The experimental results show that flowing water indeed affects the speed of light transmitted through it, that is, the mass group of water molecules in the flowing water medium drags the surrounding empty space, causing the empty space to move. Since the speed of light is constant relative to the empty space, this leads to a change in the speed of light propagating through the water.

The overall drag coefficient K of the water molecule group on the empty space can be approximately expressed as:

$$K = 1 - \frac{1}{n^2}$$

Let the velocity of the water molecule group, that is, the velocity of the flowing water, be V , then the velocity of the dragged 虚空 space V_1 is:

$$V_1 = KV = \left(1 - \frac{1}{n^2}\right)V$$

Let the speed of light in still water be C_n , where C_n is the overall speed of light propagation relative to the stationary虚空 space in water. Since the water drags the虚空 space with it when flowing, and the speed of light relative to the虚空 space remains constant, the speed of light in flowing water C_1 will also change. This change follows the Galilean transformation:

$$C_1 = C_n + \left(\mathbf{1} - \frac{\mathbf{1}}{\mathbf{n}^2}\right)V$$

This is also the case of the speed of light detected in Fizeau's flowing water experiment, consistent with the prediction of the spatial drag principle.

(3) Factors Influencing the Drag Effect

The specific formula for the dragging effect of a mass body on empty space still requires more precise experiments and verification, as there are many factors influencing the dragging effect, and there may even be undiscovered influencing factors.

The overall dragging effect of a mass group on empty space includes not only the dragging effect of individual mass bodies but also the collective dragging effect of the mass group. These effects are related to the mass group's overall ability to distort empty space, which may specifically depend on factors such as the total mass, density, and distribution of the mass group.

Taking everyday material substances as an example, besides their mass and density affecting the overall dragging capability of the empty space, their internal molecular distribution forms, structures, including different crystal structures, will also influence the overall distortion and dragging effect on the empty space.

In the previous analysis, it was assumed that mass bodies and mass groups cause distortion and dragging effects on the stationary empty space. If the empty space itself is in motion or has been distorted and dragged due to surrounding other mass bodies, the situation becomes more complex and requires further research.

In general, we can simplify the assumption that the worse the light and heat transmission properties of a material substance are, the stronger its shielding effect on electric and magnetic fields, and thus the stronger its dragging effect on the empty space, with a larger dragging coefficient. The influence of the refractive index n of water and other media on dragging capability discussed earlier is also an overall simplified reflection of the above factors.

4.9 The Relationship Between Spatial Electrodynamics and Special Relativity

(1) The Concept of Empty Space
Special relativity holds that empty space is not an objective entity and does not possess motion or mechanical properties. Space Relativity posits that empty space is an objective entity, composed of an unknown substance, and I

We can consider it as a whole, which has relative motion and dynamic characteristics.

(2) The principle of the invariance of the speed of light

Special relativity states that the speed of light is constant relative to any reference frame and any different observer. Spatial relativity posits that the speed of light is constant relative to the space it is in, and relative to different observers or different moving spaces, there will be different relative speeds.

(3) Principle of Relativity

Special relativity states: all motion is relative, and two entities in relative motion can arbitrarily assume each other's velocity as long as it conforms to the sum of relative velocities.

Spatial relativity posits: light, or electromagnetic waves, and other parts of micro-quantum entities have a constant velocity relative to the void space. Therefore, when dealing with quantum factors such as light and electromagnetic waves, one must consider the entity's velocity relative to the void space it resides in, and cannot arbitrarily assume its velocity. This velocity can be approximately regarded as the "absolute velocity" within this void space.
(4) The Relativity of Simultaneity
Special relativity holds that different reference frames in motion will have different judgments on the simultaneity of two events.
Spatial relativity posits that since there are countless observers in each reference frame, at the moment an event occurs, there can be an observer present at the same location who can confirm the time of occurrence immediately. Therefore, each inertial frame can determine the simultaneity of events without any discrepancy. Even with a single observer, one can analyze according to the rules to eliminate observational errors and make a judgment on simultaneity.

(5) The View of the Speed Limit of Light

Special relativity holds that the speed of light in a uniform, flat vacuum space, denoted as C, is the maximum speed in the universe and cannot be exceeded.

Spatial Relativity posits that the speed of light C is the speed at which light propagates relative to the surrounding vacuum space. Observers in different states of motion can have relative velocities exceeding C, and different spacecraft can accelerate through the internal drive model

Speed can exceed the speed of light. The speed of light is not the maximum speed in the universe.

(6) The View of Physical Laws and Observed Phenomena

Special relativity holds that physical laws take the same form in all inertial frames, and the propagation of light according to Maxwell's equations should also remain consistent, hence the speed of light is the same in different reference frames.

Spatial relativity posits that physical laws are objective and do not change in any reference frame, but directly observed phenomena may differ because phenomena can manifest the essence or even distort it. Therefore, it is necessary to use reasoning and research to look beyond phenomena and understand the essence. For instance, the speed of light propagation in a vacuum is constant and conforms to Maxwell's equations, which represents the essence. However, the directly observed results in different reference frames or by different observers may vary in speed, which is the observed phenomenon. The difference in observed phenomena does not negate the unity of the physical law's essence. Forcing consistency in observed phenomena across different inertial frames confuses the distinction between the physical essence and the observed phenomena.

(7) The Relationship Between Matter and Void Space

Special relativity holds that empty space is not an objective reality, does not have dynamic effects, and does not influence light or any other matter moving within it.

Space relativity posits that empty space is an objective reality, possessing dynamic effects such as density and relative motion, which influence the propagation of light and the movement of matter within it. Matter can also affect the surrounding empty space through effects like distortion and dragging, forming mutual mechanical interactions.

In short, spatial relativity argues that the description of the speed of light and spacetime in special relativity is incorrect, and it is an erroneous theory. The phenomena described, such as time dilation and length contraction, are not physical realities or observable effects; they are merely mathematical descriptions. This mathematical system distorts the understanding and description of the physical world. The so-called effects of time dilation and length contraction are simply the results of this distorted mathematical formula. That is, if the speed of light is to remain constant, time, length, and other parameters must change to maintain the consistency of the mathematical system.

Even according to this calculation system, it should be that clocks run slow and rulers expand, not contract, and this error leads to

Special relativity could not be compatible with the gravitational formula, and had to continue developing into general relativity, moving further and further away.
The Lorentz transformations and the mass-energy equation used in special relativity were previously derived by other scholars through classical mechanics. Although special relativity provided distorted and incorrect interpretations of these theories, it does not affect their validity in specific contexts. This is also the reason why special relativity can be effective in certain fields.

5 Introduction to Spatial Wave Dynamics [8][9[10]

5.1 Fundamental Properties of the Void Space

Previously, in sections on the basic principles of spatial relativity, such as Principle 2 (Principle of Spatial Objectivity) and Principle 6 (Principle of Spatial Relativity), we analyzed that the empty space is an objective physical reality and is in motion. The empty space is like an invisible superfluid ocean, in which all stars, matter, and particles interact and move.

Based on real-world experience, we can further assume that the motion of empty space can include different modes such as flow and wave.
(1) Empty Space and Wave Particles
Because there is currently a lack of scientific observational data on the detailed composition of empty space's fundamental particles, for the sake of discussion, we do not consider the essential nature of the basic particles existing in empty space. Instead, we can generally refer to them as the particles of space waves.

The motion of the void space can interact with all matter, particles, etc., producing various physical effects, which fundamentally occur through wave particles, so we can consider wave particles to have mass.

The basic properties of bosons include:
(1) Materiality: That is, the boson is material.

(2) Volume: That is, it has three-dimensional dimensions.
(3) Mass: That particles have mass and therefore possess momentum and energy.
(4) Velocity: Waves are in motion, and individual wave particles may momentarily or temporarily be at rest (relative to their spatial position), but their normal state is to be in motion.
(5) Particulate nature: That is, wave particles can be considered as particles, whose motion is linear. However, the interaction among a large number of wave particles can lead to complex motion patterns such as wave-like behavior in their collective movement.
(6) Microviscosity: The particles have smooth surfaces and very low viscosity between them, with interactions mainly involving momentum and energy exchange through collisions of the mass bodies.

In summary, they can be simply characterized as particles with mass, extremely low viscosity, and nearly ideal rigid bodies, whose essential features are possessing mass and linear motion.

(2) The Movement Pattern of Void Space and Wave Particles

Vacuum space is composed of wave particles, and overall it can be considered as a superfluid formed by an infinite number of wave particles. Therefore, vacuum space also has mass, and thus possesses density and density gradient variations, and exhibits different movements such as flow and wave.

The flow of vacuum space involves complex movements such as laminar flow and vortices.
The waves in vacuum space can be referred to as spatial waves. Due to the superfluid properties of vacuum space, spatial waves are primarily longitudinal in nature. During wave motion, the movement of wave particles is not strictly linear, meaning they do not vibrate in a single direction. Instead, they vibrate in various directions. Because the amplitude of vibrations in orthogonal directions differs, the movement pattern of all wave particles during wave propagation is generally elliptical spiral motion. When the short axis of the ellipse is sufficiently small, it can be approximated as linear vibration. When the short and long axes of the ellipse are approximately equal, it can be approximated as circular spiral motion.

The movement patterns of all other microscopic particles in spatial waves are also similar.

(3) Spatial wave velocity

Space waves interact and move in coordination with all particles, including photons. Therefore, it can be considered that the propagation speed of space waves is at least the speed of light, but this is not the speed of the wave particles; the speed of the wave particles should be greater than the speed of space waves.

All particles exist in the void space, so their motion is coordinated with the motion of space waves and is affected by them, such as acceleration or deceleration. Therefore, if a particle is accelerated by the motion of space, then its

The maximum speed of motion is also the speed of light, that is, the speed of spatial waves, unless particles are temporarily accelerated by external forces to exceed the speed of light.

5.2 The Essence of Fundamental Particles

(1) Soliton

During the process of fluctuation and flow, vacuum space can form a solitary wave, or soliton, under certain conditions (such as being subjected to external forces, etc.).

A soliton is a stable coherent structure formed by a large number of wave particles, which maintains its shape during motion. When multiple solitons meet, they still retain their individual shapes and satisfy the laws of momentum and energy conservation. Overall, it behaves as if it were a single particle, which we can call a soliton particle [12] .

Coherent structures in solitons include but are not limited to:

(1) High-speed spin

A soliton formed by the excitation of countless wave particles itself possesses high-speed spin, with surface spin velocities reaching even beyond the speed of light. This enables the soliton to generate significant rotational kinetic energy, rigidity, closure, and stability, allowing it to manifest and maintain a distinct particle-like nature.

(2) Vortex Structure

Solitons can self-organize into more smaller clusters, each exhibiting high-speed vortex motion. These different numbers of vortices are distributed within the solitons and on their surfaces. The high-speed vortex structures allow these small wave particle structures to maintain their stability and rigidity, while also giving the soliton as a whole additional rigidity, closure, and stability. Through interaction with external wave particles, they can form spatial wave motion, excite different force fields, and enable particles to acquire various properties such as different charges.

(3) Vortex Ring Structure

A large number of wave particles can also maintain structural stability and rigidity through high-speed vortex ring structures.

Each soliton contains at least one of these coherent structures, allowing it to exhibit and maintain strong particle-like properties for a long time, and interact with the external vacuum space.

Stable coherent structures may have other more forms, requiring detailed theoretical exploration and experimental verification.

Because the stability of different coherent structures or the same coherent structures varies, some solitary particles will re-collapse or disperse into wave particles and merge into the vacuum space after a period of time or upon encountering external disturbances. However, some coherent structures have higher stability and can remain stable for a long time even when facing partial external disturbances. These solitary particles have become what we normally consider as particles, and we can call them micro-particles.

All the fundamental particles we know, such as photons, neutrinos, and even electrons and protons, can be considered as types of these micro-particles, or as stable coherent structures called solitons formed based on micro-particles. That is, particles.

The microscopic particles we have already recognized are just a part of these massive micro-particles and particle types. Therefore, all the fundamental particles we discussed, from a macroscopic perspective, are formed by the excitation of empty space, and at the same time have a close and complex interaction with empty space. Hence, when discussing fundamental particles, their relationship with empty space is inseparable.

(2) The mass of fundamental particles

All fundamental particles are ultimately composed of bosons with different motion patterns and mechanical configurations. Since bosons have mass, all particles also have mass. It can be said that all particles that can undergo mutual mechanical interactions, including fermions and bosons such as photons, have mass.

In other words, mass is one of the most essential characteristics of particles or matter.
The source of matter's mass is the boson.

(3) Motion of Fundamental Particles and Wave-Particle Duality

Basic particles (hereinafter referred to as particles) move in the void space, so they are affected by void space fluctuations (with

The influence of the wave's power, direction, and speed (hereinafter referred to as spatial waves).
From the perspective of manifestation, particles can be similar to solitons, moving together with and interacting with spatial waves. This interaction endows particle motion with wave-like properties. That is, when a particle moves on its own or interacts with other particles, it follows a linear motion pattern. However, due to its interaction and coordination with spatial waves, its motion exhibits wave characteristics. This combination is manifested as wave-particle duality.

It can be said that the wave-particle duality of particles arises from the interaction between their particle nature and the wave nature of spatial waves, resulting from nonlinear interactions between particles and spatial waves.

The essence of the material world is particle-like, while wave-like behavior is a manifestation of the collective motion of particles. When a single particle undergoing linear motion encounters a group of particles exhibiting overall wave-like motion, the motion of the single particle will be influenced and exhibit wave-like properties. This is the fundamental nature of wave-particle duality.

(4) Particle Charge

The essence of electric fields, magnetic fields, and various particle fields is the different modes of motion and excitation in the vacuum space.
Fields are realized through spatial waves, which serve as the carrier for field excitation and propagation.
The electric charge of a particle is a characteristic manifested through the wave field of spatial waves, i.e., the electric and magnetic fields. The fundamental interaction between wave particles and particles at the microscopic level is the mass force. Particles are formed by the accumulation of different motion modes and mechanical configurations of wave particles, and the different charge characteristics of particles arise from these distinct motion modes and configurations of wave particles.

That is, the different motion modes of wave particles within a particle excite the surrounding vacuum space to form different wave fields, which manifest as the electric and magnetic fields of the charge.

Different electric field strengths result in different charge magnitudes of particles, so the charge of a particle is essentially the result of the coherence mode and motion intensity of its internal wave particles.

5.3 Unified Source of Force

Mass is a fundamental property of all particles, including space wave particles.

The mass and charge of particles are different aggregation patterns, structures, and movements of wave particles, forming various properties of matter and exerting effects on the surrounding space. These effects are categorized into gravitational fields, electric fields, magnetic fields, etc.

In other words, the interaction forces between substances, including gravity, electric force, magnetic force, and even strong and weak forces, are essentially different manifestations of the mass interaction forces between particles or between particles and empty space.

(1) Mass Force and Spatial Pressure

There was mass first, then force. Mass and motion are the essence, and various forces, including gravity, are phenomena resulting from the collision of mass bodies due to motion, which is known as mass force.

Mass force conforms to the principles of momentum and energy, that is, it is the result of collisions between wave particles and particle bodies that satisfy the conservation of momentum and energy.

Macroscopically, the generation mechanism of various forces is the result of the sum of the mass forces exerted by countless wave particles around a mass body. The overall macroscopic manifestation of the sum of these countless mass forces around a mass body can be regarded as the pressure of the void space, also known as spatial pressure.

(2) Electric Field Force

(1) The Generation of an Electric Field

As analyzed above, particles are formed by the stacking of different motion patterns and mechanical configurations of wave particles. These different motion patterns and configurations of wave particles give rise to the different electric charge characteristics of particles, distinguishing them as positive charge, negative charge, and neutral charge.

Also because of these different charge characteristics, charges excite the surrounding empty space, producing different electric field motions, specifically manifested as the difference in the motion patterns of electric fields around positive and negative charges.

The coherent structures that are stable in solitons, which we have analyzed above, include spin, vortices, and vortex rings. Among these, the small vortex stable structures formed by a large number of wave particles aggregating in solitons are distributed on the soliton surface, and they will increase the soliton's

In addition to rigidity and stability, this high-speed vortex motion also interacts with the external empty space, causing the external empty space to generate vortex or spiral waves.

This spiral fluctuation of the void space radiates outward in particles, and its macroscopic manifestation is the electric field E.

Whether it is a positive or negative charge, the electric field produced is a helical wave field in the void space, differing only in the helical motion pattern of the wave particles.

Observing from the direction facing the charge, the vortex of wave particles on the surface of a positive charge rotates clockwise, so the spatial wave electric field produced in the vacuum space by its disturbance also rotates in a clockwise spiral (i.e., left-handed spiral wave). The vortex of wave particles on the surface of a negative charge rotates counterclockwise, so the spatial wave electric field produced in the vacuum space by its disturbance rotates in a counterclockwise spiral (i.e., right-handed spiral wave), as shown specifically in Figure 5-1.

正电荷激发的电场（空间波）为顺时针螺旋波动（左手螺旋，面向电荷观察）

负电荷激发的电场（空间波）为逆时针螺旋波动（右手螺旋，面向电荷观察）

Figure 18: Figure 5-1 Direction of helical field of positive and negative charges

There are numerous vortex structures of wave particles on the surfaces of positive and negative charges, each vortex structure exciting a spatial wave spiral fluctuation in the vacuum space, extending outward and representing a field line in form. The more vortex structures a charge has, the more spatial spiral waves are excited, resulting in more field lines in form, which indicates a stronger electric field strength E and more charge carried by the particle. That is:

$Q \propto n(5.1)$

$E \propto n(5.2)$

Where, Q is the particle's charge, and n is the number of wave particle vortex structures on the particle's surface.
(2) Generation of Electric Field Force

When two charges meet, the spatial helical waves they excite produce different forces due to their opposite directions, and this force is known as the electric field force.

When two opposite charges excite spatial waves, the waves meet with the same helical wave direction, causing them to merge and create a spatial negative pressure between the two charges, resulting in an attractive force. The closer the distance, the stronger the helical wave intensity, so the attraction is stronger. That is:

$$F_1 \propto \frac{Q}{r^2} \propto \frac{n}{r^2} (5.3)$$

The electric field force F produced by a charge at a point is inversely proportional to the square of the distance r between the point and the charge.
When two like charges excite spatial waves, the helical wave directions are opposite when they meet, causing collisions and rebounds, which result in a spatial positive pressure between the two charges and thus produce a repulsive force. The closer the distance, the stronger the helical wave intensity, so the repulsive force F_2 becomes stronger, that is:

$$F_2 \propto \frac{Q}{r^2} \propto \frac{n}{r^2} (5.4)$$

The repulsive electric field produced by a charge at a point is inversely proportional to the square of the distance between the point and the charge, which is consistent with the electric field strength formula:

$$E = K\frac{Q}{r^2} (5.5)$$

The electric field fluctuations produced by positive and negative charges spiral in the direction shown in Figure 5-2.

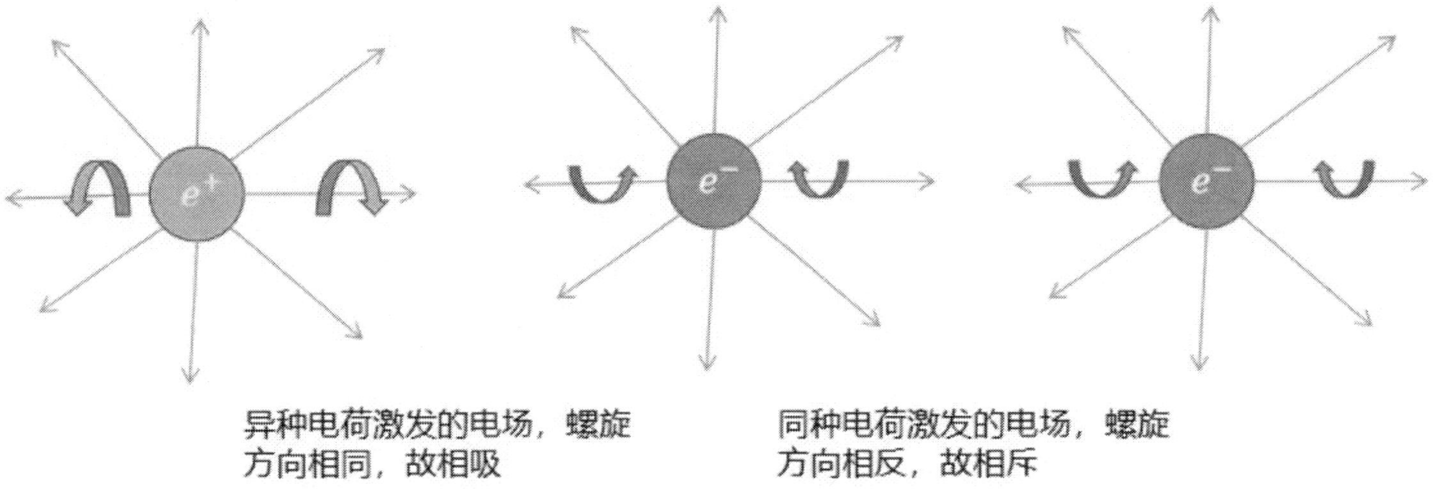

Figure 5-2 The helical direction of electric field for positive and negative charges

A more detailed wave dynamics analysis of the electric field force based on the above principle can reveal the physical meaning and specific calculation formula behind the constant K in the electric field strength formula.
(3) Electric Field Breathing Model
The electric field mechanics model discussed above can be further upgraded into an ideal breathing model. That is, while positive charges form high-speed vortex energy waves outward through the surface vortex structure, they also eject a stream of space particles. At the same time, they absorb surrounding particle streams from the void space through the gaps between vortex structures with a slower force, forming the positive charge breathing effect.

Negative charges rapidly inhale vortex energy fluctuations inward through the motion of the vortex structures on the surface, and slowly expel disordered spatial particle streams outward through the gaps between vortex structures, forming negative charge respiration.

That is, on top of the above vortex model, there exists a further breathing model, where the synthesis of spatial wave vortex fluctuations and spatial flow forms a complete and powerful electric field force.

Two positive charges meet, creating a relative jet of spatial flow between them. The spatial wave vortices rotate in opposite directions, resulting in a repulsive force.

Two negative charges meet, creating a relative absorption space flow between them, with opposite directions of space wave vortices, thus also producing a repulsive force.

Positive charges and negative charges meet, creating a flow of matching breathing space between them. The spatial wave vortices have the same direction, so they produce an attraction. The form of electric field lines for repulsion and attraction between different charges is shown in Figure 5-3.

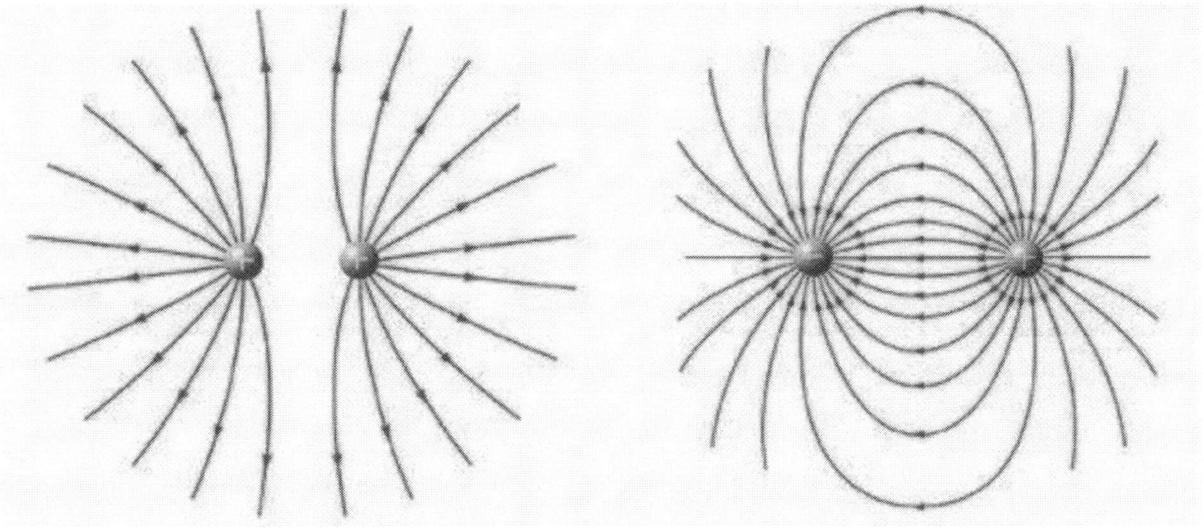

Figure 19: Electric field line distribution between two charges showing repulsion and attraction as in Figure 5-3

(3) Magnetic Force

(1) Generation of Magnetic Field
Charge excites spatial waves in the form of spiral fluctuations, which are known as electric fields.
When a charge moves, it causes three types of changes:
a. The movement of a charge in empty space excites spatial vortex ring fluctuations, which propagate outward inertially.
b. Because the charge leaves, the surrounding spiral spatial fluctuations lose their continuous driving force and synchronize during the formation of the vortex ring, resulting in a spiral vortex ring.
c. The spin of a charge excites a circular flow in the surrounding space, leading to the generation of a vortex ring with a circular flow, which merges with the helical motion of the vortex ring to form a spatial helical flow that precesses along the direction of the vortex ring.

This outwardly diffusing spiral precession vortex ring spatial flow caused by the movement of charges is the magnetic field B we observe. See Figure 5-4 for an illustration.

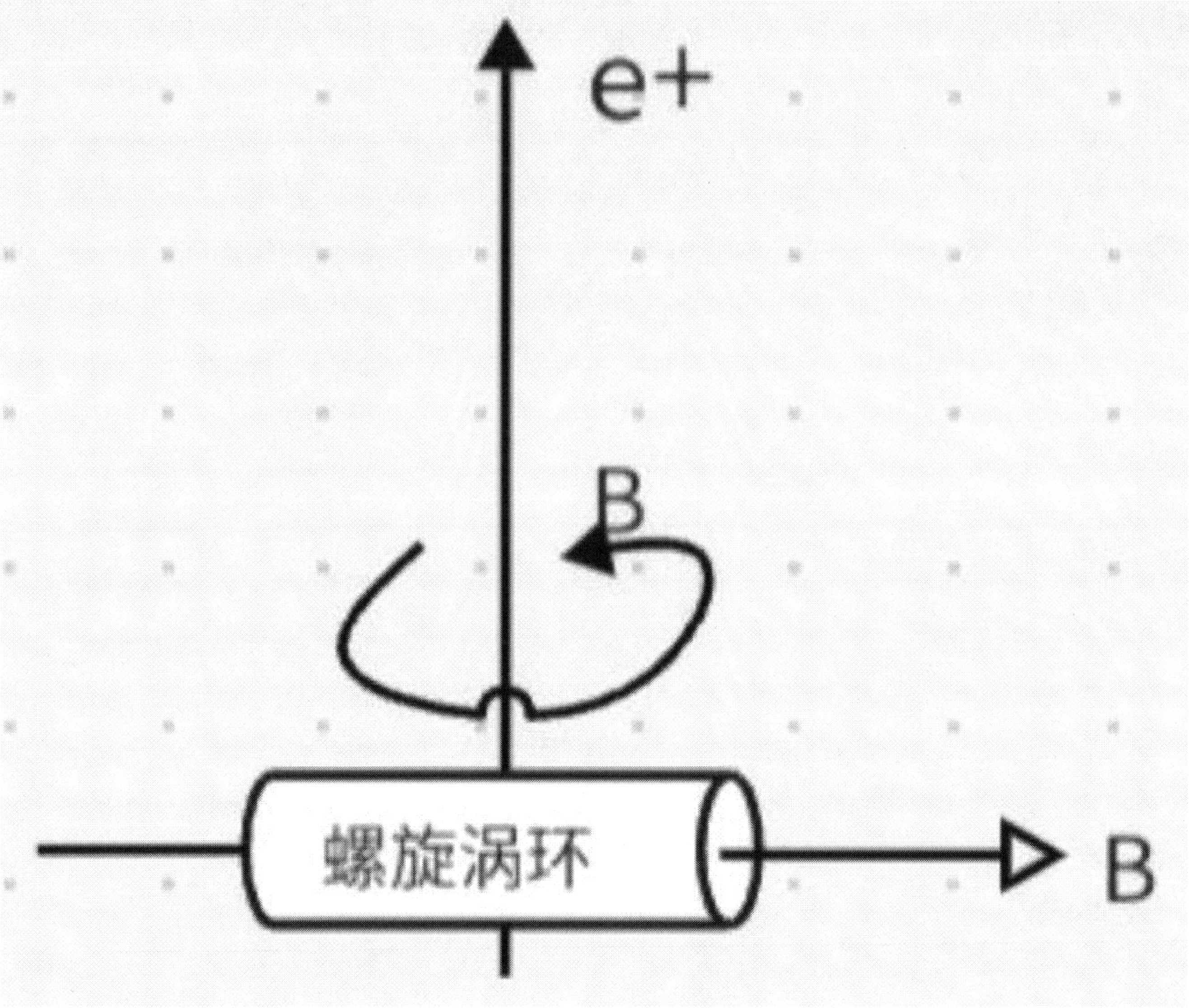

Magnetic field line vortex ring spiral direction (as viewed facing the magnetic field lines)

Figure 5-4 Spiral Vortex Ring Spatial Wave and Magnetic Field

Each magnetic field line is a closed spiral precessing vortex ring.
The spin direction of the charge affects the spiral direction of the vortex ring flow, ultimately resulting in the spiral direction of the vortex ring being counterclockwise when viewed towards the magnetic field lines.

Since the electric field is a spatial spiral wave continuously excited by charges, while the magnetic field is a spatial spiral wave no longer being excited, the force exerted by the electric field will be much greater than that of the magnetic field.

The electric field can be considered as a source-based, open magnetic field, while the magnetic field is a source-free, closed electric field.
The relationship between the electric field E and the magnetic field B follows the formula:

$$B = \frac{E}{C}(5.6)$$

where C is the speed of light.
The magnitude of the magnetic field B produced by a moving charge Q is related to the number of magnetic force vortex rings. The larger the charge quantity Q, the more magnetic force lines are generated; the farther the charge is from the point, the fewer magnetic force lines are present per unit spherical volume, meaning the magnetic field B decreases inversely with the square of the distance r. The greater the charge's velocity v, the larger the energy of the vortex ring, so the magnetic field B is proportional to the charge's velocity v.

$$B \propto \frac{Qv}{r^2} (5.7)$$

This agrees with the experimentally verified formula for magnetic field induction:

$$B = \frac{\mu_0}{4\pi} \frac{Q\vec{v} \times \vec{r}}{r^3}$$

(2) Generation of the Lorentz force in a magnetic field
When a charge is stationary in a magnetic field, the surrounding magnetic spiral vortex rings interact with the electric field fluctuations it generates, resulting in attraction or repulsion. However, since the magnetic spiral is a sourceless field without continuous charge excitation, the limited number of spiral vortex magnetic lines has a very small effect on the charge's electric field and the charge itself, making it unobservable macroscopically and negligible in calculations. See Figure 5-5 for illustration.

Figure 20: Figure 5-5 Stationary Charge in a Magnetic Field

But when a charge moves in a magnetic field, it also generates a magnetic field around itself, which interacts with the surrounding magnetic field B_0.

From the direction of entry, both sides have the same vorticity and precession direction on the charge side (such as the left), resulting in a fusion. On the other side (such as the right), the vorticity and precession directions are opposite, leading to repulsion. This allows the magnetic field lines B generated by the charge to expand outward, merging with the receding vorticity field lines of the magnetic field B_0. This change causes the magnetic field vorticity on the charge side (right side) to become sparse, creating a spatial negative pressure, which in turn generates a spatial pressure toward the right. This pressure is the Lorentz force.

The magnetic field directions generated by positive and negative charges due to their motion are opposite, so the direction of the negative pressure produced by the above mechanism is also opposite, resulting in the Lorentz force direction being opposite. As shown in Figure 5-6.

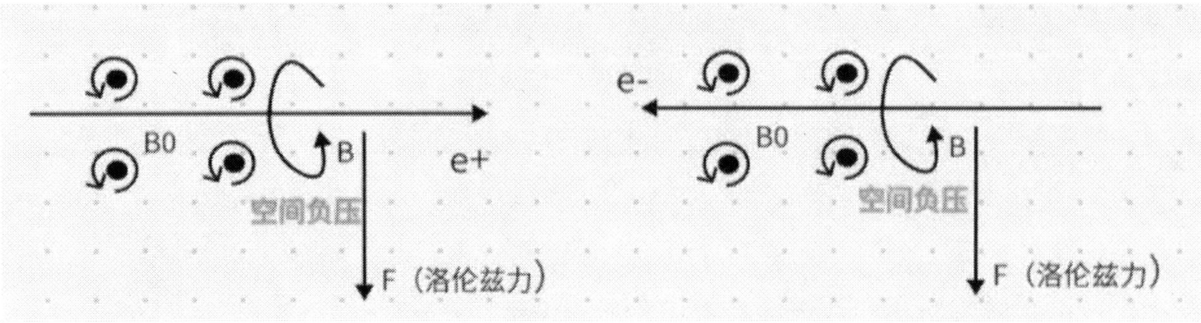

Figure 21: Figure 5-6 All Lorentz forces of positive and negative charges motion

The magnitude of the Lorentz force F is influenced by the following factors:
a. The influence of magnetic field B_0 and charge magnetic field B on the force, the larger the two magnetic fields, the greater the F, that is:

$F \propto B_0,\ F \propto B (5.9)$

b. The magnitude of the magnetic field B produced by a moving charge Q is related to the number of magnetic force vortex lines in the field. The larger the charge quantity Q, the more magnetic force lines are produced; the higher the velocity V of the charge, the greater the energy of the vortex rings, so the magnetic field B is proportional to the charge's velocity V:

$B \propto QV (5.10)$

c. The greater the velocity V of the charge Q, the more magnetic field lines it encounters and interacts with per unit time

The greater the charge velocity, the greater the sum of the Lorentz forces, that is:

$F \propto V (5.11)$

In summary, we can obtain:

$F \propto B_0 Q\ V (5.12)$

This is consistent with the experimentally verified Lorentz force formula:

$F = B_0 Q\ V (5.13)$

The above analysis is based on the case where the direction of the charge velocity is perpendicular to the magnetic field B_0. When there is an angle θ between them, the Lorentz force formula becomes:

$F = B_0 QV \sin\theta (5.14)$

(3) The Origin of the Ampère Force
When two current-carrying wires are placed side by side, the moving electrons in both wires experience the Lorentz force, resulting in a collective Lorentz force acting on the wires, which is known as the Ampère force.

The Ampère force can also be understood macroscopically as the magnetic fields generated by currents in the same direction spiral in the same direction and can merge, reducing the number of magnetic vortex rings between the two wires and creating a spatial negative pressure, thus producing an attractive force. Conversely, currents in opposite directions generate magnetic fields in opposite directions, and the magnetic vortex rings cannot merge, colliding between the two wires to create spatial positive pressure, resulting in a repulsive force. As shown in Figure 5-7.

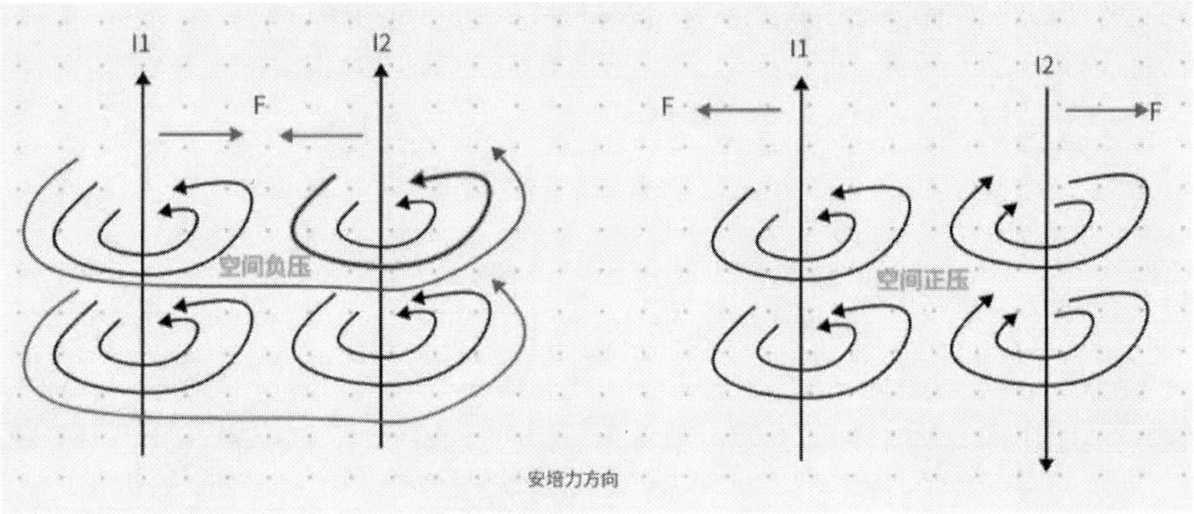

Figure 22: Figure 5-7 Schematic Illustration of the Principle of Ampère Force Generation

(4) Magnetic Force of a Magnet
The magnetic field of a magnet is the sum of an ordered magnetic field generated by countless charges inside it. All magnetic field lines converge from the N pole and flow into the S pole, forming closed spiral flow vortex rings. Between like poles, the magnetic field lines spiral in opposite directions, thus creating spatial positive pressure and repulsion. Between unlike poles, the magnetic field lines spiral in the same direction, resulting in fusion and the creation of spatial negative pressure and attraction. See Figure 5-8.

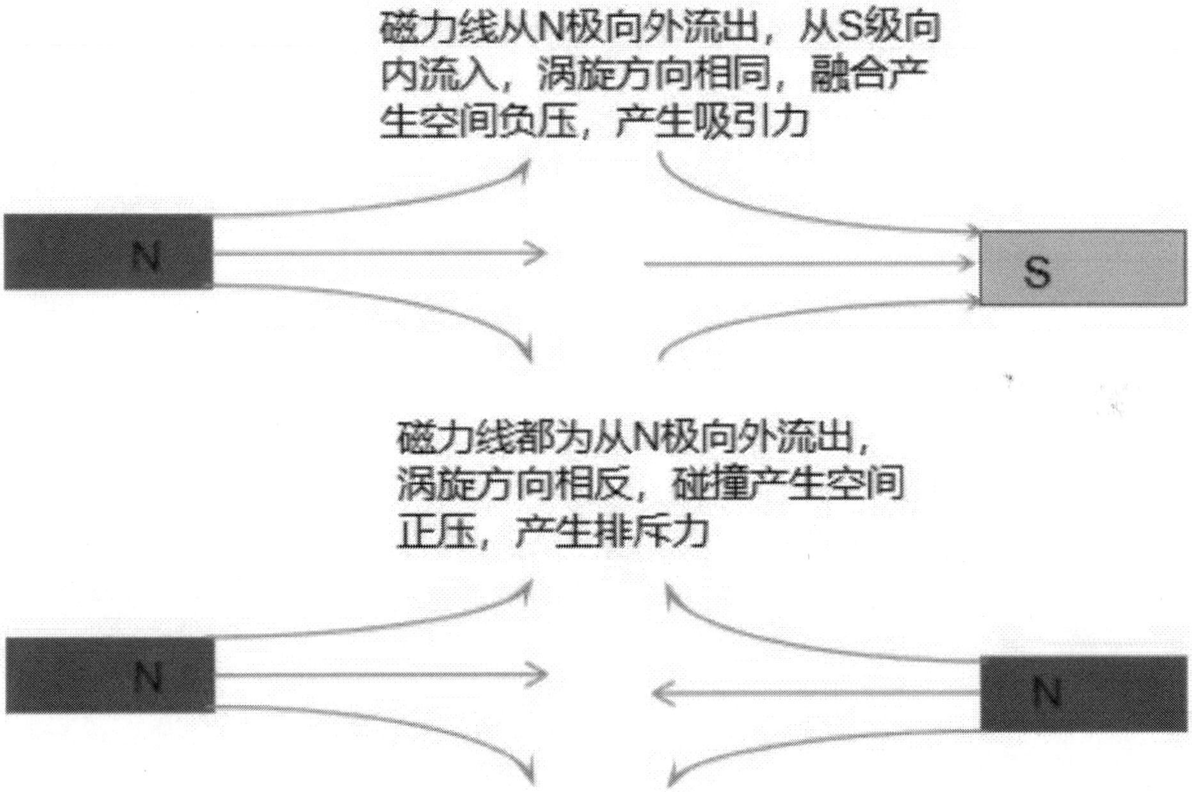

Figure 23: 1) Magnetic field lines are spatial spiral flow vortex rings, an unbounded closed field, flowing out from the N pole and into the S pole.
2) When observed in the direction of the magnetic field lines, they form a counterclockwise vortex.

Figure 5-8 Schematic illustration of the attraction and repulsion principles of a magnet's magnetic field

(5) Magnetic Monopoles
From the above analysis, it can be known that the magnetic field is a sourceless, closed-loop electric field, which is a spatial vortex flow forming vortex rings. The magnetic force is the effect produced by the closed-loop flow of spatial wave vortices, so magnetic monopoles do not exist.
(6) Magneto-optical Rotation Effect
As we analyzed above, the magnetic field is a closed spiral flow vortex ring, which has vortices along the direction of the magnetic field. On the other hand, photons themselves are particle-like, and they exhibit spiral fluctuations due to the influence of spatial wave oscillations in the vacuum, which is also the reason for their wave-particle duality.

Therefore, when photons or polarized beams move along the direction of the magnetic field, the vortex flow of the magnetic field itself affects the polarization direction of the photons, causing the polarization to deflect, which is known as the magneto-optical rotation phenomenon.

The greater the magnetic field strength B, and the longer the distance L that light travels along the magnetic field, the larger the change in the polarization angle θ caused by the influence of the magnetic field's vortex flow, that is,

$\theta \propto \mathrm{BL}(5.15)$

This has been found to have many instances in experiments, such as the Faraday effect.

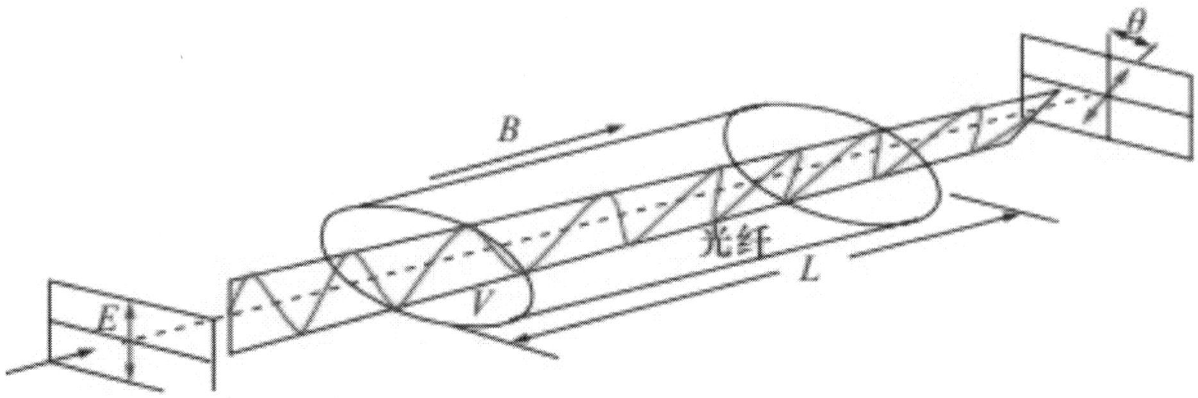

Figure 24: Figure 5-9 Schematic diagram of the Faraday effect for linearly polarized light propagating along an optical fiber

As shown in Figure 5-9, the Faraday effect [13] is a magneto-optical effect. When plane-polarized light is incident on a magnetized medium along the direction of light propagation, the polarization plane rotates. In 1845, Faraday first observed the magneto-optical effect, and first revealed the interaction between light and magnetic fields.

After a large number of experiments, it is found that the Faraday effect has $\theta = \mathrm{VLB}$ a regular pattern. θ represents the Faraday rotation angle, L is the distance the light travels through the medium, B denotes the magnetic induction strength, and V is called the Verdet constant, a physical quantity that characterizes the strength of the Faraday effect in the medium and is related to the medium's properties. This formula agrees with our previous speculation.

According to the principles of magnetic fields, when the direction of light propagation is reversed, the left and right directions of Faraday rotation will be swapped, which has also been confirmed by experiments and aligns with our prediction.

According to the principles of magnetic fields, the spatial vortex current loops are counterclockwise vortices of magnetic field lines. Therefore, when the direction of light propagation is parallel to the magnetic field direction, and the Verdet constant is positive, the polarization vector rotates counterclockwise (CCW) around the optical axis.

When the incident light is in the opposite direction to the magnetic field, the rotation becomes clockwise (CW). This is consistent with our theoretical prediction of the counterclockwise vortex of the magnetic field.

According to the principles of magnetic fields, the magneto-optic effect is essentially due to the influence of the magnetic field's own vortex flow on the light wave. Therefore, this phenomenon can exist in all media, although the different paramagnetic properties of various materials will result in different enhancements of the magnetic field, leading to different strengths of the optical rotation and different variations in the deflection angle θ .

In actual experiments, Faraday discovered the magneto-optical effect, and Verdet studied the magneto-rotation of many media, finding that the Faraday effect exists in solids, liquids, and gases. This also aligns with our prediction.
(7) Broken magnetic field and sourceless electric field
When the magnetic field is shielded or disturbed, some magnetic field lines break away from the spiral vortex ring structure, becoming broken magnetic fields or open-loop magnetic field lines. This broken magnetic field is no longer in the complete spiral vortex ring form, but instead becomes a spatial wave propagating in the void space, forming a fluctuation in the void space. The magnetic effect of this void space fluctuation is very weak, but it still possesses a certain degree of magnetic potential effect.

In 1959, Aharonov (Y. Aharonov) and Bohm (D. Bohm) published a paper which proposed that, in the space where an electron moves, the phase of the electron's wave function is affected by the electromagnetic potentials in the space, regardless of whether an electromagnetic field exists. This is the famous AB effect [14] , as shown in Figure 5-10. This effect has been experimentally confirmed and is consistent with our previous prediction.

Aharonov - Bohm Effect

- A solenoid with the current is placed just behind the slits
- Uniform magnetic field is inside the solenoid
- Magnetic field outside is equal to zero
- But vector potential of the magnetic field is non - zero outside!

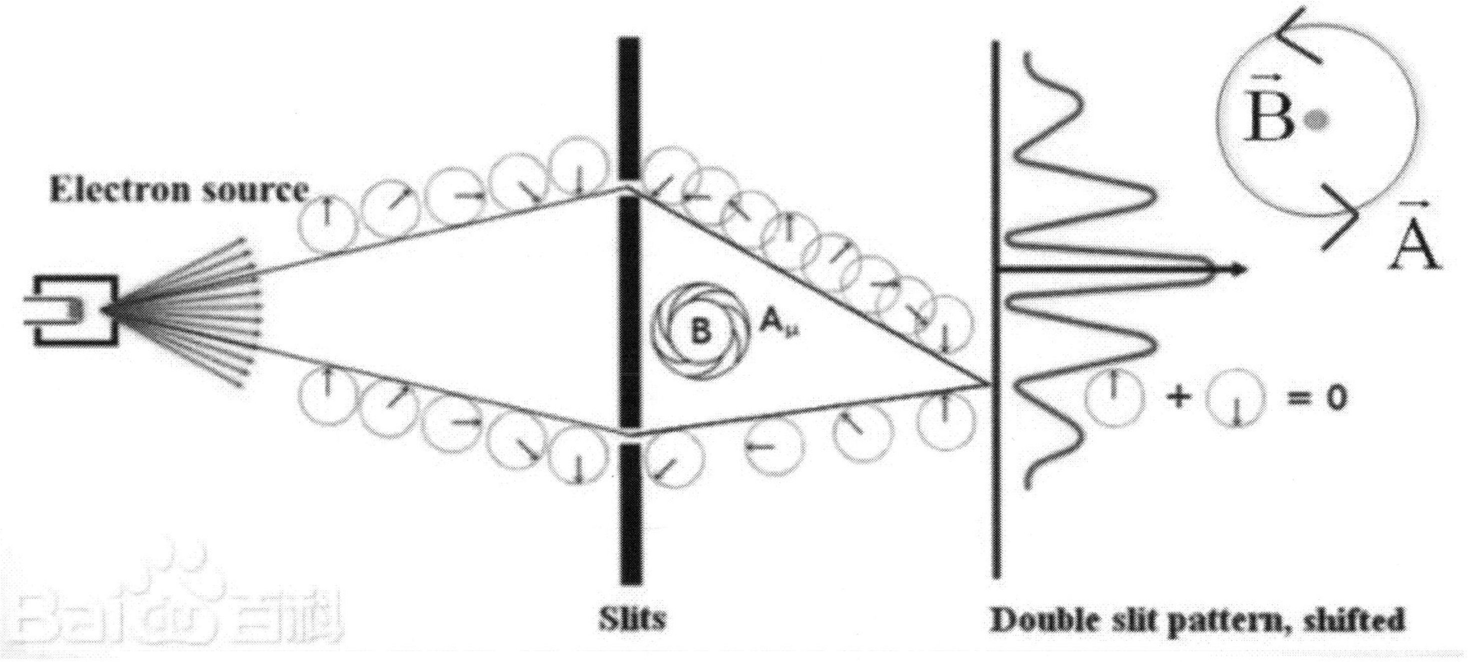

Figure 5-10 Schematic diagram of the AB effect

On the other hand, broken open-loop magnetic field lines are equivalent to sourceless electric field lines. At the same time, under conditions of shielding and interference, some electric field lines may also break, becoming spatial fluctuations propagating in the void space: broken sourceless electric field lines. The electric field effect of broken electric field lines becomes extremely weak, but there is still a certain electric potential effect.

In 1984, Aharonov and Casher predicted the AC effect based on the duality of electricity and magnetism, and this prediction was experimentally verified, which also aligns with our speculation.
(4) Gravity
As we analyzed earlier, all mass bodies in the void space are composed of the most basic wave particles stacked in different coherent structures and motion patterns. The superfluid properties of void space determine that there are no additional gravitational or repulsive forces between wave particles, only forces resulting from their mutual motion and collisions.

So, a mass body formed by an infinite number of waves, including various particles, even an infinite number of particles stacked again

Accumulated celestial bodies, matter, etc., themselves do not exist with mutual universal gravitation. The so-called universal gravitation is the result of the interaction between a mass body and its surrounding empty space, or more precisely, the macroscopic manifestation of the force arising from the interaction with countless wave particles in the surrounding empty space.

(1) Distortion of mass bodies and empty space

The mass body itself is composed of the most basic wave particles stacked, even multiple times, and is the high-energy motion and aggregation of wave particles, so the mass density of the mass body is much greater than that of the surrounding ordinary vacuum space. There is also a large amount of vacuum space within the mass body, or in other words, a large number of wave particles, such as inside atoms where there is a vast amount of vacuum space. These vacuum spaces are also filled with wave particles. Due to the intense thermal motion of particles within the mass body, it causes the internal vacuum space to also exhibit a higher thermal motion effect, which results in the density of this part of the vacuum space being less than that of the surrounding vacuum space outside the mass body.

At the same time, since the mass body itself also undergoes thermal motion, it creates disturbances in the surrounding empty space, causing the empty space around it to be excited into thermal motion as well, thereby reducing the density of the empty space.

As the distance from the mass body increases, the ability of the mass body's thermal motion to excite the empty space weakens, leading to a decrease in the thermal motion of the empty space and a gradual recovery and increase in its density. This change in the density of the empty space forms a density gradient, that is:

$\rho_m < \rho_0 < \rho_r$

Where: ρ_m is the density of the empty space inside the mass body, ρ_0 is the density of the empty space outside the surface of the mass body, and ρ_r is the density of the empty space at a distance r from the surface of the mass body.

Or:

$$\rho_r \propto \frac{r^2}{M}$$

Moreover:

$$\rho_r < \rho_c$$

Where: M is the mass of the mass body, r is the distance from the surface of the mass body, and ρ_c is the standard density of empty space.
The density gradient of this void space, when viewed macroscopically, manifests as a distortion in the void space around a mass body, meaning the density of the void space decreases toward the vicinity of the mass body.

This distortion becomes more pronounced as the mass of the mass body increases. For example, all planets, stars, and even galaxies exert an increasingly powerful converging and distorting effect on the surrounding void space.

Let the density of the void space at a distance r from the mass body with mass M be ρ_r , then:

$$\rho_r = \rho_0 + K_\rho \frac{M}{r^2}$$

Equation 5.16 is the equation for the distortion of the void space by a mass body, where ρ_0 is the density of the void space on the surface of the mass body, and K_ρ is the distortion coefficient.
(2) Drag of mass bodies on empty space
The wave particles have mass and can exert force through collisions, resulting in the exchange of momentum and energy. Therefore, when a mass body moves, it interacts with the surrounding empty space's wave particles, pushing and dragging them. The isolated particles, micro-particles, and fundamental particles formed around the mass body can also be considered as rigid mass bodies, so they will also experience the same effect and be influenced by the movement of the mass body.

The motion of these mass bodies further exerts influence on surrounding waves, particles, etc., so from a macroscopic overall perspective, it appears as the motion of mass bodies interacting with the surrounding empty space, creating a dragging and leading effect on the empty space, causing the surrounding empty space to move. This is the dragging effect of mass bodies on the surrounding empty space.

As the distance from the mass increases, the dragging effect of the mass on the vacuum space gradually weakens. Therefore, macroscopically, there is a velocity gradient in the vacuum space around the mass, with the velocity increasing as it gets closer to the mass, and the vacuum space closest to the mass's surface reaching synchronous motion

Motion, relative motion velocity is zero; the farther away from the mass body, the slower the motion speed of the void space, and the greater the relative motion velocity with respect to the mass body, eventually completely losing the dragging effect and achieving relative stillness with the distant uniform void space.

This dragging effect includes all movements of the mass body, such as vibration, self-rotation, orbital motion, etc. The greater the mass of the mass body, the larger the range of its influence on the surrounding void space and the stronger the dragging capability. For example, all planets, stars, and even galaxies will form increasingly powerful dragging effects on the surrounding void space.

Let the velocity of the mass body M be V , and the dragging motion velocity of the void space at a distance r from its center be v_1 , then:

$$v_1 = K_v \frac{M}{r} V$$

Equation 5.17 is the equation for the dragging of a mass body on empty space, where K_v is the dragging coefficient.
(3) Drag of the void space on mass bodies
a. Static pressure of space
Overall, the void space can be considered as a superfluid with mass and density, so when a mass body is at rest in the void space, it is constantly subjected to pressure from all directions of the void space, which is referred to as spatial pressure.

Just as we are in the atmosphere and constantly experience pressure from all directions, but have become accustomed to it and are unable to perceive it.

Because there is also a large amount of void inside a mass body, there will be the existence of void space fluid (or a large number of wave particles, etc.), which can connect to the void space outside the mass body and flow freely, and can be called free void space V_c . The part of the void space inside the mass body that is twisted and dragged by mass particles and cannot flow freely can be called fixed mass space V_m。

Let the total volume of the mass body be V, then:

$V = V_m + V_c(5.18)$

The pressure exerted by the void space on a mass body is equal to the mass of the void space it displaces, so the magnitude of the spatial static pressure F experienced by the mass body is related to its volume V, the density of the external void space ρ_c, and the density of the internal free space V_c and free void space ρ_0. That is:

$\mathrm{F} = \rho_c V_m + (\rho_c - \rho_0) V_c \ (5.19)$

Since the density of the vacuum space around the surface of a mass body is not much different from that inside, the spatial pressure can generally be simplified as:

$\mathrm{F} = \rho_c V_m (5.19 - 1)$

Equations 5.19 and 5.19-1 are the equations for the static pressure of the vacuum space on a relatively stationary mass body.
b. Drag effect of spatial motion
When a mass object moves through the void space, it experiences a force of motion caused by the void space, which is also referred to as spatial pressure. This is a resultant force formed by a dynamic pressure difference in the space.

The rotation of the Sun causes the surrounding void space to produce a vortex motion, which in turn further drives the motion of surrounding planets, resulting in the seven planets orbiting the Sun. This makes the seven planets become the Sun's satellites.

The direction of the vortex motion in the surrounding void space should be the same as the direction of the Sun's rotation, meaning the orbital direction of the seven planets is also the same as the Sun's rotational direction.

Not only stars, but also the rotation of planets can produce similar effects. The rotation of Earth causes the surrounding empty space to form a vortex movement, which in turn drives the Moon to orbit around Earth. Therefore, the direction of the Moon's revolution around Earth is the same as the direction of Earth's rotation.

Not only Earth, but also all the satellites of the planets in the solar system revolve around their respective planets in the same direction as the planets' rotation. This is also true for all other stars and planets.

When the scope is expanded to the galaxy, the massive central mass (mass cluster) of the galaxy rotates, causing the surrounding vast space to form a vortex, which further drives all the stars in the vicinity to rotate

The orbital motion around the galactic center is in the same direction as the rotation of the galactic center's mass group. These also align with the theoretical predictions. It can be said that the majority of satellites and celestial bodies orbit the central planet or mass group in the same direction as the rotation of the central planet or mass group. The formation of this orbital direction is influenced by the rotation direction of the nebula during the birth of the galaxy, as well as the continuous dragging effect of the rotation of the central planet or mass group on the surrounding empty space after its formation. The latter ensures the stability and continuity of the orbital direction of satellites and galaxies.

Due to factors such as accidental collisions and external forces, the orbital direction of some celestial bodies may differ from the rotational direction of the central mass body. However, once the external forces disappear, this orbital direction will not be sustainable and will gradually be corrected under the influence of spatial motion forces.
c. Spatial Motion Forces
Let the density of the void space be ρ_c, the velocity be $\overrightarrow{v_c}$, the volume of the mass body be V, the fixed mass volume inside the mass body be V_m, the cross-sectional area in the direction of motion be A, and the velocity be $\overrightarrow{v_m}$, then the drag force of the void space on the mass body is F_c:

$$F_c = K_F \rho_c \frac{V_m}{V} A \left(\overrightarrow{v_c} - \overrightarrow{v_m} \right)$$

The coefficient of force in the void space, denoted as K_F, is related to the shape of the mass body and its relative velocity with respect to the void space ($\overrightarrow{v_c} - \overrightarrow{v_m}$).

Considering that the void space can generally be regarded as a superfluid, the force exerted by the void space on a mass body moving at velocity v can also be equivalently expressed as:

$$F_l = K_l \rho_c \frac{V_m}{V} A v^2$$

where K_l is the force coefficient, related to the shape and other properties of the mass body.
5.20 and 5.21 are the equations for the force of motion in the empty space.
Due to the superfluidity and superpermeability of empty space, a large amount of empty space within a mass body can flow freely and be permeated by external empty space, so its force equation differs from that of other ordinary macroscopic fluid equations.

When the velocity of empty space v_c is greater than the velocity of the mass body v_m , empty space exerts a drag force on the mass body. When the velocity of empty space v_c is less than the velocity of the mass body v_m , it exerts a resistive force on the mass body.

Because the motion interaction between the mass body and empty space is relative, when empty space itself is in motion, the mass body outside does not experience any active force being applied. The mass body can simply start moving on its own due to the uneven spatial pressure. Just like an object stationary in the air can be moved by a strong wind.

(4) Origin of Gravity

a. Microscopic Motion of Mass Bodies

Observing from a macroscopic perspective, a mass body such as a sphere appears to be stationary on the ground with no motion. However, when observed at the microscopic level, the mass body is actually undergoing intense vibrations, and the various particles within it are engaged in a large number of complex, high-energy movements.

If a glass of water is placed there, its surface appears still, but in reality, various complex flows and waves are occurring internally, including dense vibrations of water molecules. These disordered movements subject the particles in the water to complex mechanical forces, causing the particles to exhibit irregular motion, which is the Poisson motion we can observe.

The mass body itself is a concentrated entity of high-energy particles. From a microscopic perspective, all mass bodies undergo similar complex and intense movements, releasing significant amounts of kinetic energy. This will strongly perturb and excite the surrounding vacuum space, resulting in high-energy flows, waves, and vortices in the vacuum, which are forms of thermal motion.

The fluctuations in vacuum space are mainly longitudinal waves. When the thermal motion energy of the vacuum increases, it causes changes in the pressure exerted on the mass body. This is the root cause of the gravitational phenomena appearing around the mass body.

b. Static Gravity

When the empty space between two relatively stationary mass bodies is excited by the two mass bodies, the wave speed and energy of this portion of empty space are relatively higher. According to Bernoulli's equation, this will cause the two

A slight spatial negative pressure is generated between the spheres, creating an attractive force that pulls them closer to each other.

This attraction is the static gravitational force between two massive bodies.

According to equation 5.16, it can be inferred that the intensity of the fluctuation in the surrounding vacuum space, the energy, and the mass of a mass body are proportional to its mass and inversely proportional to the square of the distance from the mass body. Let the masses of two mass bodies be M and m, and the distance between their centers be r, then the gravitational force F between them is:

$$F = K\frac{Mm}{r^2}$$

K is the gravitational coefficient between two mass bodies, which conforms to the gravitational formula we are familiar with:

$$F = G\frac{Mm}{r^2}$$

Therefore, gravity is not a real force emitted by mass bodies; it is a virtual force, representing the pressure difference in space between two or more mass bodies.

In fact, equation 5.22 is merely a simplified hypothetical equation. By considering the interaction and energy relationship between the void space and mass bodies, and incorporating the Bernoulli equation from fluid dynamics to account for the space pressure difference, we can derive a detailed expression for the gravitational coefficient K and the complete gravitational formula. When the gravitational formula is simplified under certain conditions, it can yield equation 5.22. Through this derivation, we can also obtain the detailed mathematical expression behind the gravitational constant G and its corresponding physical meaning, which is related to the interaction and energy exchange between mass bodies and the void space.

c. Centripetal Motion Gravity

When a small mass m orbits a large mass M , m is subject to an additional spatial pressure directed towards the large mass M, in addition to the static gravitational force between the large mass M and itself. Let the orbital radius of m be r, and its velocity v, then according to equations 5.21 and 5.22:

$$F_j + F_1 = K\frac{Mm}{r^2} + K_1\rho_c\frac{V_m}{V}Av^2 = \frac{mv^2}{r}$$

According to the gravitational formula, we have:

$$K\frac{Mm}{r^2} + K_1\rho_c\frac{V_m}{V}Av^2 = G_1\frac{Mm}{r^2}$$

When the force of the void space on the mass body m is not considered, the above formula simplifies to:

$$K\frac{Mm}{r^2} = G\frac{Mm}{r^2}$$

The gravitational constant G_1 differs in the two cases, as the motion of an object orbiting a mass body exhibits a different gravitational effect compared to an object at rest relative to it.

In summary, we can conclude that gravitons do not exist, and gravity is not a force emitted by mass bodies. Its essence is the spatial pressure of the void space, and it is a macroscopic phenomenon caused by the spatial pressure difference between two or more mass bodies.

(5) Range Gravitational Field

The output of vibrational energy from a mass body causes the surrounding void space to have enhanced wave capabilities and reduced spatial pressure. Therefore, a spatial pressure difference gradient field is formed around the mass body, and the macroscopic manifestation of this spatial pressure difference gradient field is a gravitational field pointing toward the mass body.

Such a spatial pressure difference gradient field exists around two mass bodies, leading to further changes in the spatial pressure gradient and further reduction in spatial pressure, thus forming a gravitational field between the two mass bodies.

The gravitational field between these two mass bodies not only affects the two mass bodies but also all mass bodies in the edge areas of the low-pressure region of the entire space. This is known as the range gravitational field, as shown in Figure 5-11.

In both the inner gravitational field and the outer gravitational field regions, the third mass m3 is subject to the gravitational forces from the two masses m1 and m2, as well as an additional gravitational force directed toward the center of the inner gravitational field. This additional gravitational force is caused by the gradient of the spatial pressure minimum existing in the outer and inner gravitational fields.

Figure 25: Outer Gravitational Field

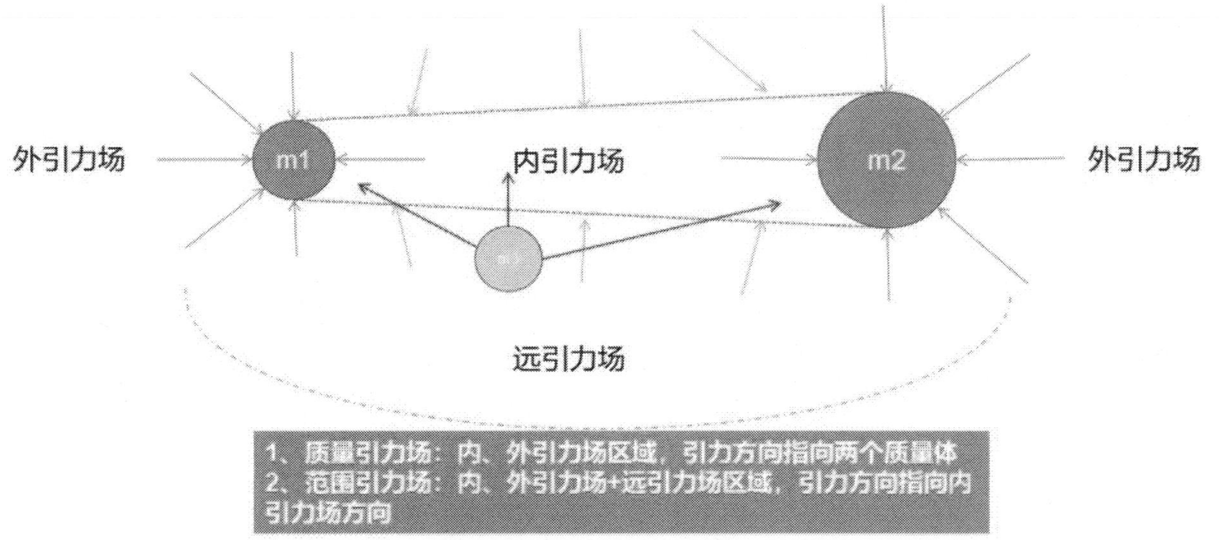

The Gravitational Field of the Void Space Range

The addition of a third mass body m3 will further strengthen both the distant gravitational field and the internal gravitational field. The traditional concept of gravity merely considers that mass bodies form linear gravitational forces pointing towards each other, whereas in the range gravitational field, there will also exist additional range gravitational forces pointing in the direction of the internal gravitational field.

The more mass aggregates, the stronger the gravitational field effect over a larger range, and the overall gravitational effect exhibited will be significantly greater than the linear gravitational effect between individual mass bodies.

In modern astronomical observations, the motion of large galaxies and galaxy clusters such as the Milky Way has been observed to exhibit a powerful converging force far exceeding the normal gravitational interaction between stars, suggesting the presence of an additional gravitational effect. For example, in 1933, astrophysicist Fritz Zwicky used spectroscopic redshift to measure the velocities of individual galaxies within the Coma Cluster relative to the cluster itself. Using the virial theorem, he found that the velocity dispersion of the galaxies was too high to be explained by the gravitational pull from the visible mass of the galaxies within the cluster alone.

Figure 26: Figure 5-12 Gravitational field range for the formation of galaxy mass clusters

The additional gravitational effects in this puzzling galaxy originate from this range gravitational field effect. Due to the existence of the range gravitational field, the almost non-viscous superfluid properties of the void space combine with all the matter in the galaxy to form a massive fluid that appears to have "viscosity," producing a powerful converging force stronger than classical gravitational effects, as shown in Figure 5-12.

(5) Other forces, Casimir effect

Electric field force, magnetic field force, and gravitational force are all macroscopic forces that manifest from spatial wave fluctuations on a macro scale. These three forces are all expressions of spatial pressure differences under different modes of spatial wave motion, and at the microscopic level, they are all forces resulting from the mass effects of spatial wave particles.

In addition to these macroscopic forces, there are also numerous various forces acting at the microscopic level. Different fundamental particles, as well as yet-to-be-observed micro-particles and solitary particles, all interact with each other through various forces. The essence of these forces is also the result of mass effects under different modes of motion.

All forces are essentially mass forces, that is, the forces resulting from momentum and kinetic energy transfer due to collisions between particles

The exchange and transfer of results only manifest as a pressure difference in space on a macro scale.
As in the Casimir effect, it is essentially a macroscopic spatial pressure difference.
Vacuum space is a superfluid, with all the quanta in space vibrating and moving constantly, forming an overall spatial wave motion in the vacuum. In other words, vacuum space is always filled with various spatial fluctuations, which are mainly longitudinal waves and also include various complex flow movements.

When two metal plates gradually approach each other, about a few hundred nanometers apart, the distance between them becomes close to or reaches the wavelength region of the mainstream long-wave in space waves. Therefore, the long-wave of space waves will gradually be excluded, and the fluctuation energy and fluctuation speed of space waves will increase, leading to a greater spatial negative pressure between the two metal plates, and thus causing an additional attraction between them. This is the Casimir effect.

Even according to the equation for generating "gravity" through spatial pressure, the normal spatial pressure difference at this moment might increase significantly (a more detailed derivation of the spatial pressure coefficient is needed to confirm this).

In addition to the Casimir effect, many other mechanical phenomena can also be explained by the same principle.

5.4 Anti-gravity Principle

Anti-gravity has always been an ideal propulsion that humans have relentlessly pursued. Through the above analysis, we have understood the principles and nature of gravity. To achieve anti-gravity, we must also consider the formation mechanism of gravity.

Gravity is a manifestation of the macroscopic spatial pressure gradient field caused by the thermal motion of mass bodies leading to the movement and collision of countless particles in the surrounding vacuum space. It is a virtual force, not a real existing force. Therefore, there are several ways to achieve anti-gravity.

(1) Creating reverse spatial pressure: reactionless engine
Gravity is a spatial pressure difference pointing toward massive objects (such as Earth). To create anti-gravity, one would need to produce a reverse spatial pressure difference to counteract the gravitational effect. For example, through special propulsion

The system generates strong spatial flows toward Earth, which in turn can create reverse spatial pressure on the propulsion system and aircraft, thereby counteracting gravity and forming the driving force for the aircraft's movement.

Considering the equations we analyzed earlier, such as the equation for the distortion of empty space by a mass body (Equation 5.16), the equation for the dragging of empty space by a mass body (Equation 5.17), the equation for the static pressure of empty space on a relatively stationary mass body (Equation 5.19), and the equation for the force of motion of empty space (Equation 5.21), we can perform dynamic design based on these equations to generate reverse space flow, create reverse space pressure, and achieve the goal of anti-gravity.

As with all mass jet engines, they actually cause the flow of empty space, but this unoptimized propulsion has a minimal effect on the movement of empty space and cannot generate effective spatial thrust.

For example, as we analyzed earlier, a magnetic field is a type of spatial vortex flow field. Therefore, if we create a strong, directed toward Earth, non-closed instantaneous magnetic field spatial flow, i.e., an instantaneous open-loop magnetic field spatial flow, it would also produce additional spatial thrust that can counteract gravity.

Another method is to create a powerful spatial vortex to produce a unidirectional jet, which would result in the maximum spatial flow and generate strong spatial pressure. This can be achieved through methods such as ultra-high-speed vortex of high-density fluids, to create such spatial vortices and control the direction of spatial jet.

Because spatial vortex jet differs from traditional engines, it does not require traditional propellants such as fuel, and can be considered as a propellantless engine, which is an important propulsion method for future space travel.
(2) Manufacturing special materials or equipment to eliminate the factors that generate gravity
According to the formation principles of spatial pressure and gravity, as described by equations such as 5.19, 5.21, and 5.22, gravity is formed due to the high-energy fluctuations created by a mass body in the surrounding empty space, resulting in a spatial negative pressure between two mass bodies.

Therefore, to eliminate gravity, one can focus on canceling this spatial negative pressure. If special materials can be manufactured that reduce or prevent the transmission of fluctuation energy to the external empty space, then

the spatial negative pressure between objects and the Earth can be reduced, thereby decreasing the pressure. If the ability to weaken the fluctuation energy in the external empty space is further enhanced, it can weaken or even completely cancel the spatial negative pressure formed by the Earth or other mass bodies. If the weakening capability of the external empty space fluctuation energy is further improved, it can create a reverse spatial pressure, which is anti-gravity, and serve as the driving force for spacecraft.

Of course, such an effect can also be achieved by using specially designed equipment in conjunction.
(3) Manufacturing spatial shielding and spatial bubbles
Since the spatial pressure that creates a gravitational effect acts on the spacecraft through the void space, another way to counteract gravity is to create shielding and isolation of the void space, separating it from the spacecraft. In this way, the spacecraft would no longer experience this spatial pressure, and thus would not feel the gravitational effect.

For instance, creating a sudden spatial bubble through a spatial wave burst could temporarily shield the spacecraft from all forces transmitted through the void space, including gravitational effects, thereby achieving anti-gravity and even generating spatial propulsion.

In the deep sea, some submarines or torpedoes can significantly increase their speed by creating such a bubble, and this principle is also applicable to aircraft flying in the void space.

However, the flow and fluctuation speed of void space is extremely fast, likely at the speed of light, so a very high density of continuous space bubble creation is required to maintain a long-term space negative pressure, achieving anti-gravity and space propulsion power.

When flying in a space bubble, since there is no force from the void space, the aircraft no longer experiences inertial resistance caused by the void space's forces, allowing its speed to instantly reach a very high level. Inside the aircraft, through the design of artificial gravity fields, pilots can also be free from the limitations of acceleration inertial forces. In ordinary aircraft, pilots can already feel the limit of 9 Gs, but in a space bubble, there is no such restriction because there is no inertial force to resist. Therefore, even a small force can cause significant acceleration of the aircraft, and even with large acceleration, it will not harm the aircraft or the people inside.

(4) Enhance the gravitational pull of other massive bodies

Like spacecraft on Earth, they are subject to Earth's gravity and the counteracting gravity of the Moon, but because the distance to the Moon is too great, the gravitational force is very weak. By applying the principle of gravity, the gravitational force between the spacecraft and the Moon can be enhanced, thereby achieving the effect of canceling gravity or even generating gravitational propulsion.

Considering gravity is caused by the spatial negative pressure resulting from the increase in spatial wave motion energy, it is possible to increase the spatial energy output towards the Moon through methods such as high-energy electromagnetic waves disturbance, thereby enhancing the spatial wave energy between the Earth and the Moon, increasing the spatial negative pressure, and thus increasing the gravitational force between them, achieving the effect of counteracting Earth's gravity.

In summary, these are several approaches to creating anti-gravity based on the principles of spatial motion. Of course, in actual engineering implementation, more influencing and interfering factors need to be considered.

The numerous reports of UFOs and flying saucers around the world suggest that if such powerful alien vehicles truly exist, their propulsion system is very likely based on this spatial power. Because this spatial power can generate strong spatial flows, it would cause significant electromagnetic interference, leading to the failure of all electrical devices, communication systems, lighting, etc., and the lights would go out. After the UFO passes, the systems would return to normal. If such intense electromagnetic interference phenomena exist in many UFO reports, it may indicate a real alien UFO event.

5.5 Gravitational and Inertial Mass

According to equation 5.19-1, a stationary mass will experience a spatial pressure F from the surrounding vacuum space:

$$F = \rho_c V_m$$

ρ_c is the density of the surrounding vacuum space where the mass body is located, V_m is the fixed mass space volume inside the mass body.

Two mass bodies m and M are relatively stationary in their respective vacuum spaces due to the vacuum space fluctuations between them

Energy叠加, enhancement, creates spatial negative pressure, resulting in a spatial pressure difference pointing towards each other.
We examine the small mass body m, where the spatial pressure in the direction opposite to the large mass body M is normal F_b , and the spatial pressure in the direction facing the large mass body M is F_a , then the "gravitational" force F experienced by it from the large mass body is:

$$F = F_b - F_a$$

The acceleration a exhibited by the mass body is:

$$a = \frac{F}{m} = \frac{F_b - F_a}{m} = \frac{\rho_b V_m - \rho_a V_m}{m}$$

Where ρ_b、ρ_a represents the density of the void space in the direction opposite and facing the large mass body M for the mass body m . By observing the acceleration a and the gravitational force F, we can obtain the gravitational mass of the mass body:

$$m = \frac{\rho_b V_m - \rho_a V_m}{a}$$

Gravitational mass is the mass observed based on the ability to change the motion of a mass body through gravity.
We now consider that when a mass body m is at rest in the void space, it experiences spatial pressure from all directions F_o which is balanced. If an external force F 1 is applied to m, it will cause an imbalance in the forces acting on m, resulting in accelerated motion:

$$a = \frac{F_1 - F_c}{m} = \frac{F_1 - \rho_c V_m}{m}$$

Where ρ_c is the density of the void space surrounding the mass body m .
By observing the acceleration a and the external force F, we can determine the inertial mass of the mass body:

$$m = \frac{F_1 - \rho_c V_m}{a}$$

Inertial mass is the mass observed based on the external force's effect on the mass body's ability to change its motion.
can be considered as the same mass, as the gravitational mass is the mass manifested under the drive of spatial pressure, and the inertial mass is the mass manifested as resistance from spatial pressure.

The essence is the same, both representing the intensity of the action of empty space on mass bodies.
The magnitude of gravitational mass and inertial mass is related to the density of the vacuum space where the mass body is located ρ_c and the fixed mass space volume within the mass body V_m .

5.6 Rest Mass and Moving Mass

The mass exhibited by a mass body when it transitions from a state of rest to motion under the influence of a force, such as gravitational mass and inertial mass analyzed above, can all be considered as rest mass.

According to equation 5.21, when a mass body is in relative motion within the void space, it experiences a force exerted by the void space:

$$F_l = K_l \rho_c \frac{V_m}{V} A v^2$$

As we can see, as the speed of an object increases, the force exerted by the vacuum space on it increases exponentially with the square of the speed. If this is equated to the mass effect during the object's motion, the observed result is that the object's mass increases with the increase in its velocity. This is known as the relativistic mass of the object.

Additionally, in the previous section on spatial electrodynamics, "Achieving Superluminal Principle," we analyzed that when particles or mass bodies are accelerated by external forces, the acceleration effect decreases as their speed increases, even approaching the speed limit of the external force, until it tends toward zero.

For example, if a particle is accelerated by electromagnetic force, since the speed limit of electromagnetic force is the speed of light, the acceleration effect on the particle diminishes as its speed increases and approaches the speed of light, eventually tending toward zero.

If this effect is equated to the mass effect during the object's motion, then the observed result is also that the mass of the mass body increases with the increase in its velocity.

under the two effects of the upper limit of acceleration force velocity and the change of spatial resistance, the acceleration by external force

A mass body will exhibit an increasingly significant mass increase effect as its velocity increases.
Therefore, the rest mass and moving mass of matter are consistent, and the magnitude of the mass is determined by the quantity of mass particles such as wave particles it contains, regardless of whether the matter is at rest or in motion (unless the matter absorbs or loses some mass particles when moving; assuming this does not occur, the mass of matter remains unchanged).

The mass effect of this substance is merely an observational equivalence, resulting from changes in the velocity of the mass body, which cause variations in external accelerating force and spatial resistance effects.

This also aligns with the basic principle of our spatial relativity 7: the principle of conservation of mass-energy, which states that the mass of a material system in a closed system remains constant, neither increasing nor decreasing out of nowhere, that is:

$$\mathrm{M}_g = \mathrm{M}_i = \mathrm{M}_s = \mathrm{M}_v$$

The gravitational mass M_g , inertial mass M_i , rest mass M_s , and moving mass M_v of a mass body are equal.

5.7 Galaxy Gravity and Dark Matter

According to the equations of mass body's distortion of the void space (Equation 5.16), the mass body's dragging of the void space (Equation 5.17), and other related equations, the larger the mass of a mass body, the greater its ability to distort and drag the surrounding void space. The denser the surrounding void space, the greater the total mass that the void space follows as the mass body moves.

At the same time, according to the equations of force in the void space motion (Equation 5.20, Equation 5.21), the movement of void space will also produce a dragging effect on other mass bodies within it.

A large galaxy contains a vast number of stars, planets, and other massive bodies, which create a very significant distortion and dragging effect on the surrounding void space. These dragged void spaces further pull these mass bodies and other mass bodies, eventually forming a vortex mass complex.

This vortex mass complex not only has a massive amount of mass from various stars and observable matter, but the overall mass of the void space that cannot be directly observed is also extremely large. Under the pressure from the most massive body at the galaxy center, the mass clusters, and the surrounding void space with the highest density and mass, the entire galaxy and the surrounding void space are pulled into motion.

Especially the superfluid properties of the void space make the entire galaxy exhibit distinct characteristics of vortices and swirling flows.

This is why all galaxies have a spiral shape.
The massive mass and force of the void space play a significant role here; otherwise, without considering the effect of the void space, it would be impossible to achieve such a strong pulling force and form such a standard vortex structure.

The spiral structures of all galaxies are essentially the result and manifestation of the superfluid vortex motion in the void space.

In the physics community, since astronomer Jacobus Kapteyn proposed in 1922 that the existence of invisible matter around stars could be inferred indirectly through the motion of stellar systems, numerous scientists and studies have found substantial evidence, such as (1) the velocity dispersion of galaxies within galaxy clusters is too high to be explained by the gravitational pull from the visible galaxies alone, indicating the presence of a large amount of dark matter with a mass at least 100 times that of the visible galaxies. (2) The rotational velocity of stars in the outer regions of galaxies is much higher than expected by Kepler's law, corresponding to a larger mass-to-light ratio. This suggests that there may be a significant amount of dark matter in these galaxies. (3) The dark matter in our own galaxy cluster is about ten times more massive than the visible matter. (4) The velocity of stars in the outer regions of galaxies remains constant over a considerable range. This implies that there may be a substantial amount of invisible matter not only concentrated in the core of galaxies, but also distributed throughout their outer regions, with a mass far exceeding the total mass of luminous stars. (5) The gravitational lensing effect observed when looking at background galaxy clusters, the temperature distribution of hot gas in galaxies and galaxy clusters, and the anisotropy of the cosmic microwave background radiation.

All of these have led the physics community to conclude that dark matter, which we are currently unable to observe, is universally present in galaxies throughout the universe.

This aligns with the predictions of our theory, and dark matter can be considered as the void space we have discussed

In this context, we consider the empty space as a whole and a black box, including the wave particles we have defined in our discussion. We have not yet reached a conclusion on how many different substances dark matter is composed of, but overall, they can be equivalent to the entities of the empty space we have mentioned.

Another source of the additional gravity beyond the gravitational influence of the observable mass in a galaxy is the range gravitational field effect discussed in our previous gravity chapter.

5.8 Gravitational Waves, Gravitational Lenses

Gravity itself is a macroscopic manifestation of spatial pressure difference, essentially spatial pressure, which is also transmitted through fluctuations in the vacuum space. Therefore, what is called a gravitational wave is the wave motion of the spatial wave.

We have analyzed before that the propagation speed of spatial wave fluctuations is no less than the speed of light. Assuming the propagation speed of the spatial wave is the speed of light, then the propagation speed of the so-called gravitational wave is also the speed of light.

That is, when two massive bodies, such as the Earth and the Moon, move relative to each other, the speed at which the change in their gravitational interaction propagates is also the speed of light.

In our previous overview of spatial electrodynamics, we discussed that massive bodies distort the surrounding empty space, creating density gradient changes. These density gradient changes cause a lensing effect in the empty space around the massive bodies. Since all electromagnetic waves and gravitational effects propagate through empty space as wave phenomena, they will all experience refraction when passing through the spatial lens around a massive body.

For gravity, this spatial lensing phenomenon is known as gravitational lensing, which refers to the refraction effect experienced by gravitational waves as they pass through a spatial lens. Its essence is a spatial wave lens, and it exists both in macroscopic cosmic space and microscopic particle space, following the same refraction laws as the light lenses we use and observe in daily life.

The refractive index n is determined by the density of the vacuum space within the lens region and the condition of the density interface.

5.9 Redshift and Blueshift of Light

The redshift and blueshift of light waves mainly occur in the following situations.
(1) Change in spatial density
Photons and spatial waves move together, forming the light wave we observe. The energy of a light wave with a fixed frequency in a unit volume is fixed. When passing through regions of different vacuum space densities, its propagation speed changes. Since the frequency remains unchanged, the wavelength changes, which is the phenomenon we observe as redshift and blueshift.

For example, when a light wave enters a region of higher vacuum space density, its wavelength becomes longer and its speed increases, resulting in redshift. When it enters a region of lower vacuum space density, its wavelength becomes shorter and its speed decreases, resulting in blueshift.

The gravitational redshift we refer to also falls into this category. When starlight passes near a massive planet, because the vacuum space density around the planet decreases, it experiences blueshift upon entering and redshift upon leaving the area.

Light emitted from stars and other celestial bodies becomes longer in wavelength as it travels away from them, due to the increasing density of empty space with distance from the star. This phenomenon is what we observe as "gravitational redshift."

According to equation 5.16:

$$\rho_r = \rho_0 + K_\rho \frac{M}{r^2}$$

Where ρ_0 is the density of the void space at the surface of the mass body, ρ_r is the density of the void space at a distance r from the mass body, and K_ρ is the distortion coefficient.

The density of the void space around a star, ρ_c , is related to the star's mass M and the distance r from its center of mass, while the change in wavelength of light is directly related to the change in void space density, $\Delta\rho_c$, so we can have:

$$\Delta\lambda = \mathrm{K}\Delta\rho_r = \mathrm{K}\left(\rho_{r2} - \rho_{r1}\right) = \mathrm{K}K_\rho M\left(\frac{1}{r_2^2} - \frac{1}{r_1^2}\right)$$

Equation 5.32 is the gravitational redshift formula.

Where: $\Delta\lambda$ is the change in wavelength of light traveling from r_1 to r_2 , K is the redshift coefficient, r_1、r_2 is the distance from the star's center to the starting and ending points of the observed light, ρ_{r2} is the density of the vacuum space at distance r_2 , and ρ_{r1} is the density of the vacuum space at distance r_1 .

The above gravitational redshift formula should be equivalent to the currently tested and verified gravitational formula (Equation 5.33):

$$\mathrm{z} = \frac{\mathrm{KM}}{\mathrm{C^2r}}$$

Among them:
Z: is the redshift, the ratio of the increase in the observed wavelength of light to the original wavelength.
K : is the Newtonian gravitational constant (commonly denoted as (G)).
M: is the mass of the star from which the light escapes.
c: is the speed of light in a vacuum.
r : is the radial distance from the center of the celestial body.
In the entire universe space, the density of the void space is uneven, so the light from distant galaxies and stars undergoes many redshifts or blueshifts during its propagation to the solar system and Earth, and what we observe is the final result.
(2) Doppler effect caused by relative motion
Because of the relative motion between the light source and the observer, the Doppler effect occurs, and we can observe blueshift or redshift of light.

When a light source moves closer to an observer, the observed wavelength of the light becomes shorter, resulting in a blue shift. Conversely, when the light source moves away from the observer, the observed wavelength becomes longer, resulting in a red shift.

the same reasoning, when an observer moves away from a light source, they will observe the light waves emitted by the source to have a longer wavelength, resulting in redshift. Conversely, when the observer moves closer to the light source, they will observe the light waves to have a shorter wavelength

The light wavelength becomes shorter, resulting in a blue shift.
Of course, all of these are assumptions made when the relative motion between the observer and the light source is below the speed of light. If their relative motion exceeds the speed of light, the observed results will show more complex changes.
(3) Changes in light wave energy
Light propagates through the void space and moves together with the spatial wave's oscillation. During this process, it is affected by the energy of the spatial wave's oscillation, leading to energy loss of the light, which also causes the wavelength of the light wave to become longer, resulting in redshift.

The effect of spatial wave oscillation on photon energy is very small, and it is difficult to observe over short distances. However, after experiencing long-distance propagation in the universe's starry space, this effect becomes increasingly apparent. That is, the farther a star is from us, the more obvious the redshift phenomenon we can observe from its emitted light waves.

$$\Delta\lambda = \mathrm{K_eD}$$

Equation 5.34 is the energy loss redshift formula, where K_e is the energy loss coefficient of the light wave, and D is the distance over which the light wave propagates.

As shown by the formula, the farther a star is from us, the more its emitted light's energy is lost, so the greater the redshift, which is consistent with our observations of the universe.

In fact, similar theories and research have already been developed in the academic community. For example, regarding the issue that the James Webb Space Telescope's observations confirm the universe's age to be much greater than what the current cosmological models predict, Guta proposed a new extended model by modifying the metric and adding a new "coupling constant," recalculating the cosmological equations, and incorporating the photon fatigue hypothesis. The final prediction results matched the observations from the James Webb Space Telescope, solving the problem of "impossible early galaxies." This also proves the rationality of our light wave energy loss redshift theory (Equation 5.34) or the photon fatigue hypothesis.

(4) Redshift Limit Silence

It can be inferred that when the distance of light waves is sufficiently far and their energy decreases to a sufficiently low level, photons will return to a non-excited state and merge into the vacuum space waves, no longer being observable electromagnetic waves. That is,

Redshift of light waves $\Delta\lambda$ has a limit $\Delta\lambda_{max}$:

$$\Delta\lambda < \Delta\lambda_{max} \quad (5.35)$$

There may be a lower limit to the frequency of light waves, and when this lower limit is reached, they will return to a non-excited state and no longer be observable electromagnetic waves. That is, the frequency of light waves λ has a lower limit λ_{min} :

$$\lambda > \lambda_{min}$$

Equations 5.35 and 5.36 are the redshift limit formulas.

According to the above equation, it can be determined that when starlight travels far enough in the universe, it will fall into a non-excited state due to exceeding the redshift limit, and thus no longer be observable electromagnetic waves. This is referred to as the redshift limit silence.

In 1823, German astronomer Heinrich Olbers proposed the famous Olbers' paradox. If the light from countless stars in the universe has accumulated for over十多亿 years, why hasn't it filled the entire universe with observable light waves? The underlying reason is the existence of the redshift limit principle. A large amount of starlight, due to long-distance propagation, exhausts its energy and falls into the redshift limit silence. This observation also aligns with our prediction.

5.10 About the Big Bang Theory [19]

(1) Hubble Redshift

The Big Bang theory is the mainstream theory of the origin of the universe today, and one of its bases is the Hubble redshift law.

In the 1920s, astronomer Edwin Hubble studied the observations made by Vesto Slipher. He noticed that the colors of distant galaxies were slightly redder than those of nearby galaxies. Hubble carefully measured this reddening and made a graph. He found that this reddening (redshift) was systematic; the farther a galaxy was from us, the redder it appeared. Therefore, based on the Doppler effect, he deduced that these galaxies were moving away from us at greater speeds. The elongation of light waves was a result of the universe's expansion, and on this basis, he proposed the Big Bang theory of the universe's origin.

In fact, according to our above formula for the redshift caused by light wave energy loss (Equation 5.34), the farther a galaxy is from us, the greater the redshift of the light it emits. This is because the farther the galaxy, the longer the distance the light wave has to travel, resulting in greater energy loss. The decrease in energy leads to longer wavelengths, thus causing a larger redshift.

The cause of the Hubble redshift may actually be related to the energy loss of light waves, rather than the result of distant galaxies moving away from us at higher speeds. Therefore, the Big Bang theory derived from this basis is worth re-examining and considering its rationality.

(2) Cosmic Microwave Background Radiation

Cosmic background radiation is an isotropic microwave radiation coming from the background of the universe, also known as microwave background radiation. In the early 1960s, American scientists Penzias and R. W. Wilson built a highly sensitive horn-type receiving antenna system to improve satellite communications, but they detected an unavoidable background noise at centimeter wavelengths. These microwave noises from the universe, with wavelengths ranging from 0.3 cm to 75 cm, are blackbody radiation with a temperature near 2.7 K, commonly referred to as 3 K background radiation, which is the cosmic microwave background radiation.

The cosmic microwave background radiation is also considered one of the foundational pieces of evidence for the Big Bang theory.
However, if we consider the properties of the void space from the perspective of spatial relativity, void space is a superfluid that permeates the entire universe space, itself existing with endless fluctuations and flows, thus possessing energy. These fluctuations would cause a kind of overall, disordered excitation in void space, resulting in a weak radiation. Given the property of void space permeating the entire universe and its extremely low temperature, this radiation should also be isotropic, consistent with the characteristics of blackbody radiation observed at a temperature near 2.7 K.

Therefore, this cosmic microwave background radiation may precisely represent the overall spontaneous radiation of the void space that permeates the entire universe space.

In other words, the 2.7 K cosmic background radiation indicates that the average temperature of the superfluid void space throughout the universe is 2.7 K, and this background radiation is the macroscopic manifestation of that.

(3) Helium Abundance

Helium abundance refers to the relative content of helium isotopes 4He in the universe. According to cosmological observations, hydrogen elements account for 25% , and helium elements account for 75% in the universe. This abundance is considered one of the evidence for the Big Bang theory.

But as we analyzed earlier, when the vacuum space is excited, various fundamental particles are generated. It can be considered that all matter and fundamental particles originate from the results of the vacuum space being excited by different energy levels. Hydrogen and helium elements are also consistent with this point.

Vacuum space, as an ocean of wave particles弥漫 throughout the universe, possesses an overall average energy fluctuation. It can be excited by the energy of surrounding mass bodies to generate new fluctuations, which in turn excite vacuum space and give rise to new particles. Among all the particles, the most fundamental are hydrogen and helium elements. Hydrogen is relatively less stable than helium, which leads to the current abundance contrast between hydrogen and helium in the universe.

Similar to the previously discussed stellar light redshift and microwave background radiation, the uniformity of vacuum space across the entire universe results in the uniform distribution of hydrogen and helium elements throughout the cosmos. This uniformity can serve as evidence that these three observational phenomena are all outcomes of the cosmic isotropy of vacuum space.

Therefore, it is necessary to re-examine the validity of the Big Bang origin theory supported by the above three observational evidences.

5.11 Black Holes

(1) Spatial Vortex Black Holes
The light or electromagnetic waves propagate through the void space, as discussed in the basic principles of spatial relativity 4: the new light-speed invariance principle. The propagation speed of light is constant at C in the uniform void space where it propagates, but it will vary relative to other void spaces or observers in relative motion, that is, light

The speed conforms to the principle of relativity.
Moreover, in different density vacuum spaces, the speed of light varies; the higher the density of the vacuum space, the faster the speed of light within it.

Due to the superfluid properties of vacuum space, there are numerous complex waves and flows in vacuum space, from the microscopic particle world to the macroscopic universe. There are also a lot of vortex and vortex rings and other motion states.

If there exists a high-density, ultra-high-energy space vortex flow in the universe, and the flow speed of the vortex reaches or exceeds the speed of light within it, then the light in this vortex flow cannot escape. Observers from outside would see a black circular hole in space, which is the black hole we observe.

So, black holes in the universe are an observed phenomenon when the vortex of the void space reaches the speed of light. Light is merely a transverse wave fluctuation within the void space, and in such intense vortex flow, it cannot be transmitted outward. Moreover, assuming there are more vortex movements within this spatial vortex, they would enhance the containment of internal fluctuations, similar to a rigid soliton formed by an ultra-large coherent structure.

The black holes we observe in the universe are all regions where light is blocked by these spatial vortices.
The void space converges toward the black hole's center along with the black hole's spatial vortex, and will form a jet perpendicular to the black hole's vortex plane at the center.
(2) Binary Black Holes and Spatial Black Hole Pairs
Black holes are spatial vortices caused by massive energy disturbances or the rotation of large mass objects or groups of mass objects. Therefore, when two black holes meet, they will pull and consume each other due to their respective spatial vortices, eventually merging into a single black hole.

In certain situations, a large disturbance may also trigger two spatial vortices simultaneously, forming two black holes with opposite vortex directions, i.e., one clockwise and one counterclockwise when observed from the black hole.

The vortexes of two black holes form the same black hole vortex singularity, and the void space flows drawn into the vortexes on both sides also converge here to form the black hole jet outward.

(3) Discussion on Mass Singular Black Holes

According to our analysis of gravity, its essence is the macroscopic manifestation of spatial pressure, not a force emitted by mass itself, but influenced by the momentum of mass particles. Therefore, spatial pressure, spatial density, and the density of the mass body itself will all have limits and will not increase infinitely. Gravity also has a limit, so a mass body will not collapse into a so-called mass singularity due to "infinite" gravity, and thus a black hole centered on a mass singularity will not form.

On September 28, 2014, Laura Mersini-Houghton, a theoretical physicist at the University of North Carolina at Chapel Hill, published an article on the ArXiv online physics database, stating that she had mathematically proven that "black holes do not exist." Through mathematical analysis, she argued that when a star dies and collapses, it would emit the radiation proposed by Hawking. During this process, the star itself would lose a significant portion of its mass, and the remaining density of the dead star would be insufficient to form a black hole.

This statement also aligns with our prediction.

5.12 Perihelion Precession of Mercury

We have previously analyzed that massive celestial bodies will cause distortion and dragging in the surrounding empty space. This distortion leads to a density gradient in the empty space around the mass body, with the density decreasing as it approaches the surface of the mass body and increasing as it moves away, eventually approaching the uniform density of the surrounding empty space.

However, this is only a consideration under ideal conditions. In reality, especially around massive, high-energy objects such as stars, the temperature changes in the vicinity of the star may cause the density gradient of the surrounding empty space to no longer change smoothly, but instead may exhibit some fluctuations.

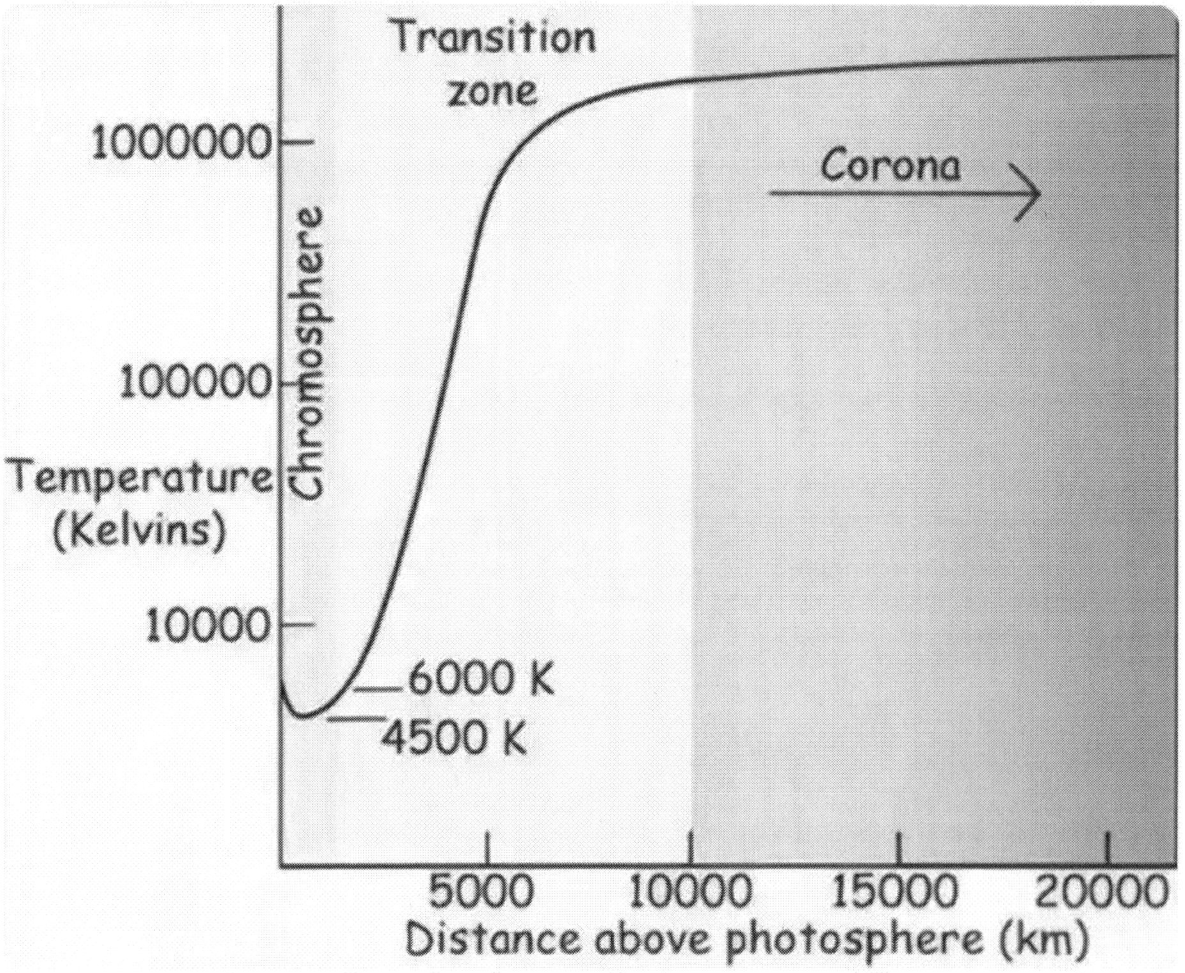

Figure 27: Figure 5-13 Temperature variation with height in the solar atmosphere

As the density of the void space normally increases with distance from the star's surface, but since the temperature abnormally increases during the process of moving away from the star's surface, as shown in Figure 5-13, and temperature directly affects the density of the void space, this non-linear fluctuation and change in temperature will cause the density of the void space to also fluctuate and change non-linearly, resulting in the density of the void space around the star first increasing, then decreasing, and then increasing again, as shown in Figure 5-14.

When other mass bodies pass through this range of void space density variation, the spatial pressure (gravity), drag force of the void space, and other related forces will all change.

Figure 28: Void space density

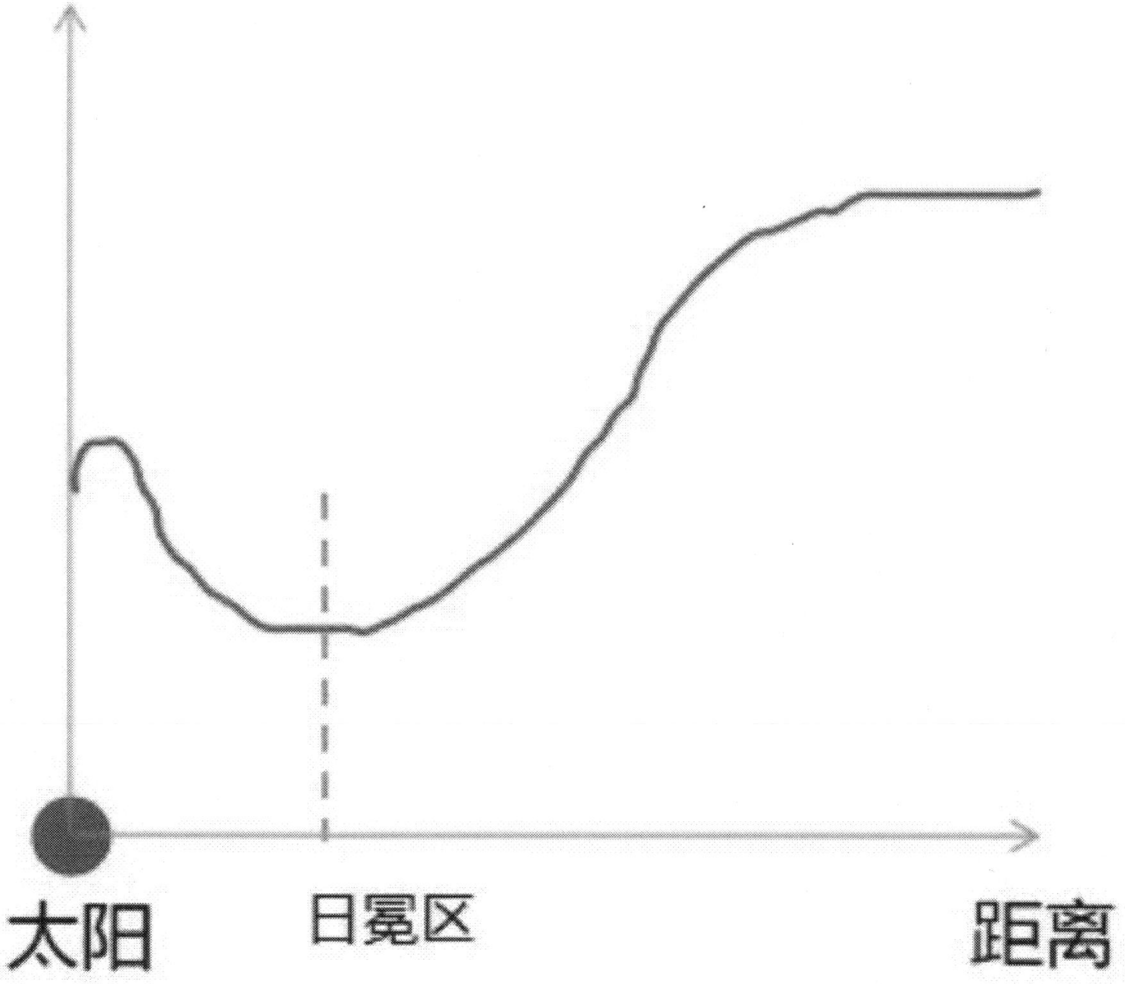

Figure 5-14 Density fluctuations of the vacuum space around the Sun

In the solar system, Mercury is the closest planet to the Sun. When its orbital path approaches the Sun, it may encounter the range of these vacuum space density changes, causing variations in the "gravitational" force it experiences and the dragging effect of the vacuum space, which results in changes to its orbit. This phenomenon is known as the precession of Mercury's perihelion.

Other planets, due to their greater distance from the Sun, experience a smaller effect of this kind, and the phenomenon is not as pronounced as with Mercury.

5.13 Analysis of Bose-Einstein condensate

(1) Bose-Einstein condensate

A Bose-Einstein condensate is a gaseous, superfluid state of matter that boson atoms exhibit when cooled to near absolute zero. At this point, a large number of bosons behave as if they have "condensed" into the same state, which is known as the Bose-Einstein condensate.

In 1995, Wolfgang Ketterle from MIT and Eric Cornell and Carl Wieman from the University of Colorado at Boulder first achieved the Bose-Einstein condensate using gaseous aluminum atoms at $170\text{nK}\ (1.7 \times 10^{-7}\ \text{K})$. In this state, nearly all the atoms gather into the lowest energy quantum state, forming a macroscopic quantum state.

The explanation in quantum mechanics is that according to the de Broglie matter wave theory, as temperature decreases, the speed of particle motion slows down, and the wavelength of the matter wave becomes longer. When the temperature is low enough, the de Broglie wavelength of the atoms becomes comparable to the distance between them. At this point, the matter waves interact to reach a completely identical state, and the properties can be described by a single atomic wave function. When the temperature reaches absolute zero, thermal motion disappears, and the atoms are in an ideal Bose-Einstein condensate state.

(2) Principle of Thermal Motion

When modern theories consider the thermal motion of molecules, atoms, etc., they mainly take into account the thermal motion of molecules, atoms, etc. caused by radiation

Radiation or conduction mechanisms cause the absorption of energy, increasing the motion speed, thereby leading to increased thermal motion, which is manifested as a rise in temperature.

In fact, the interaction between the vacuum space and molecules also affects thermal motion. When light or other radiation comes in, it brings along synchronized fluctuations of the vacuum space, which also carry energy. These fluctuations can enhance the vacuum space vibrations in the region where the molecules are located, thereby influencing all molecules and particles within the entire vacuum space range and causing their motion to intensify, that is, thermal motion becomes more vigorous.

When the fluctuation energy of the vacuum space decreases, it also affects all molecules and particles within its range, causing their motion to weaken and resulting in a decrease in thermal motion, which is manifested as a drop in temperature.

Therefore, the thermal motion of molecules and other microscopic particles is closely related to the movement of the vacuum space itself. When the fluctuation energy of the vacuum space approaches zero, it can have an effect similar to freezing on all microscopic particles within it, making their energy and thermal motion approach zero. Smaller mass particles are more significantly affected. This is also one of the reasons for the formation of the Bose-Einstein condensate at the microscopic level.

(3) Formation Mechanism of Bose-Einstein Condensate

Quantum theory explains the formation mechanism of Bose-Einstein condensate, but does not explain why the thermal motion of fundamental particles slows down as temperature decreases. As analyzed above, the movement of matter is closely related to and interacts with the vacuum space, especially for microscopic particles with small mass, where this interaction is more pronounced. Characteristics such as wave-particle duality of particles are also the results of the influence of vacuum space's fluctuations on particle motion.

When the temperature decreases, the fluctuations of the vacuum space are suppressed. When the temperature approaches absolute zero, the fluctuations of the vacuum space also drop to a very low level, similar to being frozen or in a state of thermal stillness. In this case, the motion of all fundamental particles within it is affected, which is why the thermal motion slows down. This effect is most noticeable for the particles with the smallest mass, leading to the behavior of bosons.

The bosons, along with the vacuum space, have almost fallen into a state of thermal equilibrium, and due to their extremely low viscosity, they become completely integrated with the vacuum space, exhibiting superfluid properties that are almost non-viscous as the vacuum space itself. It is precisely this characteristic that allows them to achieve a unified state, forming the Bose-Einstein condensate.

This principle also applies in fields such as superconductivity and power transmission, where the influence and role of vacuum space must be considered in order to fully understand the underlying mechanisms and make practical use of them.

5.14 The Relationship Between Spatial Wave Dynamics and General Relativity

General relativity analyzes the gravitational effect from the perspective of spatial geometric structure. It exists structurally in contradiction with quantum mechanics in terms of mathematical systems, making it non-quantizable, unable to clarify the essence of spatial distortion, and incapable of integrating with quantum mechanics to achieve the unification of all mechanics.

Space Relativity analyzes the entity formation of empty space, providing a common origin for all forces such as gravity and electromagnetism. There is mass first, then force; all forces are manifestations of mass forces. On a macro scale, gravity and electromagnetism are expressions of spatial pressure differences in empty space under different motion modes, and on a micro scale, they are also manifestations of mass forces between various fundamental particles under different motion modes.

Space Relativity clarifies the essential meaning of space distortion in general relativity, which is the distortion of mass bodies on surrounding empty space, that is, the effect of mass bodies on the density of surrounding empty space, leading to changes in density gradients. It solves the problem that the spacetime structure system of general relativity cannot be quantized and exists in structural contradictions with quantum mechanics, achieving a unified theoretical framework for cosmological mechanics.

Space Relativity is suitable for macrocosmic mechanics such as gravity in the universe, and also applicable to micro-particle mechanics. It and general relativity are two theoretical systems for the macro gravity theory and cosmological theory, so many theoretical prediction results are consistent. However, due to differences in theoretical foundations and understanding of the essence of the universe,

there are also differences between them in many aspects.

Space Relativity is a correction and improvement of General Relativity in a broader field, capable of more reasonable and simple explanations of the mechanical laws and phenomena in the macrocosmic universe and the microcosmic particle world.

5.15 The Relationship Between Spatial Wave Dynamics and Fluid Mechanics [20]

Space Relativity treats the void space as an objective entity similar to a superfluid, like dark matter, so many established theories from fluid mechanics are used in its theoretical analysis. However, due to the special characteristics of void space resembling a superfluid and not considering gravity, Space Relativity differs from the existing fluid mechanics theories in the real world.

Considering that void space can be regarded as an objective entity of a superfluid, utilizing the well-established theoretical accumulation in fluid mechanics is a good methodological approach for studying the relative motion and interaction between mass bodies and void space in both macroscopic universe and microscopic particle fields, by drawing on observations and understanding of everyday fluid phenomena.

Through the study of the superfluid characteristics of cosmic void space, we can also better refine the existing classical fluid mechanics theory system and deepen our understanding.

An Introduction to Classical Dynamics in Space

Assuming an object is stationary relative to the observer, we examine its motion laws.

6.1 When no external forces are acting on an object

When no external forces act on an object, there is only the interaction between the object and the empty space.
When an object is in relative static equilibrium with the void space, the object will remain in a stationary state.
When an object is in relative motion with the void space, it will experience a force from the void space, initiating accelerated motion in the same direction as the void space's movement relative to the object.

Assuming the void space is in a one-dimensional uniform flow state, when an object accelerates to the same speed as the void space, i.e., achieving relative static equilibrium with the void space, the force exerted by the void space on the object will drop to zero, and the object will begin to move at a constant velocity together with the void space.

In this process, the stationary observer will observe that the small ball starts accelerating from rest relative to themselves and then enters a state of uniform motion.

If the void space is in a two-dimensional or three-dimensional flow state, the movement of objects will also follow the void space to gradually produce motion in the same direction.

6.2 When external forces are applied to an object

When an external force F acts on an object, the object's motion state will be the result of the combined effect of the external force and the force exerted by the void space.

When an object is in relative static equilibrium with the void space, it will begin to accelerate under the external force, but it will simultaneously experience a reverse spatial resistance from the void space F_0 , so its acceleration a will be determined by the difference between the external force F and the spatial resistance F_0 , that is:

$$a = \frac{F - F_0}{m} \quad (6.1)$$

Equation 6.1 is the acceleration formula of an object in empty space, where m is the mass of the object.
According to equation 5.21, as the speed of an object increases, the resistance it encounters from the void space will also increase, with the resistance being proportional to the square of the object's velocity. Therefore, as the object's speed increases, the resistance it experiences increases, leading to a decrease in its acceleration.

When the void space is at rest, the object's acceleration reduces to zero as the externally applied force F balances the resistance force from the void space, and it begins to move at a constant speed under the combined effect of the external force and the resistance from the void space.

When the void space is in a flowing state, objects will experience different motion conditions due to the difference in direction between external forces and the void space's forces.

6.3 Action and Reaction Force Situations

No matter what motion state the void space is in, the law of action and reaction of classical mechanics remains unchanged, that is, the force exerted by a third party on an object is equal in magnitude and opposite in direction to the reaction force exerted by the object on the third party.

6.4 The Relationship Between Classical Dynamics in Void Space and Classical Mechanics

When the force exerted by the void space on an object is not considered, or when the object has a large mass and density and is in a low-speed motion state, making the force from the void space negligible, the motion laws of the object can be approximately described by the Newtonian laws of classical mechanics.

When considering the force exerted by the vacuum space on macroscopic objects, treating the vacuum space as an ideal non-viscous superfluid and analyzing it through the mature classical fluid dynamics theory is a good methodological approach.

When considering the force exerted by the vacuum space on microscopic particles, because the mass of microscopic particles is very small, they are significantly affected by vacuum space fluctuations, it is also necessary to take into account the quantum dynamic effects and interactions of the vacuum space.

7 Introduction to Quantum Dynamics of Space [21][22]

We have analyzed in the basic principles of space relativity 2: the principle of objective space, that the vacuum space is consistent at both macroscopic and microscopic levels. Whether at the macroscopic cosmic level or the microscopic particle and quantum level, the essence of the vacuum space is consistent and the same, and it is an overall superfluid composed of the fundamental mass wave particles. Therefore, the vacuum space has the same properties at both macroscopic and microscopic levels and interacts with mass bodies in the same way.

Macroscopic planets, matter, and microscopic particles, quanta, etc., are all composed of the most fundamental mass entities called wave particles, which aggregate into different coherent results and motion patterns. All the mass of substances comes from wave particles, so the numerous mechanical interactions and laws, phenomena analyzed in the previous sections on space electrodynamics and space wave dynamics, also exist in the microscopic quantum world.

Of course, because the masses of various particles and quanta in the microscopic particle world are very small, they will interact with the empty space in different ways and exhibit some phenomena that do not exist or are difficult to observe at the macroscopic level.

7.1 Wave-Particle Duality

(1) Origin of Wave-Particle Duality
The essence of the wave-particle duality exhibited by microscopic particles has been analyzed in the fourth chapter of the basic principles of spatial relativity: the new constancy of light speed, and in the chapter on the fundamental nature of particles in the general introduction to spatial wave dynamics.

All matter, including fundamental particles, is essentially particulate, and its motion is also particulate linear motion. However, since all matter and particles move within the environment of empty space, which is a superfluid with a large number of fluctuations and complex movements, all matter and particles moving in it will be affected by these fluctuations.

The influence of fluctuations and movements in the void space causes particles that propagate linearly to be disturbed, forming the characteristics of transverse wave fluctuations. The flow and movement of void space can alter the direction and speed of particle propagation.

Because microscopic particles have relatively small mass, they are more significantly affected by the fluctuations and flows of void space, thus exhibiting more pronounced wave-like properties and acquiring the characteristics of transverse wave fluctuations. As a result, particles display wave-particle duality in their overall behavior.

It can be said that macroscopic matter, even celestial bodies, are also influenced by the fluctuations of void space. Therefore, their motion is not strictly linear but also possesses transverse wave fluctuation characteristics, thereby having a certain degree of "wave-particle duality." However, due to the large mass of macroscopic matter and celestial bodies, the effect of void space fluctuations is minimal, making this duality difficult to observe at the macroscopic level.

Thus, wave-particle duality is a common motion characteristic of all particles, matter, celestial bodies, etc. Its essence is the result of the interaction between the linear motion of particles and matter and the fluctuation motion of void space, forming a composite motion.

This synthetic motion can explain the origin of de Broglie matter waves in quantum mechanics and is consistent with the predictions of the de Broglie matter wave theory.
(2) Principle of Minimum Energy Path Selection
When a large number of particles move together and have complex interactions, they eventually evolve into a motion pattern with minimum energy consumption and continue to move according to this pattern. This is one of the reasons for the emergence of large-scale phenomena in nature. Essentially, the emergence of all patterns is a dynamic equilibrium state formed by the interactions among all subsystems within the system. This dynamic equilibrium allows the system to operate with minimal resistance and to maintain stability and continuity for a longer period.

The principle of least action in physics is also the result and manifestation of this dynamic equilibrium, meaning that a physical system will choose the path that results in an extreme value of the action (usually a local minimum) when evolving from one state to another.

Various fundamental particles are affected by the fluctuations of the void space when moving through it, leading them to eventually produce synchronized transverse wave oscillations. This is also a manifestation of the principle of least action, as this collaborative motion pattern consumes the least amount of energy for the subsequent movement of the particles, resulting in the minimal resistance.

This is the principle of minimum energy consumption path selection for particles moving in void space.

(3) Particle's speed and direction

Particles propagate through void space, and their speed and direction change due to the influence of void space fluctuations.

For example, when spatial waves interfere, particles moving within them will choose the path of least resistance in their linear motion direction, because the particle itself has transverse wave motion, so it will choose the direction where wave propagation still exists, even the direction with the maximum wave propagation energy. That is, the interference of spatial waves affects the direction selection of the particle's wave motion.

This process is also a manifestation of the principle of minimum energy consumption path selection.
In simple terms, it is the particle that determines the direction and speed of its linear motion, while the wave motion of spatial waves determines the speed and direction of the particle's wave movement.

Through the above analysis, we can better understand the wave-particle duality of particles and use it to explain the classic double-slit interference experiment in quantum mechanics.

(4) Polarization and energy of particles [23]

Elliptical wave motion is the most general form of motion for all transverse wave phenomena, because the wave effects in the void space cause the wave nature of various fundamental particles to manifest as elliptical waves. This is the polarization property of waves we are familiar with: elliptical polarization.

As the elliptical short radius varies, the wave motion of the particle exhibits approximate linear polarization and circular polarization.

Because the particle's motion is a synthesis of its linear motion and the wave motion of the void space, the energy it manifests is also a synthesis of the energy from its own linear motion and the wave energy of the void space.

Let the energy of the particle's linear motion be E_1 , and the energy of the void space's wave motion be E_0 , then the total energy E manifested by the particle is:

$$\mathrm{E} = \mathrm{E}_1 + \mathrm{E}_0$$

Equation 7.1 is the particle's apparent total energy equation.
Because the vacuum space has complex fluctuations, its fluctuation energy E_0 changes dynamically, possibly increasing or decreasing at any time, oscillating around the average energy, the total energy of particles' microscopic behavior also exhibits fluctuation changes.

Through this principle, the quantum tunneling phenomenon that occurs偶然 in the quantum microscopic world can be explained.

7.2 Double Slit Interference

In quantum mechanics, the double-slit experiment is a demonstration of the wave-like and particle-like behavior of microscopic entities such as photons or electrons. The double-slit experiment is a "two-path experiment," and quantum theory suggests that microscopic objects can simultaneously pass through both paths or through either path from the initial point to the final point. The path difference causes the quantum state that describes the physical behavior of the microscopic object to undergo a phase shift, thus resulting in an interference phenomenon.

The double-slit experiment, also known as the Young's double-slit experiment, is specifically designed to demonstrate the interference behavior of light waves. If the light beam is considered to move from the light source to the detection screen as particles, the number of particles arriving at any position on the detection screen should be equal to the sum of the number of particles that passed through the left slit and those that passed through the right slit previously. The irradiance at any position on the detection screen, when both slits are open, should be equal to the sum of the irradiance when only the left slit is closed and when only the right slit is closed. However, when both slits are open, the result is not simply the sum; some areas on the detection screen appear brighter while others appear darker. This pattern can only be explained by the wave theory of light through constructive and destructive interference, not by the simple particle theory of light.

Addition of quantities, as shown in Figure 7-1.

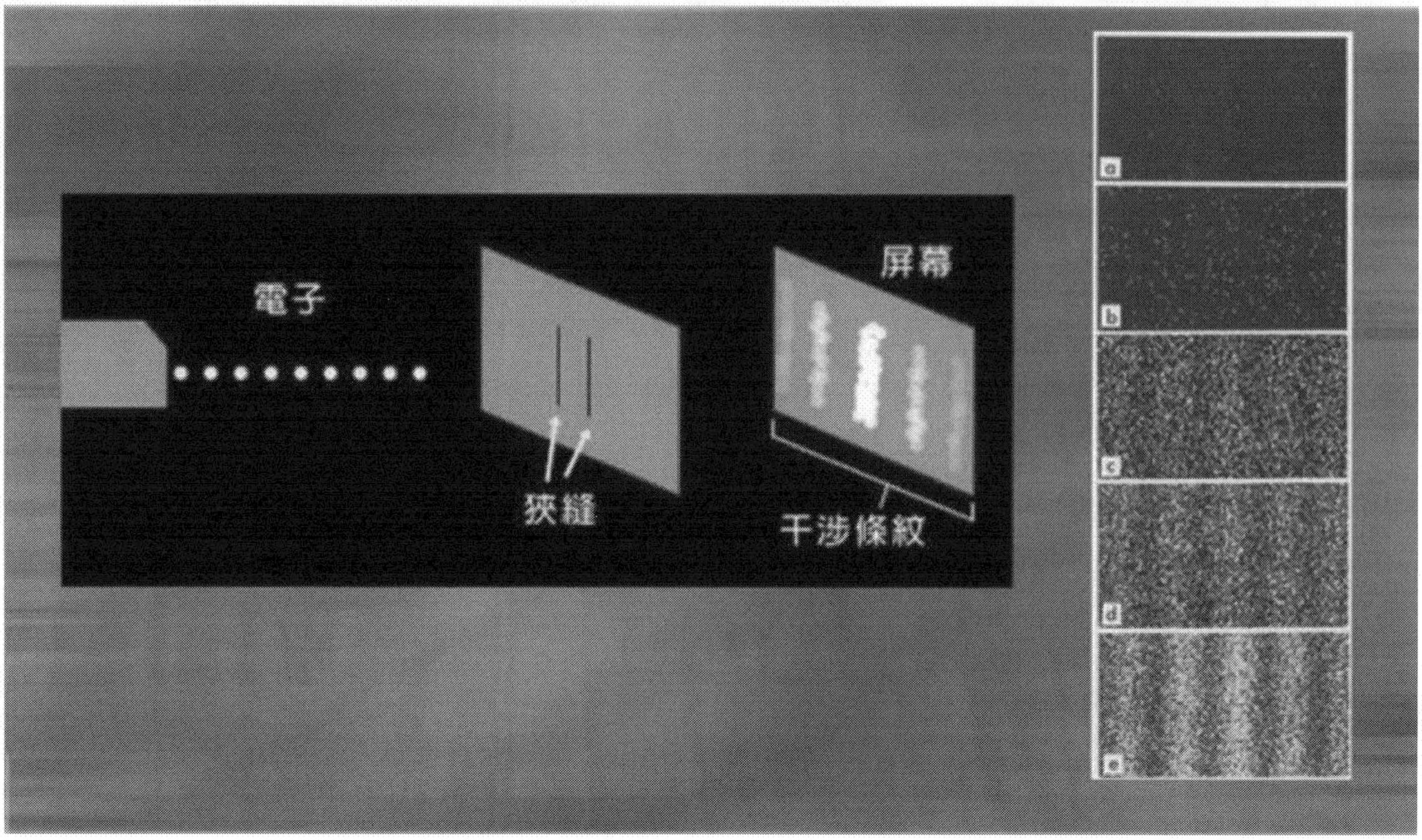

Figure 7-1 Young's Double-Slit Experiment

We have analyzed before that the essence of wave-particle duality is the superposition of a particle's linear motion and the wave motion of the empty space, forming a composite motion.

When we excite a particle, such as a photon, it simultaneously triggers fluctuations in the vacuum space. The photon moves together with the vacuum space fluctuations, forming a combined wave-particle motion pattern.

When the photon and the vacuum space fluctuations reach the double slits, the photon passes through only one slit, while the vacuum space fluctuations pass through both slits at the same time and produce interference after the slits.

Due to the appearance of space wave interference, the principle of minimum energy path selection causes the photon to choose the path with the least resistance, which is to continue moving in the direction of the fluctuation. Eventually, it hits the final fluorescent screen.

A large number of photons accumulate on the screen along this path, ultimately forming the interference pattern. The key points throughout the process are:
(1) When a single photon is emitted, it simultaneously excites fluctuations in the vacuum space, and the photon travels with the spatial wave fluctuations through the double slits.
(2) Each photon passes through only one slit at a time, while the interference pattern formed by going through both slits is created by the spatial wave fluctuations that travel with the photon.
(3) After passing through the double slits, the spatial wave fluctuations interfere with each other, and the photon follows the principle of least energy consumption, choosing the direction where the fluctuations exist to continue forward.

These few simple principles form the interference fringes we observe on the screen.
When we observe the double-slit interference through instruments, the act of observation disturbs the spatial wave formed after the double slits, destroying the interference state and restoring the disordered state. In this case, the movement of photons no longer is influenced by the spatial wave interference state, returning to a basic linear state, which causes the interference pattern to disappear.

There is no unexplainable or mysterious force involved in the entire process.
The double-slit interference experiment is one of the most mysterious and confusing phenomena in quantum mechanics, casting a shadow of mysticism over the entire field. The famous physicist Richard Feynman once stated in his book "The Feynman Lectures on Physics" that the quantum phenomena demonstrated by the double-slit experiment cannot, absolutely cannot, be explained by any classical means, and it embodies the core ideas of quantum mechanics.

But the facts are clearly not like that. Everything can be clearly explained by the wave-particle duality principle, which is the combination of classical particles and spatial waves.

There are no mysteries in the universe; any observed so-called mysteries can be reasonably understood and explained through analysis that looks beyond the surface. The mystical and metaphysical views and interpretations imposed on quantum mechanics due to phenomena such as wave-particle duality, double-slit interference, and quantum entanglement are clearly inappropriate.

7.3 Probability and Randomness in Quantum Theory

(1) Probability in Quantum Theory

The randomness in quantum theory is one of the core characteristics that distinguish it from classical physics. The state of a quantum system is described by a wave function (probability amplitude), and its modulus squared gives the probability of finding a particle at a particular position. When

When measuring a quantum system, the wave function randomly "collapses" to a particular eigenstate (such as a definite value of position or momentum). Along with other principles like the uncertainty principle, these features give quantum theory its mysterious randomness and uncertainty.

For this incomprehensible randomness, many scholars including Einstein and Schrödinger have had intense debates. Schrödinger even designed a hypothetical experiment called "Schrödinger's cat" to question it, and surprisingly, Schrödinger's cat has become a milestone in the theory of quantum uncertainty.

(2) Theoretical categories from different perspectives

In fact, the description of the physical world can exist in various perspectives. Different perspectives can lead to different theoretical systems, which may differ significantly in their descriptions but can be completely compatible in essence.

Just as we analyzed in the basic principles of spatial relativity 1: the principle of objective physical laws, the underlying physical laws of the world are the essence, while the phenomena we directly observe are the appearances. Phenomena can reflect the essence or even distort it. Therefore, observers need to use rational analysis to look beyond the appearances and understand the essence.

Theoretical systems formed by the description and analysis of the same physical laws can differ in type due to different perspectives:
(1) Principle-based descriptive theory: such as classical mechanics and many other theories we are familiar with.
(2) Statistical Description Theory: Such as quantum mechanics, a theory that describes phenomena in terms of randomness in many aspects.
(3) Geometric Mathematical Description Theory: Such as general relativity, which describes the universe and gravity from the perspective of geometric structures.

As long as these theories can well and reasonably self-consistently explain the phenomena we observe, they are suitable tools. Although we are more accustomed to principle-based descriptions in our thinking, which allow us to understand the causes and mechanisms between each interaction more accurately, other theories are clearly also powerful tools in application.

For example, statistical description theory, our daily statistics itself is an essential analytical tool.

All the textual outputs and image generation provided by current AI large models are also results based on statistical theory. This demonstrates the correctness of statistical theory.

The correctness of statistical theory, in turn, better illustrates that there are profound laws at work behind the physical facts it describes. That is, statistical theory makes predictions based on the statistics of phenomena and outcomes. Behind every statistical theory description, there must be a principle-based theory explanation.

(3) Origin of Probability and Uncertainty

Our understanding of probability and uncertainty in the physical world originates from the limitations of our observational methods and data acquisition capabilities. In the endless sea of motion within empty space, there are numerous interactions, and we are unable to accurately observe and obtain precise information about these interactions. Therefore, we can only analyze them from the perspective of probability and statistics and make predictions based on their macroscopic manifestations.

As in our coin toss experiment, the probability of which side lands up is actually influenced by various factors such as the force, angle, air flow, and environmental vibrations during each toss. These interfering factors cause the experimental results to appear random, but the underlying physical principles still make the outcomes follow statistical laws. If we can precisely control all these factors to a certain level of accuracy, we can ensure that the coin lands with the same side up every time, eliminating uncertainty and randomness.

The randomness in quantum theory follows the same principle. Microscopic particles are greatly influenced by the surrounding vacuum space, which is a superfluid composed of countless wave particles, exhibiting complex movements and producing intricate, dense interactions with the particles. Without the means to observe or obtain data, we can only provide theoretical analysis and predictions through macroscopic statistical descriptions.

(4) How to explain the randomness in quantum mechanics

A large number of uncontrollable factors interacting with each other will result in randomness in observations, which is a fundamental aspect of quantum mechanics

The randomness of particle motion arises from two sources: one is its mathematical description through statistical systems, and the other comes from the unknowability of the motion of ultra-microscopic void space and wave particles, as well as their effects on microscopic particles.

Void space is a superfluid composed of countless wave particles, which can interact with fundamental particles through wave-like motion. By analyzing and understanding the hydrodynamics of the superfluid void space, we can explain and comprehend probabilistic phenomena in quantum mechanics, thereby obtaining a principle-based theoretical description underlying the statistical nature of quantum mechanics.

(5) The Principle-Based Explanation of the Probability Cloud of Electrons Orbiting the Atomic Nucleus

In the overview section of spatial electrodynamics, we analyzed a long-standing puzzle in physics: electrons undergoing acceleration emit electromagnetic waves, but why do electrons orbiting the nucleus in circular motion not radiate electromagnetic waves? If electrons did radiate electromagnetic waves, they would quickly lose energy and fall into the nucleus, which is clearly not the case in reality.

To resolve this difficulty, quantum mechanics describes the electron's trajectory as a probability cloud, meaning it randomly appears at some point around the nucleus. The probability of its appearance is described by its wave function, which is also one of the sources of probabilistic nature in quantum theory.

In the overview of spatial electrodynamics, we have analyzed that according to the basic principle 2 of spatial relativity: the principle of spatial objectivity, and basic principle 6: the principle of spatial relativity, the vacuum space is an objective reality. Electric charges only experience acceleration relative to the vacuum space they are in, and electromagnetic waves are radiated in this vacuum space only when their speed, acceleration, amplitude, etc., reach the threshold of the vacuum space.

Therefore, electrons orbit the atomic nucleus without radiating electromagnetic waves, and there are two influencing factors:

(1) The movement of electrons has not reached the threshold to emit electromagnetic waves. Although we observe macroscopically that accelerated motion of charges can emit electromagnetic waves, the macroscopic distance is extremely large for microscopic particles. When considering the tiny region around an atomic nucleus, the acceleration of electrons may not meet the condition to emit electromagnetic waves, that is, it may not reach the threshold for emitting electromagnetic waves in the vacuum space.

(2) The vacuum space around the atomic nucleus itself exists in motion. The high-speed spin and movement of the atomic nucleus can distort and drag the surrounding vacuum space, causing it to move as well. Electrons moving in this moving vacuum space may have a reduced relative velocity with respect to the vacuum space, or even exist in a state of synchronous motion with the vacuum space, i.e., relative rest. Under such conditions, it is obviously impossible to emit electromagnetic waves.

For example, when a charge is stationary on the Earth's surface, the Earth is in relative motion around the Sun, experiencing accelerated movement, but the charge does not emit electromagnetic waves, which is due to this reason.

So, through the analysis of the relationship between electronic excitation and electromagnetic waves and relative motion with empty space, we can provide a theoretical explanation from the principle, becoming the principle-type descriptive theory underlying the statistical description theory of quantum mechanics for the uncertainty of electron motion laws.

(6) Principle-based theoretical explanation of the double-slit interference quantum phenomenon

As we have analyzed above regarding the double-slit interference, we can also better understand the classical physics principle-based theoretical explanation corresponding to the probabilistic wave function of quantum particles.

As in the double-slit interference experiment:

(1) Emitting a single photon. In quantum theory, a single photon is represented by a wave function and follows the principle of superposition. According to spatial relativity, this photon is actually a superposition of spatial wave fluctuations that are synchronized with the vacuum space, forming a combination of linear motion and wave motion, which is the wave-particle duality.

(2) Photons pass through the double slits. According to quantum theory, at this moment the photon is in a superposition state as described by the wave function, meaning it is in a combined state of passing through one slit and passing through the other slit. According to spatial relativity, at this moment the photon passes through one slit, while the synchronized spatial wave oscillation passes through both slits simultaneously.

(3) Reaching the screen. In quantum theory, photons in two superposition states have a phase difference, so after passing through two slits, interference occurs. When reaching the screen, the photon's wave function collapses due to measurement, resulting in a definite position, which appears as a bright spot on the screen. According to spatial relativity, the spatial wave oscillating in sync with the photon, after passing through the two slits, will also produce interference, and the photon will choose the path with the least energy consumption.

Principle: choose to move in the direction of spatial wave oscillation that still exists in the interference field, reach the screen, and form a light spot.

(4) After countless light spots converge, they form alternating bright and dark interference fringes. In quantum theory, the probability of a photon collapsing to a specific point on the screen is proportional to the square of the wave function at that point. According to spatial relativity, the specific point on the screen where a photon hits is the result of the photon's initial linear motion direction and the minimum energy path selection in the interference field formed by spatial wave interference after passing through the double slits.

Therefore, the principle of superposition of spatial waves and linear motion of photons in spatial relativity can well explain the probabilistic statistical description in quantum theory. It is the principle-based description theory corresponding to the statistical description theory in quantum theory. The statistical description of the wave function in quantum mechanics is equivalent to the principle description of the wave-particle duality theory of photons in spatial relativity.

In summary, many similar confusions and phenomena in the quantum realm are essentially the result of quantum masses being too small, making them more significantly affected by fluctuations in the vacuum space. Therefore, when describing quantum motion, it is necessary to take into account the interaction between the quantum and the vacuum space. Other probabilistic descriptions in quantum mechanics that are of a statistical nature can also be analyzed through this principle to find the corresponding theoretical descriptions of fundamental principles.

The two theories are equivalent in macroscopic observations.

7.4 Quantum Entanglement and Locality [24][25]

In quantum mechanics, when several particles interact with each other, and their individual properties have become integrated into the overall properties of the system, making it impossible to describe the properties of each particle separately, but only the properties of the entire system, this phenomenon is called quantum entanglement or quantum entanglement (quantum entanglement).

Two particles in an entangled state, no matter how far apart they are, will have their states instantly determined at the same time when the state of one particle is measured. This mysterious superposition and non-locality faster than light have sparked long-lasting debates among scientists such as Einstein, Schrödinger, and Bohr.

To understand this quantum entanglement feature in quantum mechanics, two aspects need to be clarified.

(1) The new locality principle

Einstein insisted that quantum mechanics should comply with locality, which aligns with most people's understanding, but Einstein's concept of locality has a problem: it limits the speed of interaction to the speed of light. This speed limit not only restricts the cognition and development of theoretical physics, but also troubled Einstein himself.

However, based on our previous analysis in the introduction to spatial electrodynamics, the speed of light is just an ordinary speed in the universe, not the maximum speed. Through the mode of self-driven acceleration, we can completely reach and even surpass the speed of light.

Therefore, we need to update the definition of locality: all physical interactions in the objective world require time and are not instantaneous.所谓 instantaneous is merely an observation result due to insufficient observational precision. That is: let the interaction time between two objects be Δt , then we have:

$$\Delta t > 0$$

This equation is the new principle of locality.
Including quantum theory and all physical theories, all interactions in them should conform to the new principle of locality.

(2) Physical Interpretation of Wave Function Collapse

In addition to the locality issue, another major controversy regarding quantum entanglement lies in whether two "entangled" particles remain connected after being separated, which relates to the physical interpretation of wave function collapse in quantum mechanics.

Einstein, Schrödinger, and others believed that the relationship between particles conforms to the principle of determinism, with their states being determined at the moment they are separated, and no further interaction occurring between them after separation. We only learn their specific state information at the time of measurement.

but the Copenhagen interpretation of quantum theory suggests that two "entangled" particles, after being separated, have their states

The state is not determined; it is only when we perform a measurement that the particle's wave function "collapses" into a specific state, simultaneously interacting instantaneously with other "entangled" particles to determine their states as well.

Based on the above view and our new locality principle, we can determine two interpretations of the collapse of the wave function in quantum mechanics:

(1) Classical interpretation

That is, the view of Einstein, Schrödinger, and others, that the two particles already have definite states when they separate, and there is no instantaneous action at a distance between them. The so-called wave function collapse is not a physical phenomenon that actually occurs.

(2) Quantum mechanics interpretation

Two "entangled" particles remain connected even after being separated, maintaining a nonlocal interaction. Our measurement causes the wave function of one particle to collapse, and due to this nonlocal interaction, the other entangled particle also collapses instantaneously, determining their respective states.

This is the traditional explanation of quantum mechanics, known as the Copenhagen interpretation. However, according to our new locality principle above, even if a particle truly collapses due to measurement, the process of the other "entangled" particle collapsing is not instantaneous. The time required for the interaction, denoted as Δt , is extremely small but should and must be greater than 0.

That is, the nonlocal interaction of quantum entanglement must conform to the new locality principle.

(3) Origin of the Entanglement Problem

The origin of the quantum entanglement problem in quantum mechanics lies in the mathematical system of the wave function used to describe quantum states. The wave function itself is a probabilistic mathematical system, so it can only provide a probabilistic distribution of the states of two "entangled" particles. The state of a quantum particle can only be determined through the act of measurement, which causes the wave function to collapse.

It can be said that the probabilistic and statistical nature of the mathematical system of quantum mechanics is the cause of quantum entanglement

Controversy over the problem.
We have discussed earlier that the laws of the physical world are objective, but their description and understanding are subjective. Therefore, different theories can be formed based on different analytical perspectives, including principle-based description theory, statistical description theory, and geometric mathematical description theory.

Among them, the statistical description theory and the geometric mathematical description theory are both theories that describe and explain the macroscopic phenomena of the physical world. Behind these theories, there must be corresponding principle-based description theories.

The statistical and mathematical description theories at the macroscopic level can provide accurate analysis and description of physical phenomena, just like our current AI can generate dialogue text according to user needs, or even create images and videos, achieving very accurate results. However, such statistical systems are based only on the statistics of phenomena and do not involve the underlying physical principles and essence. For example, the images and videos generated by AI are based on the analysis and prediction of pixel distributions formed during training, but they lack understanding of the physical meaning and significance behind each pixel. This is why distortions, or what is commonly referred to as "hallucinations," often appear in AI-generated videos and images.

Of course, as AI's intelligence level and logical capabilities improve along with the enhancement of training data, these hallucination issues will significantly decrease. However, to completely eliminate them, AI needs to have an understanding of the physical world and objective laws, meaning it must grasp the underlying physical principle-based laws corresponding to the statistical and mathematical description laws of the macroscopic phenomena.

Statistical descriptive theories in the theoretical domain (such as quantum mechanics) and geometric mathematical descriptive theories (such as general relativity) are also the same.

(4) Principle of Determinism

Einstein said that God does not play dice, while Bohr asked, how can we be sure that God does not play dice? This is a topic that cannot have a definite answer.

The issue lies in the fact that each successful theory will have a consistent mathematical description and explanatory system.

It would be difficult, if not impossible, to refute within such a self-consistent theoretical system.
However, in the process of exploring the many unknown areas of the universe and science, we must also follow several fundamental principles, and the principle of determinism should be one of them.

We have discussed in the basic principles of spatial relativity, principle 1: the objectivity of physical laws, that the world should be materialistic, and physical laws are objective and deterministic. If there are elements of mysticism or the occult in some theories or scientific fields, it can only mean that there are certain underlying physical laws that we have not yet fully understood and mastered. To find these laws and eliminate mysticism and occultism is the primary task of scientific development.

According to our previous analysis, many of the incomprehensible issues in quantum mechanics, such as wave-particle duality and the double-slit interference, actually result from its influence by special relativity, which has neglected the objective reality of empty space and its superfluid characteristics with dynamic mechanics. The description of wave-like properties of each quantum, or even all material motion, originates from the wave motion of empty space. By incorporating this underlying principle, the numerous mysterious and occult aspects in quantum mechanics can be resolved, and more complete, accurate, and deterministic explanations can be provided.

(5) Explanation of Quantum Entanglement

Above, we analyzed the classical and quantum mechanical interpretations of quantum entanglement. Through numerous experiments, the academic community has demonstrated the invalidity of Bell's inequalities, including the 2022 Nobel Prize in Physics, which was awarded to three scholars who experimentally proved Bell's inequalities. However, we still tend to believe that the essence of quantum entanglement is the classical interpretation, that is, once two entangled particles are separated, their states are already determined, and whether or not one particle is measured does not affect or change the state of the other.

The negation of experimental verification of Bell's inequalities can be considered a necessary condition rather than a sufficient condition for rejecting the classical interpretation.

7.5 Fields, Virtual Photons

In quantum theory, the most important mathematical tool is the field, which quantum theory considers as a substance

Observing reality, it is matter that possesses energy, but one cannot form a clear understanding of what kind of matter a field actually is.

The field can be understood as a wave form of the empty space, through our previous analysis. Different particles and mass bodies cause different wave and flow movements in the surrounding empty space. The region of empty space with different wave and motion forms around a particle or mass body is the essence of the field.

In quantum mechanics, the field is a mathematical analysis and description tool for the wave and flow movements of empty space. The interaction between particles and the field essentially represents the interaction between particles and the surrounding empty space: macroscopically, it is the interaction with the overall flow field and wave field of empty space; microscopically, it is the interaction between the particle and countless wave particles, or even isolated particles, in the empty space.

All material mechanical interactions, both microscopically and fundamentally, are manifestations of the interaction between particles and the surrounding empty space under different motion patterns.

Take the electromagnetic field as an example. In quantum theory, the force between electromagnetic fields is mediated by virtual photons. Since the origin of virtual photons cannot be explained, they are defined as the instantaneous excited states of the field allowed under the uncertainty principle. In reality, the essence of virtual photons is the ubiquitous wave particles, or even isolated particles, in empty space. These fundamental particles have always existed in empty space and possess various forms of motion with different energies, though they are not easily observable. When they are excited or interact with fundamental particles, their existence becomes apparent.

There are endless space fluctuations, flows, and space energy in the vast universe, which are also the sources of observable world matter movement and energy.

7.6 Quantum Tunneling

(1) Principle Explanation of Quantum Tunneling

In microscopic quantum motion, there is an effect known as quantum tunneling, where a moving particle encounters a potential barrier, and the energy of the barrier is higher than the energy of the particle, yet the particle still has a certain probability of passing through the barrier,

Particles can pass through the barrier, as if tunneling.
In our previous analysis of the wave-particle duality of particles, we have seen that all microscopic particles, due to their small mass, are significantly affected by the fluctuations of the vacuum space. The motion, velocity, and energy exhibited by microscopic particles are the synthesis of their own particle motion, velocity, and energy with the velocity and energy of the vacuum space fluctuations. This is the root cause of the wave-particle duality exhibited by particles.

That is, the energy E of a microscopic particle satisfies the particle's apparent total energy formula (Equation 7.1):

$$E = E_1 + E_0$$

Where E_0 is the particle's own energy, and E_1 is the energy of the spatial wave fluctuation.
When a particle encounters a potential barrier with higher energy than itself, it generally cannot pass through. However, due to the complex fluctuations in the vacuum space, the energy E_0 will fluctuate, meaning that E_0 will dynamically change. When the change in E_0 reaches a certain extent, making the apparent total energy E of the particle described by equation

7.1 exceed the potential barrier energy, the particle can then tunnel through the barrier, resulting in the quantum tunneling phenomenon.

Since the wave energy of spatial waves fluctuates around an equilibrium value, there is a certain probability that the energy of a particle can exceed the barrier, thus the quantum tunneling phenomenon also exhibits a probabilistic nature.
(2) Quantum activities conform to the principle of conservation of mass and energy
For the quantum tunneling phenomenon, quantum mechanics explains it through the energy-time uncertainty principle. That is, according to the uncertainty principle in quantum mechanics, particles can temporarily borrow part of the energy, allowing their total energy to exceed the energy of the barrier, thus enabling them to pass through the barrier. At the same time, in order not to violate the energy-time uncertainty principle, the particle must return the borrowed energy within the time frame specified by the principle.

This process can bypass the principle of energy conservation. Through this mechanism, particles can have a certain probability of passing through the barrier.

The uncertainty principle, which allows for temporary violation of energy conservation and borrowing energy from the vacuum, also has applications in areas such as virtual photons under electromagnetic forces, which is something quantum theory does not consider the vacuum space itself

The need to accommodate the dynamics and energy of particles, which necessitates a compromise on the conservation of energy theorem.
In fact, we have discussed in the basic principle 7 of our space relativity: the principle of mass-energy conservation, that mass conservation, energy conservation, and mass-energy conservation occur in all physical processes of a system.

Considering the vast amount of wave energy in the vacuum space, this process is inherently a normal energy exchange and interaction, and the entire process conforms to the principle of energy conservation. There is no need to "temporarily violate the principle of energy conservation" through the uncertainty principle to secretly borrow and return energy.

All quantum activities conform to the principle of mass-energy conservation, which includes mass conservation, energy conservation, and mass-energy conservation.

(3) Experimental Verification

According to the principles of quantum tunneling we have analyzed, when a quantum encounters a potential barrier higher than its own energy, if there are particle oscillations around it, the greater the energy of the particle oscillations, the higher the probability of the quantum "tunneling" through the barrier, and the higher the upper limit of the energy of the barrier it can tunnel through.

Alternatively, by emitting electromagnetic waves into the space region where the quantum is located, the higher the energy of the electromagnetic waves, the higher the probability of the quantum "tunneling" through the barrier, and the higher the upper limit of the energy of the barrier it can tunnel through.

Because particle oscillations or electromagnetic waves will both excite fluctuations in the vacuum space. When the space wave overlaps with the vacuum space fluctuations around the quantum, it leads to an instantaneous increase in the energy of the region's space wave and an increase in amplitude. According to equation 7.1, this allows the quantum to gain a larger base energy from the space. When the total energy exceeds the barrier energy, the "tunneling" effect can be achieved.

7.7 Quantum Teleportation

In quantum theory, quantum teleportation is a method of transmitting quantum states. Its principle involves two parties communicating at different locations using a pair of entangled particles to recreate an unknown quantum state of a particle at one location onto another particle at the other location.

In this process, the original object is not transferred to the recipient; it remains with the sender at all times, and only the information is transmitted

The quantum state of the original object is transmitted, and the sender may even be completely unaware of this quantum state, while the receiver transforms another material unit (such as a particle) into the exact same quantum state as the original. This is similar to the propagation of classical waves, for example, a wave traveling from point A to point B actually transfers the vibrational state from A to B, not the actual particles (such as air molecules in sound waves) from A to B. The quantum state of the original object is destroyed when the sender performs a measurement and extracts classical information, as shown in Figure 7-2.

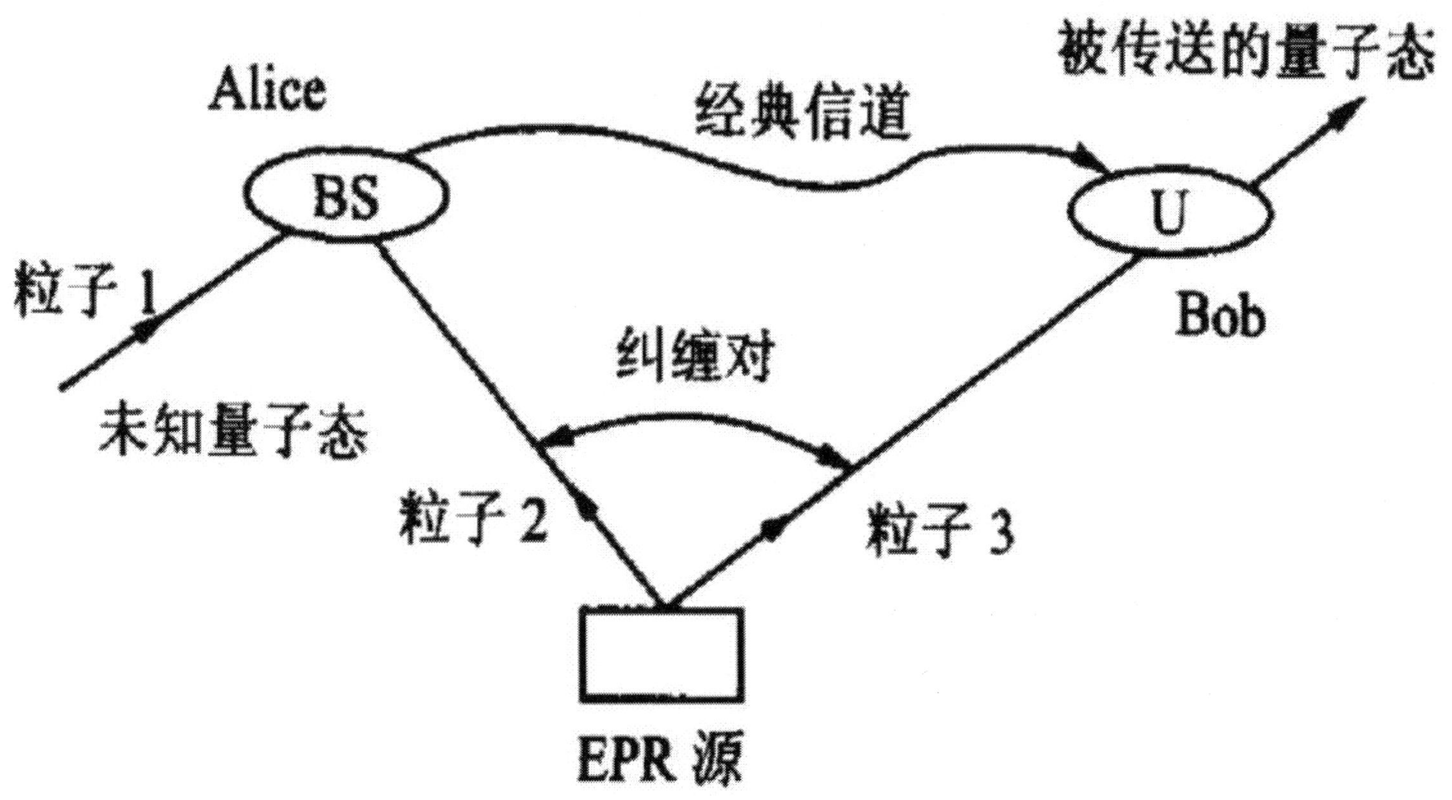

Figure 7-2 Quantum Teleportation Principle Diagram (BS represents Bell basis measurement, U represents transformation matrix)

In 1997, the Zeilinger group in Austria first experimentally verified the principle of quantum teleportation in the laboratory, and quantum teleportation has since experienced significant development.

However, as we analyzed earlier, there is no instantaneous response between the two entangled particles 2 and 3 over a distance, meaning there is no true entanglement. Their states are actually determined from the moment they are created, and there is only a correlation between them, which is also the case in this experiment.

In quantum teleportation, the two entangled particles 2 and 3 act as templates. By performing a measurement operation on the particle 1 that carries the quantum state to be transmitted and one of the entangled particles (such as particle 2), and then sending this measurement data to a distant location to adjust particle 3 accordingly, we can obtain the state of particle 1. This process can be imaginatively or simply analogized as follows: the entangled particles 2 and 3 are actually two completely identical (or with a specific correlation) clay models, while particle 1 is a sculpted model that contains the information to be transmitted. The joint measurement of particles 1 and 2 reveals how to sculpt particle 2, which is then used to recreate the state of particle 1 in particle 3.

Thus, it becomes the same sculpture model as particle 1, then send this data to particle 3, and perform the corresponding operation (carving) on particle 3 to make it the same sculpture model as particle 1.

The entanglement between particle 2 and particle 3 essentially establishes a definite relationship between them, allowing the operation on the latter to be calculated based on the former.

So, to achieve this so-called quantum teleportation, it is not necessary for particle 2 and particle 3 to be entangled particles. We can uniformly set specific forms of electrons, atoms, etc., as standard electrons, atoms, etc. The sender and receiver can each prepare standard particles according to the standard, and then perform the above process, which can also achieve the goal of teleportation. The operation to be performed on particle 3 can be obtained through corresponding transformations based on the operation data of particle 2 and the relationship between standard particles.

If standard particle 2 and standard particle 3 are completely identical, then the simplest case is to record all the operations needed to transform particle 2 into the sculpture model particle 1, and then send these operation data to the other side. The receiver can perform the corresponding operations on particle 3 based on the data, thereby obtaining a sculpture model identical to particle 1 and achieving "quantum teleportation."

The teleportation here uses the simplest "copy operation" mathematical transformation.
In current quantum teleportation, the transformation only utilizes the mathematical relationship of entanglement between particle 2 and particle 3 to obtain the operation needed for particle 3.

7.8 Identical Particles

In quantum mechanics, many particles cannot be distinguished from each other by any physical measurement or observation. These particles have identical intrinsic properties, such as mass, charge, and spin, and are therefore called identical particles.

In fact, these particles differ in properties such as mass, charge, and spin, but we cannot observe these differences in our macroscopic observations. At the extremely microscopic level of wave particles, they still exhibit clear distinctions, such as differences in the number of wave particles accumulated within the mass body, and differences in charge and other properties resulting from the interaction between the mass body and the surrounding empty space

Features will also differ due to differences in the number of interactions with bosons, but these differences are so subtle that they occur at an extremely microscopic level of bosons, which our macroscopic detection methods based on photons cannot identify.

Just as using a one-ton weight cannot detect a difference of a few milligrams between two masses.
There are no two completely identical leaves in the world, nor are there two completely identical particles; the concept of identical particles in quantum mechanics is merely an approximation from a macroscopic perspective.

7.9 Electron Spin

In quantum mechanics, electrons possess the property of spin. Initially, people thought that the spin of an electron would be similar to the spin of a planet, a physical spin. However, during actual analysis, it was found that if this were true, many difficulties would arise, one of which is that if an electron really spins, the tangential velocity at its surface would far exceed the speed of light.

This was considered to contradict the relativistic theory's assumption of the speed of light as an upper limit, so quantum theory dared not acknowledge this as a physical spin of the electron, and could only regard it as a peculiar property of the electron, as analyzed in Figure 7-3 [26] .

Electron spin is an intrinsic degree of freedom of electrons, a new physical quantity in quantum mechanics with no classical counterpart. (Such as orbital angular momentum corresponding to $\vec{r} \times \vec{p}$ in classical physics.)

Uhlenbeck and Goudsmit initially understood electron spin as the self-rotation of an electron, but this approach would face many difficulties:

1. Assuming the electron is a finite-sized sphere (denoted as r_e), with charge distributed within the sphere according to some pattern: angular momentum due to rotation $S \propto m_e r_e^2 \omega \sim \frac{\hbar}{2}$
According to Einstein's mass-energy relation: $m_e c^2 \sim \frac{e^2}{r_e}$, the classical radius of the electron can be estimated: $r_e \sim \frac{e^2}{m_e c^2}$
Electron spin angular velocity: $\omega \sim \frac{\hbar}{2m_e r_e^2}$
Tangential velocity of electron surface motion: $v \sim \omega r_e \sim \frac{\hbar}{2m_e r_e} \sim \frac{\hbar c^2}{2e^2} \sim \frac{c}{2\alpha} \sim 70c \gg c$, much greater than the speed of light, inconsistent with relativity.
2. According to quantum mechanics, if spin is assumed to come from the rotation of an electron, the Hamiltonian operator is: $H = \frac{L^2}{2I_e}$
Its eigenfunctions are Y_{lm} (fixed rotation), $\frac{1}{\sqrt{2\pi}} e^{im\varphi}$ (fixed axis rotation), l、m are all integers, which would be unable to explain the experimental fact that the electron spin is a half-integer [64] .

Figure 7-3 Spin Properties of Charge

But once we understand the relationship between the speed of light and the vacuum space, we realize that the speed of light C is not the ultimate speed in the universe, nor is it an objective physical law.

Therefore, we should not be afraid to acknowledge the physical spin of an electron due to these reasons and limitations. The spin of an electron and its charge is its own physical spin, and any particle can have spin motion. The reluctance to acknowledge the spin of an electron is also one of the many constraints imposed by relativity and the constancy of the speed of light on theoretical physics.

Of course, in a microscopic environment, its spin also involves interactions with the vacuum space, and there are some special properties that need to be considered.

7.10 Photon Mass, Momentum, and Energy

(1) Photon Momentum and Energy

All particles and matter have mass, as we analyzed earlier.
Let the mass of a photon be m, since its linear velocity in the standard density vacuum space is C, then the photon's momentum P is:

$$P = mc(7.2)$$

Photon Kinetic Energy E_1 :

$$E_1 = \frac{1}{2}mc^2(7.3)$$

Since photons are affected by spatial wave fluctuations and exhibit transverse wave motion, they possess potential energy E_2 :

$$E_2 = \frac{1}{2}mc^2(7.4)$$

So the total energy E of a photon is:

$$E = E_1 + E_2 = mc^2(7.5)$$

These topics have been discussed and explained in the basic principles of spatial relativity 7: the principle of conservation of mass and energy.
At the same time, according to the basic principle of spatial relativity 4: the new constancy of light speed, the speed of light conforms to relativity,

that is, the speed of light observed by different observers in different motion will be inconsistent.
Assuming the speed of the observer relative to the vacuum space where light propagates is V , then the observed speed of light $C1$ is:

$$C_1 = C - V(7.6)$$

In this case, the momentum and kinetic energy of the photon will also change.
Photon relativistic momentum:

$$P = m(C - V)(7.7)$$

Photon relativistic kinetic energy:

$$E_1 = \frac{1}{2}m(C - V)^2(7.8)$$

Photon Relative Potential Energy:

$$E_2 = \frac{1}{2}m(C - V)^2(7.9)$$

Photon Relative Energy:

$$E = E_1 + E_2 = m(C - V)^2(7.10)$$

(2) Photon Total Path Energy

The discussion above considers the energy of photons as wave quanta moving at a constant linear velocity (such as the speed of light C). In reality, during the wave process, photons have a helical motion, so their actual path velocity is not merely the linear speed of light.

We assume that in standard vacuum density, the linear speed of light is C, and the wavelength of motion is λ , but because of the helical motion, we assume it can be approximately considered as a circular helical motion. Thus, the total wavelength $\lambda1$ of the motion trajectory is:

$$\lambda1 = \sqrt{(2\pi\ A)^2 + \lambda^2(7.11)}$$

Where A is the amplitude of the spiral wave.

Let the frequency of light wave be f, then:

$$C = f\lambda$$

The total travel velocity of the photon C_1 is:

$$C_1 = f\lambda 1 = C\sqrt{1 + (2\pi fA)^2}(7.12)$$

The energy of the photon in this case is:

$$E = mC^2\left(1 + (2\pi fA)^2\right)(7.13)$$

When the frequency f approaches zero, it can be simplified to the normal energy formula:

$$E = mC^2$$

(3) Photoelectric Effect and Light Wave Energy
The Planck relation, discovered by Max Planck in his study of blackbody radiation, suggested that the energy of light waves should be discrete, that is, the basic energy E is:

$$E = hf$$

Where $ h $ is the Planck constant and $ f $ is the frequency of light.
Later, when Einstein explained the photoelectric effect, he proposed that photons are particle-like. Here, the basic energy represents the energy of a photon, that is, the energy of a photon with frequency f is $E = hf$.

However, we have analyzed earlier that the apparent total energy E of all fundamental quantum particles, including photons, is the sum of their own particle energy E_1 and the energy E_0 of the common motion of the vacuum space wave.

According to the apparent total energy equation of a particle (Equation 7.1):

$$E = hf = E_1 + E_0$$

That is:

$$hf = mC^2 + E_0(7.14)$$

Or:

$$hf = mC^2\left(1 + (2\pi fA)^2\right) + E_0(7.15)$$

Equations 7.14 and 7.15 are the quantum energy equations of light waves, meaning the energy of a light quantum is the sum of the particle energy of the photon itself and the energy of the spatial wave it moves with.

Not only photons but all other fundamental particles, even all matter, conform to this energy formula, meaning that macroscopic matter also follows the wave properties of matter. However, microscopic particles, due to their small mass, are significantly affected by spatial wave fluctuations, thus exhibiting obvious wave-like behavior, whereas macroscopic matter is almost unaffected, and the corresponding phenomena cannot be observed macroscopically.

7.11 The Relationship Between Spatial Quantum Dynamics and Quantum Mechanics

As a science of the microscopic quantum domain, quantum mechanics has achieved great success, which can prove the correctness of this theoretical framework. However, as a statistical description theory, it lacks clear explanations and understanding of the underlying physical principles. Moreover, because it does not consider the objective entity of empty space, which has significant dynamic effects on particles, there are many descriptions and principles in the theory that are either incomprehensible or difficult to understand.

The spatial quantum dynamics part of spatial relativity solves the problematic theoretical descriptions in quantum theory, such as probability and instantaneousness, by taking into account the influence and interaction of empty space on microscopic quantum particles, and provides an explanation of the underlying physical principles.

Space quantum dynamics and quantum mechanics are two theoretical frameworks for interpreting phenomena in the microscopic quantum domain. Quantum mechanics is a statistical description theory, while space quantum dynamics is a corresponding physical principle-based description theory. It addresses some of the shortcomings in quantum mechanics by introducing new principles of localization, providing a complete explanation of wave-particle duality and the double-slit interference, and restoring determinism to probabilistic quantum physics.

Space quantum dynamics is a partial correction and improvement of quantum mechanics.
Historically, the de Broglie-Bohm theory also made some efforts toward restoring a deterministic classical interpretation of quantum mechanics, proposing the pilot-wave theory which abandoned the explanation of wave function collapse that was difficult to understand

It can well explain the probabilistic description in quantum mechanics from the perspective of classical mechanics and perfectly accommodate all the conclusions of quantum mechanics. In this context, the so-called "pilot wave" it mentions is actually the spatial wave in the empty space as pointed out in our space relativity theory. However, Bohm did not recognize this, leading to criticism from the mainstream academic community that adding a pilot wave was unnecessary.

At the same time, the de Broglie-Bohm theory holds a non-local view of quantum entanglement, believing that the effects of entanglement occur instantaneously at a distance, which is inconsistent with our new locality principle.

It is worth noting that after 2009, French scientists such as Yves Couder used fluid droplets like silicone oil to conduct experiments, creating an analogy to single-particle quantum systems and giving the navigation wave theory a physical realization [27][28] . This experimental result has led some scientists to question the traditional probability wave interpretation. It also aligns with the principles of our spatial quantum dynamics.

8 Introduction to Spatial Particle Dynamics

8.1 Basic Principles of Spatial Particle Dynamics

(1) Motion and mechanical properties of spatial superfluids
We have discussed in our previous chapters that the universe is filled with a kind of substance similar to a superfluid [29][30] , which we are currently unable to observe. This superfluid has mass, volume, and dynamic properties, and all matter in the universe is generated through excitation of this fundamental superfluid. Like the vast ocean of emptiness in the universe, all particles, stars, and other matter move and interact within it. This is the core foundation of spatial relativity and also the basic starting point of the principles of spatial particle dynamics.

Of course, the void space we refer to here is a general term for the superfluid matter in space, not space itself.

To more clearly define the concept, we can further explain and define it:
Vacuum space: space with no matter present.
Void space: a general term for the superfluid that permeates vacuum space and possesses properties of motion and mechanics. The terms "void space superfluid" and "space superfluid" mentioned later both refer to this concept.
(2) Principles of particle stacking and evolution
By referring to the formation of all things in our world, all life initially began as a simple combination of basic amino acids and proteins. Through mutual interactions with the surrounding natural environment, these life forms gradually evolved into functionally differentiated modules. These different functional modules further developed and combined, eventually stacking together to form a rich and diverse biosphere, evolving countless life wonders.

All life appears very miraculous, but fundamentally it all originates from the simplest ammonia

The combination of amino acids and proteins began, and at first there was not even the slightest life activity.
All the matter in our world and the universe appears to be rich and colorful, very mysterious and complex, but when they are generated from the basic vacuum space superfluid, they are composed of some very simple components, or initial particles. The interactions between these particles, including their interactions with the vacuum space, further evolve and gradually form the vast and magnificent world of the universe we see today.

Just as the evolution of life, just as our integrated circuit systems, started with AND gates and NOT gates, then stacked and differentiated to form functional adders, multipliers, and registers, and eventually evolved into more complex systems such as CPUs and GPUs, our software systems also follow a similar path. They began with only 0、1 machine language, then progressed to assembly language, and further to higher-level languages, and now even to the natural language used by artificial intelligence. These systems also started with the simplest symbols, gradually stacking and differentiating to form various functional modules, and continuing to evolve into the complex and colorful software world we see on phones, computers, and other devices today.

The emergence of material particles and the evolution into the macroscopic world should also follow the same principles. The process of stacking, differentiating, and evolving from basic elements such as empty space and wave particles is the fundamental principle of space particle dynamics.
(3) The relationship between mass, motion, and various forces
There is mass before there is force. The collision between mass bodies and the exchange of momentum and kinetic energy are the sources of all macroscopic and microscopic forces.

Vacuum space and bosons have mass, volume, and dynamic characteristics, and all matter particles are aggregates of bosons in different motion patterns and coherent structural modes, so they also possess mass, volume, and dynamic characteristics.

Mass, volume, and motion are the most fundamental properties of all substances in the material world. All forces are manifestations of collision effects between mass bodies or groups of mass bodies under different motion patterns. Essentially, they are all results of collisions between mass bodies

The result of momentum and energy exchange during the collision.
All forces, including the gravitational force, electromagnetic force, strong force, weak force, and other more forces yet to be discovered, conform to this principle.
(4) Analytical methods of fluid mechanics
Coherent structures are formed in the vacuum due to energy fluctuations and excitation. Coherent structures formed with

different excitation energies will show differences. These coherent structures exhibit certain rigidity and closure due to their own motion, which is the original prototype of all particles.

The most basic coherent structure can be referred to as the fundamental soliton or fundamental solitary particle. All subsequent other particles appear as different evolutions due to varying excitation energies, as well as higher-order, more complex particle forms resulting from the fusion and combination of solitary particles.

Since the vacuum space can be considered as a superfluid, these excitations, coherent motions, and interactions between particles can all be analyzed based on fluid dynamics [31] , in accordance with the principles of fluid dynamics [32] . This is also the fundamental analytical method and model for spatial particle dynamics.

At the same time, due to the quantum nature of the basic elements called quasiparticles, all the fundamental coherent structures excited in a superfluid are quantized. That is, as long as there is no collision, there will be no mechanical interaction between the quasiparticles within these extremely microscopic coherent structures, and they will not affect each other.

8.2 Formation and Evolution of Spatial Particles

(1) Soliton
When a fluid is subjected to specific disturbances, it forms a wave with an independent coherent structure, which has a certain degree of closure and stability, and even rigidity, similar to a special kind of particle. Therefore, it is also called a "solitary wave" (Solitary Wave), or "soliton" [15][16] .

In August 1834, the British scientist and shipbuilder John Scott Russell observed a pile of water being pushed out from the bow of a moving wooden boat; when the boat suddenly stopped, this pile of water remained in place, seemingly unaffected

The shape propagates forward at a speed of approximately 13 kilometers per hour. Ten years later, at the 14th meeting of the British Association for the Advancement of Science, he published a paper titled "On Water Waves," describing this phenomenon. He referred to this peculiar moving water mass as "solitary wave" or "soliton," which is the solitary wave phenomenon discovered in the macroscopic real world.

Solitary waves can exist for a long time, moving at a certain speed until they encounter an obstacle or their energy is exhausted, at which point they disperse back into fluid molecules and disappear.

When the superfluid of the void space is disturbed, similar solitary waves can also form. Unlike the macroscopic solitary waves we observe in the real world, these microscopic solitary waves in void space possess quantum properties due to the superfluid characteristics of void space. They can be called solitons or solitary wave particles, which can exist and move stably for a long time, exhibiting properties such as enclosure, stability, and rigidity similar to those of particles.

At the same time, due to the quantum properties of superfluid quasiparticles, these quantum solitons will have better stability and longevity.

Depending on the type of perturbation, solitary particles can have different types.
(1) Vortex
When subjected to a slight one-dimensional perturbation (linear perturbation), the superfluid vacuum space excites the primary soliton: the vortex. Due to the properties of the superfluid vacuum space, which has extremely low viscosity, it generates quantum vortices, as shown in Figure 8-1.

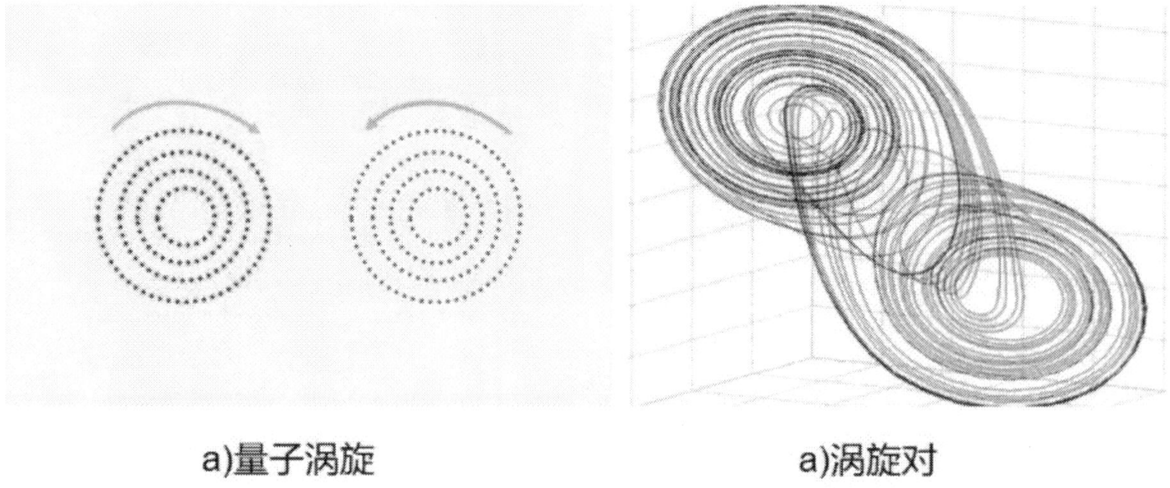

Figure 29: Figure 8-1 Quantum Vortex

Quantum vortices are the most fundamental soliton coherent structures, as the bosons have no viscosity, and the bosons within the vortex are entirely driven by momentum effects. When the vortex energy is extremely high, the vortex develops a certain degree of closure and rigidity, acquiring a level of stability, which is the characteristic of an initial rigid particle.

If a quantum vortex does not collide with other vortices in its surroundings, it can remain stable for a long time. As it moves through the surrounding space and interacts with the surrounding vacuum, its energy gradually diminishes, eventually decaying and dispersing back into the vacuum, reabsorbing into the superfluid of the vacuum space.

(2) Vortex

Vortices are the most fundamental motion and solitons in quantum solitons. When the disturbance energy in the void space increases, such as two-dimensional or three-dimensional disturbances, it further excites a large number of quantum vortices, which aggregate into a collective state composed of numerous vortices, forming a three-dimensional stacked vortex structure.

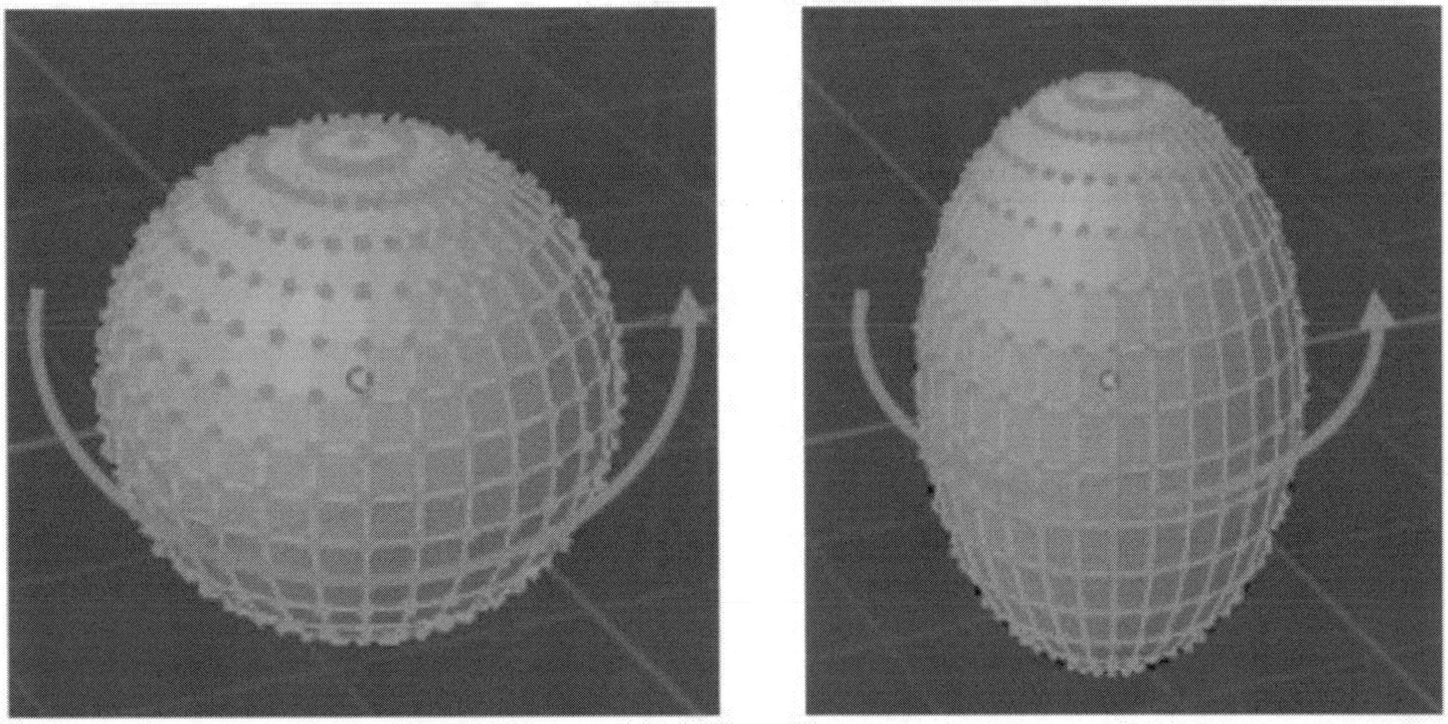

Figure 8-2 Vortex: Closed vortex matrix formed by multiple vortices

This three-dimensional stacking vortex is similar to water droplets in the macroscopic world, and we can call it a vortexon. Vortexons can be ellipsoidal or spherical and can be considered as vortex matrices formed by multiple vortices, creating a unified large vortex closed flow field. Their mass, volume, and spin speed are all significantly higher than those of the basic quantum vortices, and they also possess stronger closure, stability, and rigidity, exhibiting more distinct particle characteristics, as shown in Figure 8-2.

(3) Vortex Ring

When the disturbance energy increases, or when special two-dimensional or three-dimensional disturbances occur, the superfluid vacuum space can also excite the formation of higher-energy solitons: vortex rings.

The vortex ring is a three-dimensional upgrade of a vortex, not only involving circular flow around the vortex center on a plane, but also three-dimensional vortex motion, forming a closed vortex ring body, as shown in Figure 8-3.

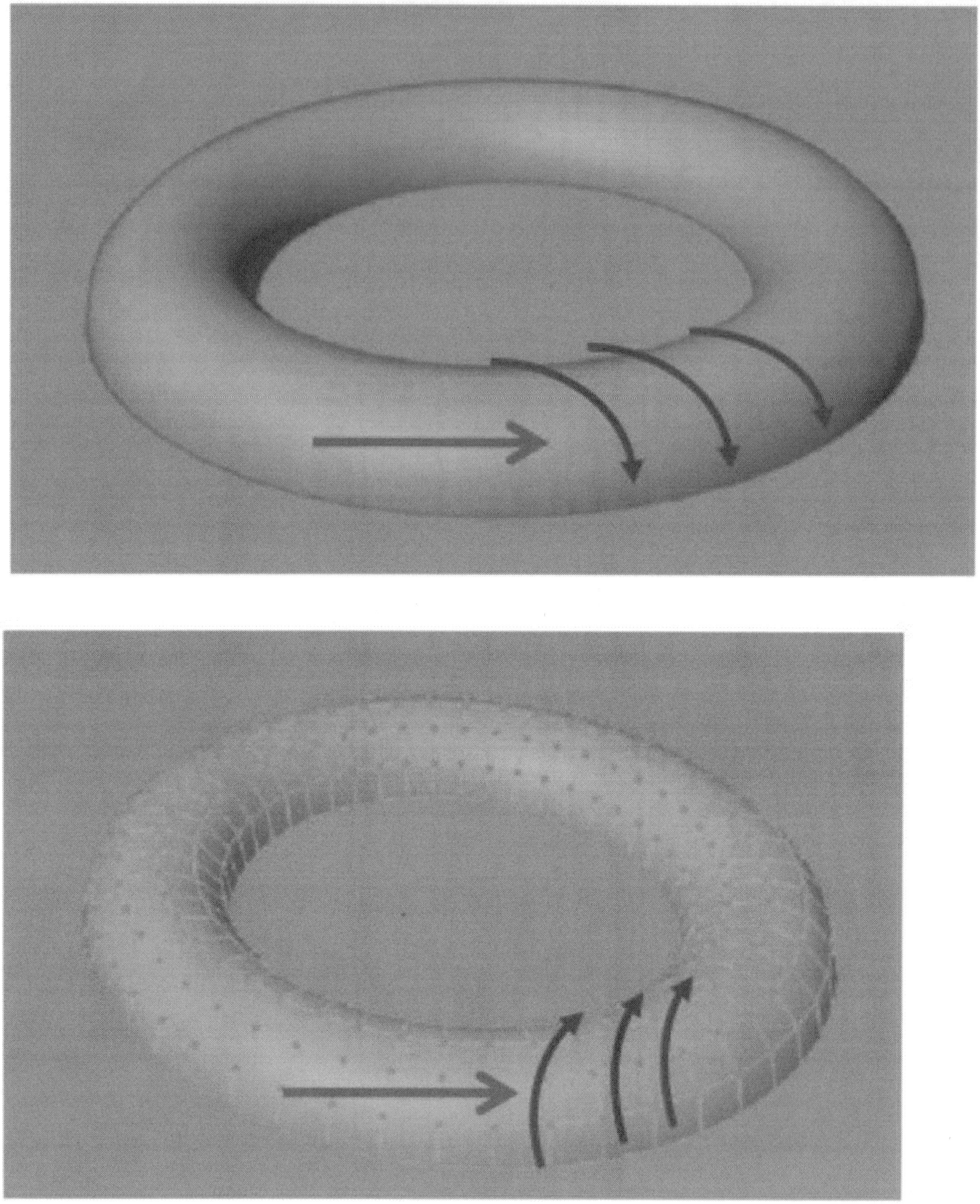

Figure 8-3 Closed Vortex Ring

The vortex ring has higher energy and therefore possesses greater rigidity, closure, and stability than a vortex, allowing it to exist and move for a longer period.

(4) Vortex ball

When the excitation intensity of the vacuum space continues to increase, it can excite the production of higher energy solitary particles: vortex balls.

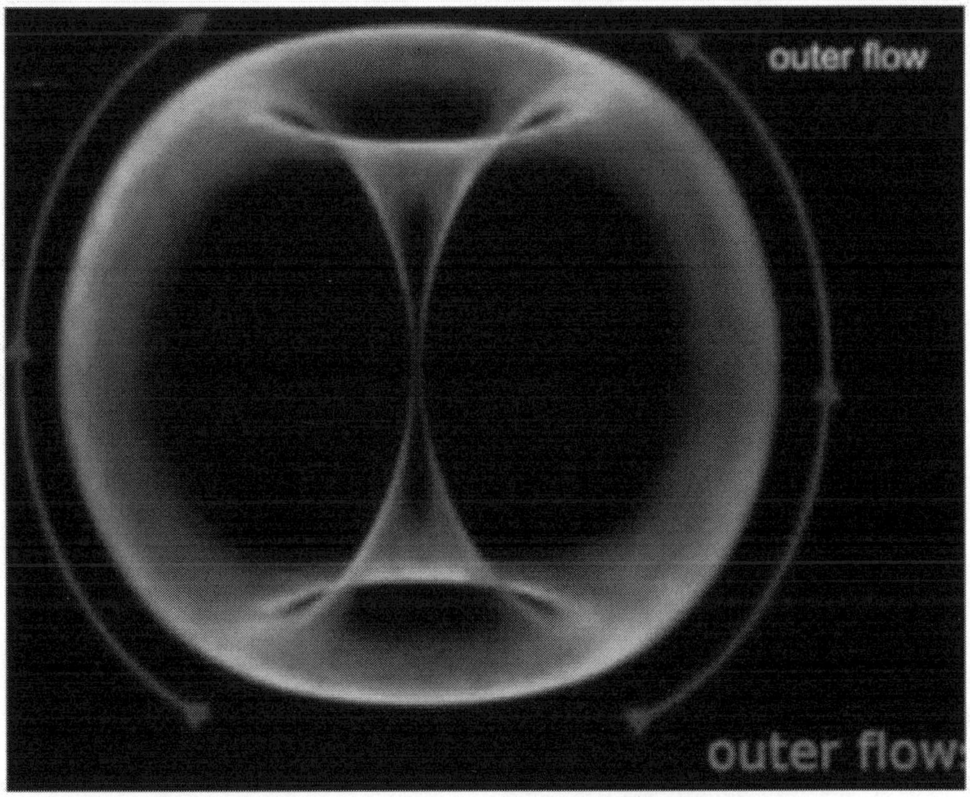
outer flow
outer flows

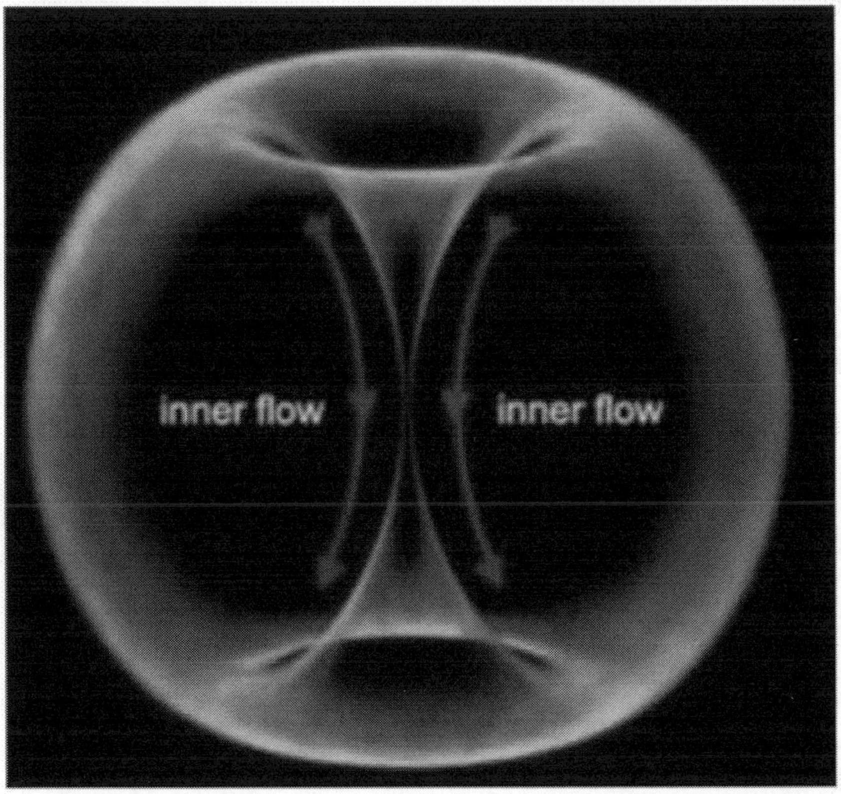
inner flow
inner flow

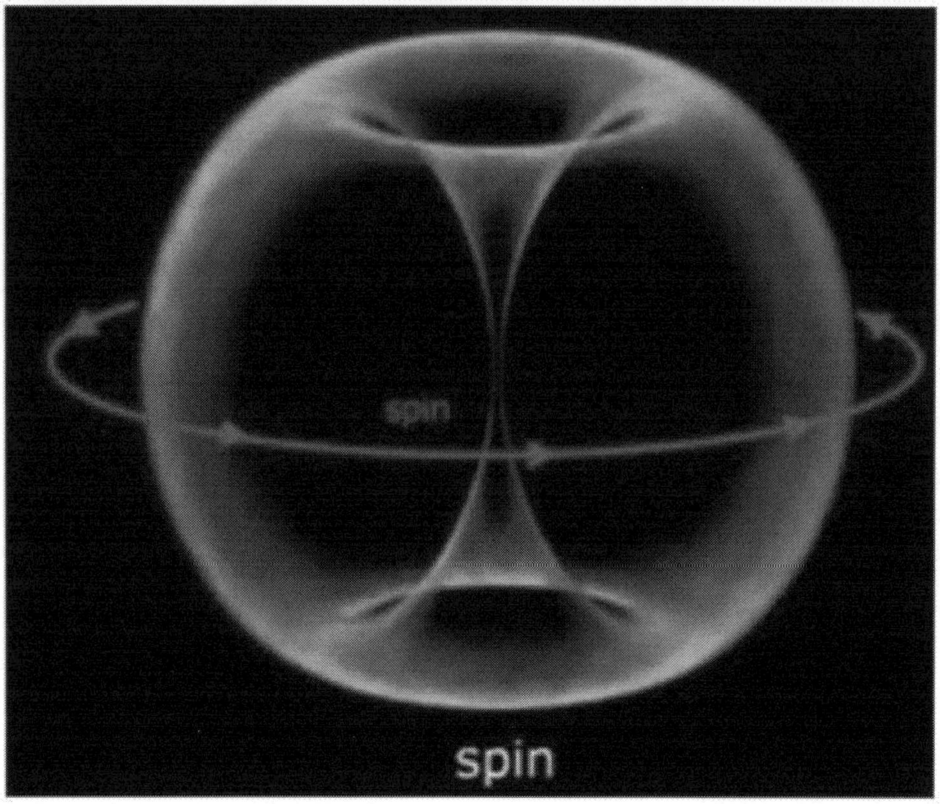

Figure 8-4: Internal and External Flow Fields of a Torus and Spin

The vortex ball has greater mass and kinetic energy, as well as a larger volume, equivalent to a three-dimensional reinforced version of the vortex ring, with external flow, internal flow, and overall spin flow, forming

A closed flow field. The vortex ball also has stronger closure, stability, and rigidity compared to the vortex ring, and can already be referred to as a micro-particle, capable of existing and moving stably for a long time, as shown in Figure 8-4.

Quantum vortices, vortex rings, and vortex balls are the most basic quantum coherent structures formed when the vacuum space is excited, and they are also the fundamental forms of all particles.

It should be noted that the interior of vortex rings and vortex balls is not entirely empty; there are also a large number of wave particles present, including small vortices, vortex particles, vortex rings, and even vortex balls. These internal particles interact with soliton particles, together forming the soliton particle body, which is an important source of the soliton's total mass, momentum, and energy.

(2) Composite Particles

If vortices, vortex particles, vortex rings, and vortex balls are the fundamental particles formed when the void space is disturbed, then as the disturbance energy in the void space continues to increase, these primary particles, in addition to increasing in mass, energy, and volume, will further interact to form more complex and powerful composite particles.

The formation patterns of these composite particles include:

(1) Nuclear vortex sphere

The solitons we discussed earlier, or vortices in the first-level particles, are hollow vortices without a core. When a vortex enters the flow loop of a larger vortex and becomes its inner core, it transforms the larger vortex into a vortex with a core, as shown in Figure 8-5.

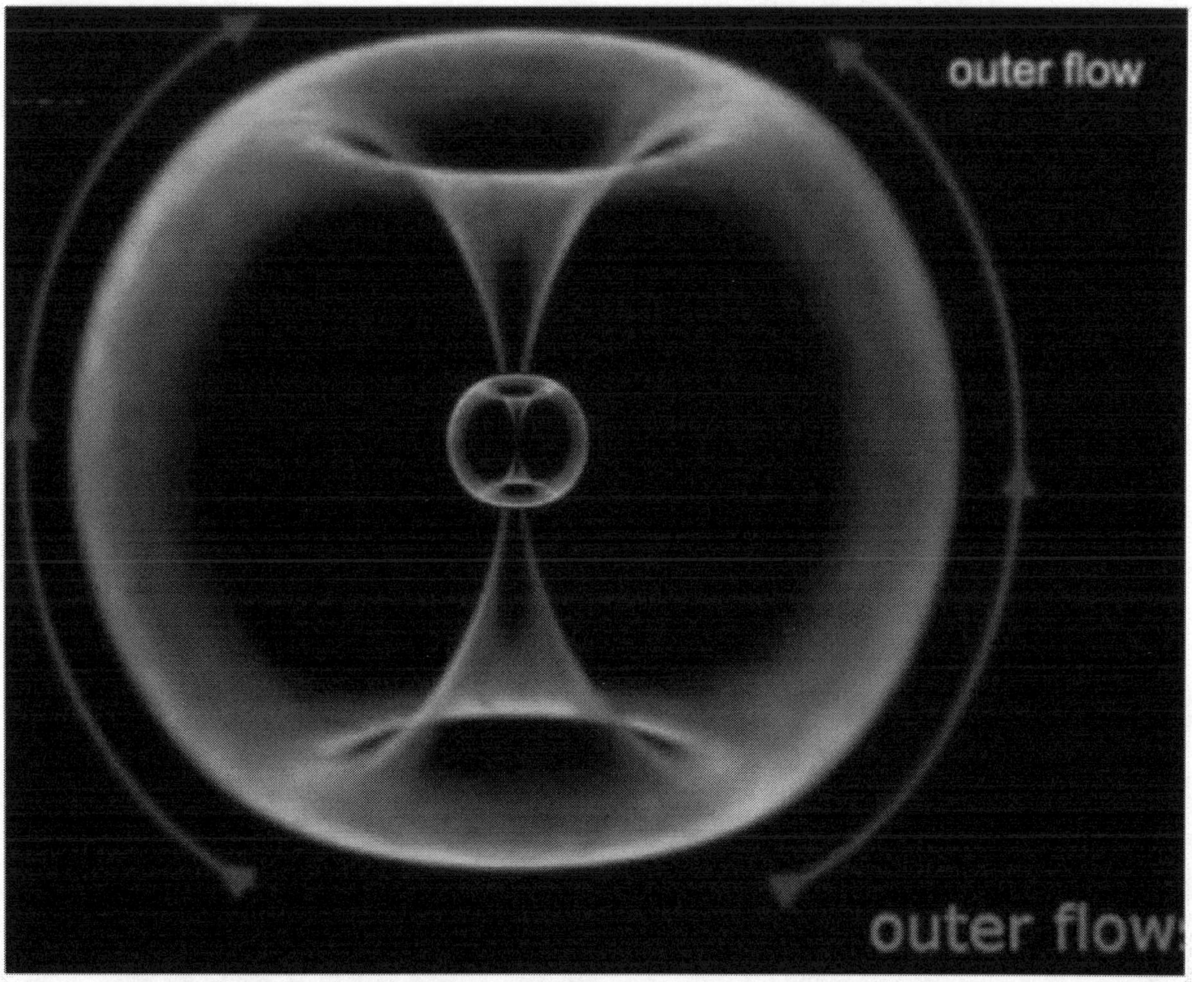

Figure 30: Nuclear Vortex Sphere 8-5

Due to the formation of the nucleus, the mass and energy of the vortex ball are greatly enhanced, and its rigidity and stability are further improved, making it a composite particle that can exist for a relatively long time.

Like vortices, there are also a large number of wave particles, even vortices, vortex particles, vortex rings, and vortex balls, existing in the internal space of composite particles, in addition to the vortex ball nucleus. All these soliton particles, together with the internal and external flow fields of the vortex ball composite particle, form a complex internal energy field, which collectively constitute the mass and energy sources of the composite particle.

At the same time, the formation of the nucleus also gives the overall vortex ball particles a more complex multi-layered flow field, creating different energy levels in the external flow field, as shown in Figure 8-6.

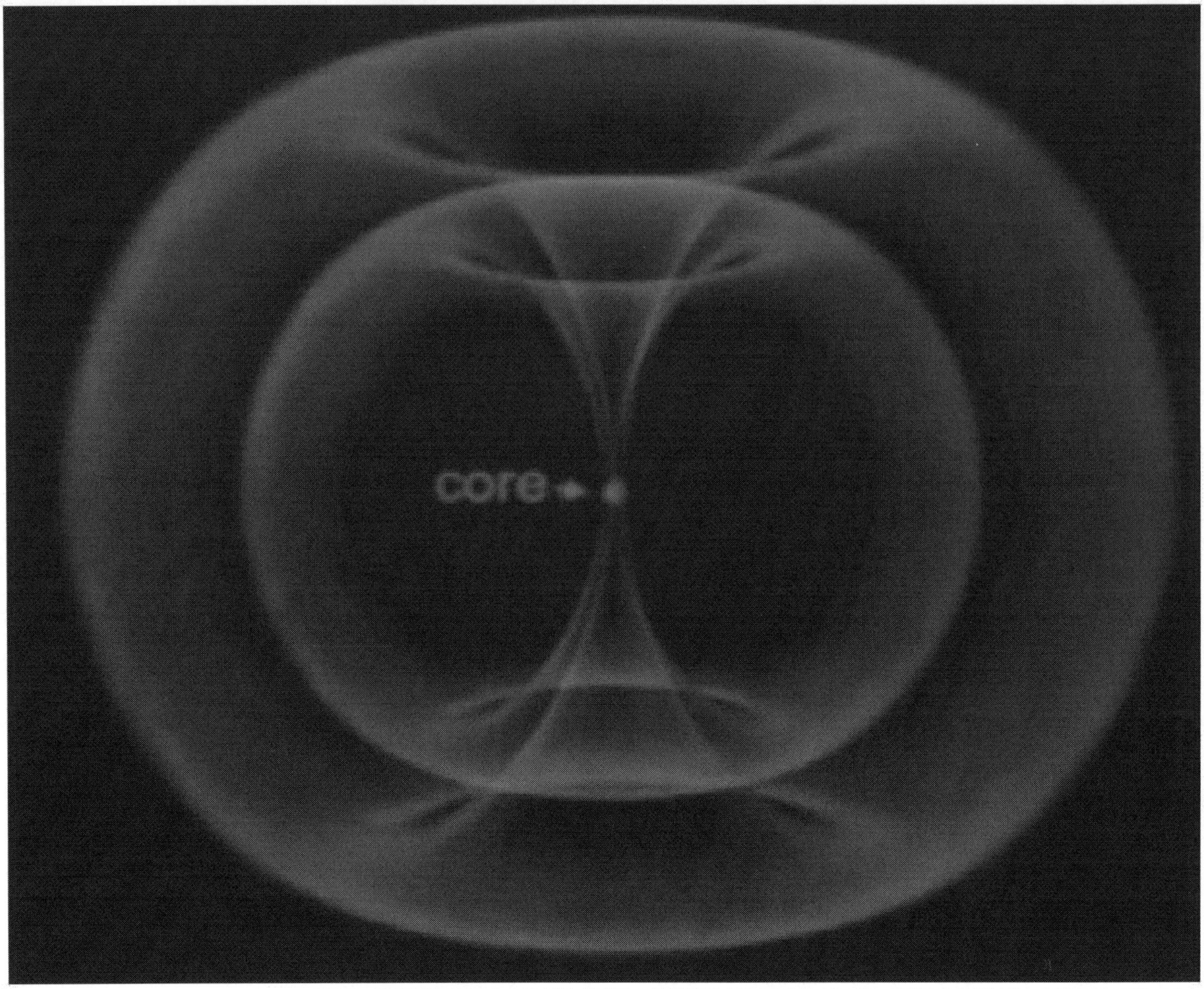

Figure 8-6 Multi-layered flow field of a nuclear vortex sphere: particle energy levels

(2) Vortex Ball Combination
Two vortex balls can be compressed and deformed under stronger external forces, eventually forming an additional outer unified field while retaining their own complete flow fields, thus merging into a new independent particle.

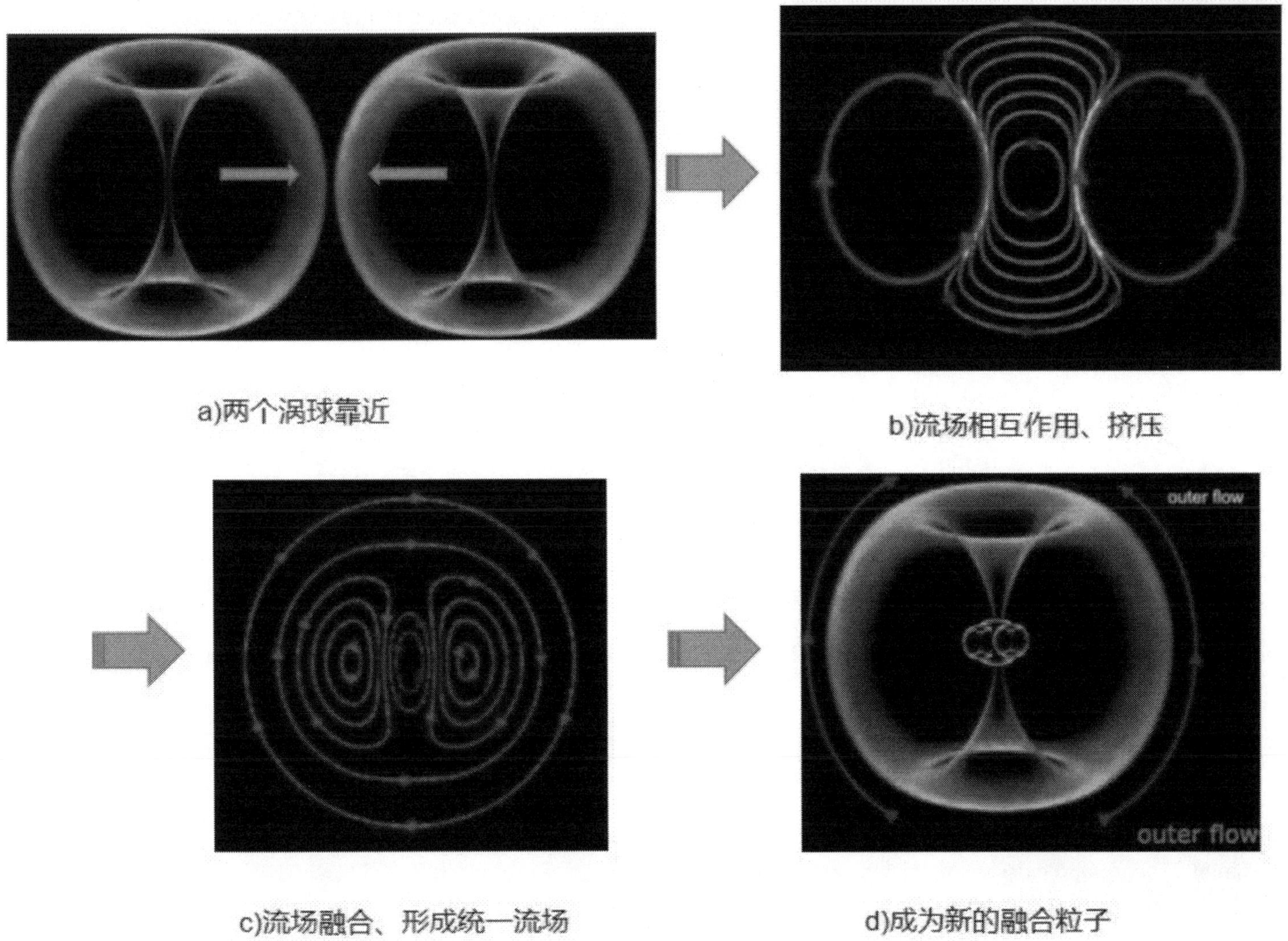

Figure 8-7 Vortex Ball Particles Fuse into a New Composite Particle

In the newly formed composite particle, a new vortex field is also generated between the two solitons, similar to a new soliton. The new soliton, together with the two merged solitons, collectively forms the new composite particle. Additionally, the presence or excitation of a large number of solitons within the composite particle leads to a significant increase in its mass and energy, as shown in Figure 8-7.

(3) Nuclear arrays, nuclear nesting

The vortex ball nucleus of a composite particle may be a single vortex ball or an array composed of multiple vortex balls. At the same time, the nucleus of a composite particle itself may also be a composite particle, meaning that the nucleus of a vortex ball may have its own vortex ball nucleus, forming a nested nucleus structure.

The existence of these nuclear arrays and nuclear nesting makes composite particles more complex high-level particles, with properties such as energy, mass, and volume continuing to increase to enormous levels.

The pattern mentioned earlier of vortex ball combinations forming new composite particles, in this new composite particle, two

The combination of vortices can actually be viewed as a nuclear array of a new particle.

(4) Vortex Ball Array Combination

When more than two vortices combine to form a new vortex composite particle, more complex vortex array combinations are created. The mass and energy of the composite particles formed by these array combinations will also be greater, with more complex internal energy flow field distributions.

By combining in the aforementioned ways, the primary particles initially excited from the void space—the solitary particles—can gradually stack together to form more complex composite particles with higher quality and energy. They can further evolve into all the fundamental particles and matter we observe, gradually constructing our complex and diverse material world.

(3) Fundamental Motion Properties of Particles

According to the above analysis, all particles, including solitary particles and composite particles, have the following motion properties [33] :
(1) Spin
All particles have spin properties. Since all primary particles: isolated particles are first generated due to spin, the composite particles formed by their combination also possess the characteristic of high-speed spin.

Spin is the fundamental reason for the particle's closure, stability, and rigidity.
(2) Spin
The vacuum space has wave-like and flowing motion characteristics. All particles are affected by the action of vacuum space during their propagation and movement, so they also exhibit wave characteristics. The smaller the mass of the particle, the greater the influence it receives, and thus the more pronounced its transverse wave motion characteristics will be. This is the origin of the material wave properties of all particles and even all matter.

Because of the characteristics of transverse wave motion, this wave will not be absolutely ideal linear motion, but will have an elliptical trajectory formed by a long axis and a short axis. Therefore, the addition of this elliptical wave characteristic during linear motion of the particle results in a gyroscopic motion pattern, where the particle exhibits elliptical gyroscopic motion along the direction of linear movement.

Motion.
For photons, this kind of spin motion is their polarization, so light polarization is generally elliptical polarization. When the short axis of the ellipse approaches zero, it can be approximated as linear polarization, and when the long and short axes approach equality, it can be approximated as circular polarization.

According to the above principle, all microscopic particles and even all matter have similar polarization characteristics, but their polarization characteristics weaken as the mass increases, leading to a reduction in the effect of their spin motion.

(3) Vortex Flow Field

According to the analysis of the principle of particle formation above, all vortex sphere particles and composite particles formed by stacking vortex sphere particles have overall internal and external closed flow fields. The external flow field of the closed flow field enters the particle from one end to become the internal flow field (which can be called the vortex flow field inlet, S pole), and exits the particle from the other end to form the external flow field (which can be called the vortex flow field outlet, N pole), as shown in Figure 8-8.

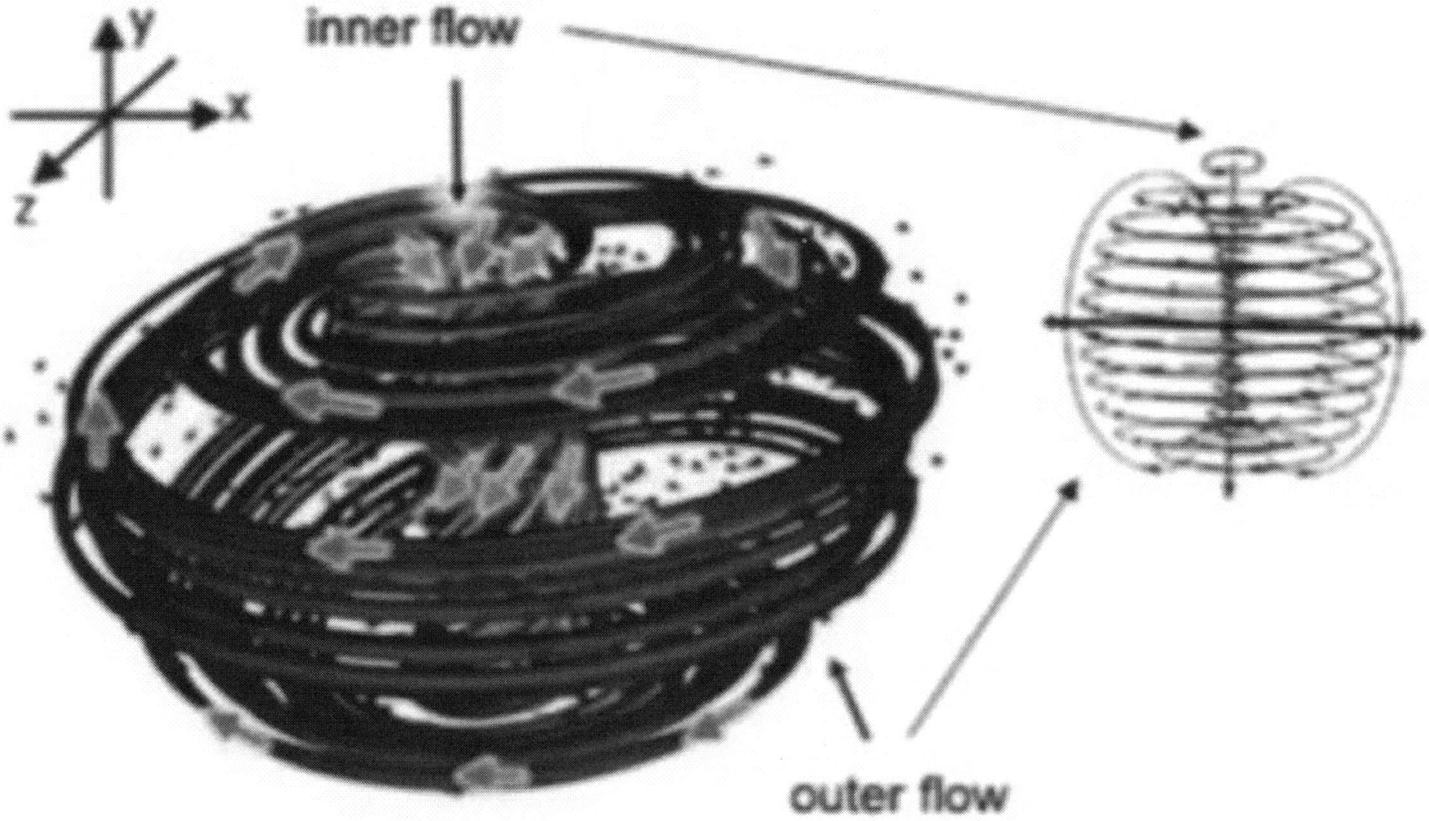

Figure 8-8: Inlet and Outlet of the Particle Vortex Flow Field

This closed vortex flow field allows particles to have a determinable spin direction. When observed from the outlet of the vortex flow field, particles can be classified into two types: clockwise spin and counterclockwise spin. This distinction enables clear differentiation among particles and provides them with the ability for more complex stacking, differentiation, evolution, and interactions. Combined with the particles' spin and orbital motion, it allows for the emergence of various new effects and interaction patterns, as shown in Figure 8-9.

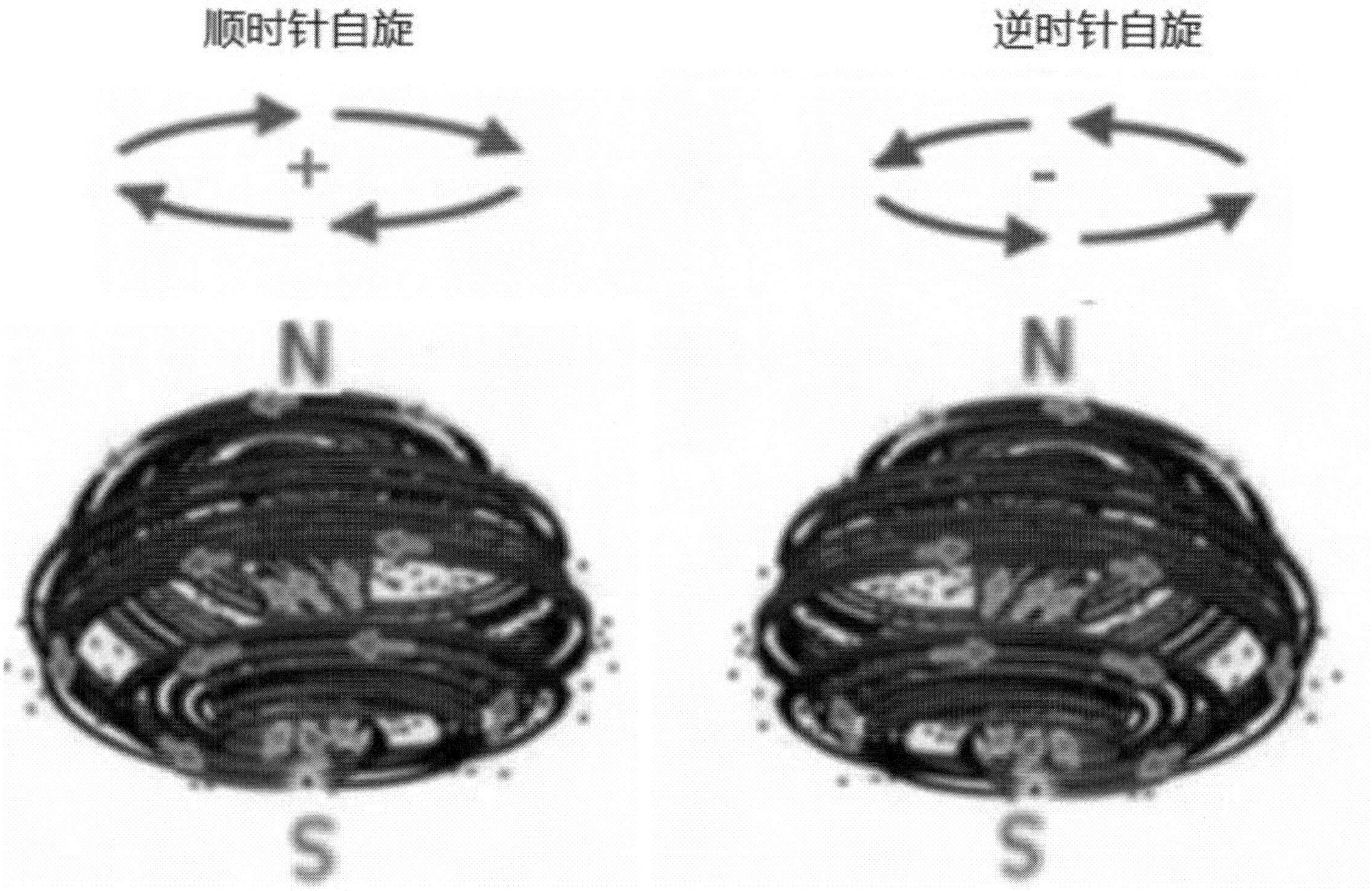

Figure 8-9 Particle Spin and Flow Field Direction

The outflow field of a particle forms a closed and rigid shell around it, and the multiple layered shells of composite particles constitute the energy levels and orbital layers of the particle, as shown in Figure 8-10.

Due to the complexity of the composite particle flow field, a particle may have multiple vortex flow field entrances and exits.

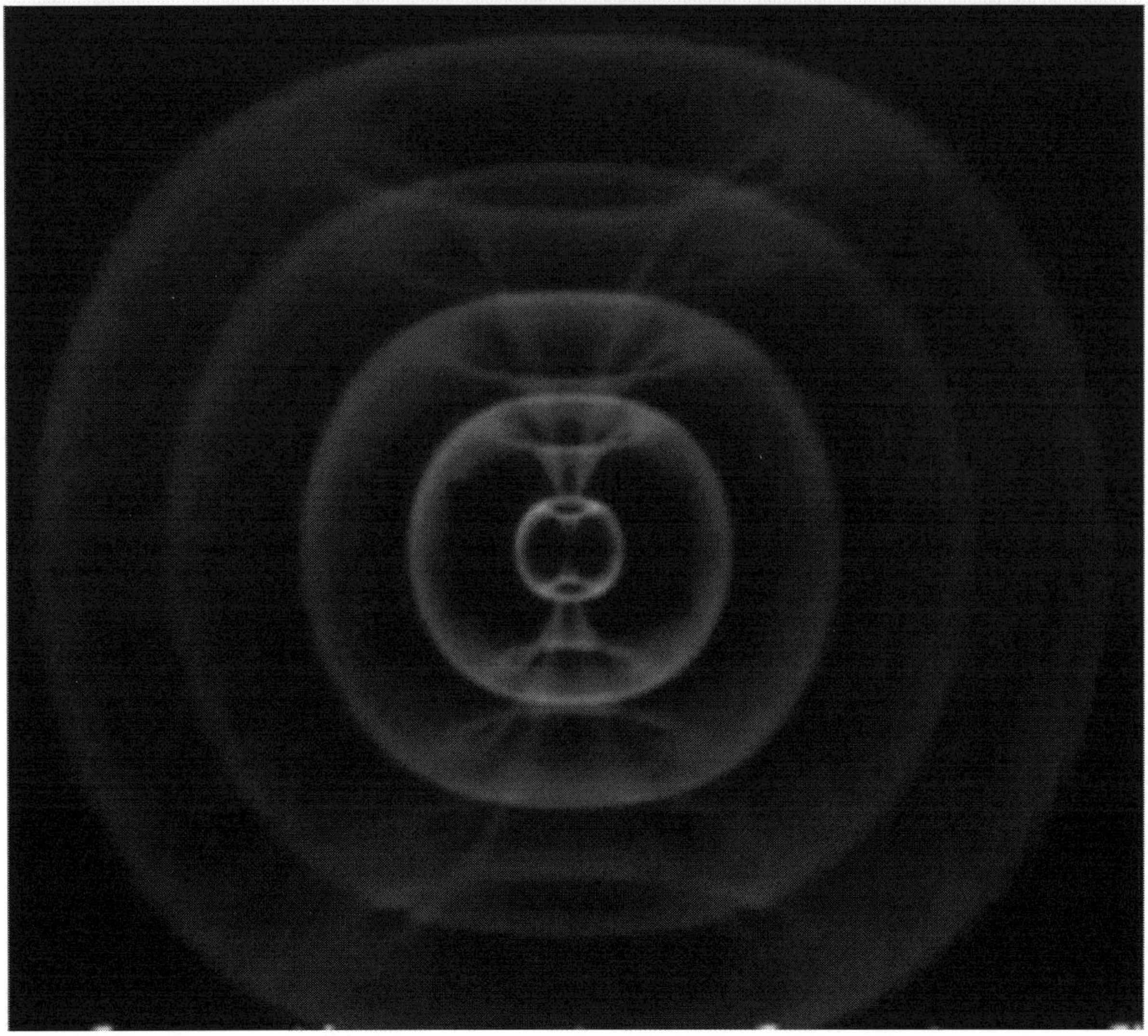

Figure 31: Figure 8-10 Multi-level and orbital structure of composite particles

(4) Vibration

The high-speed spin of a particle allows it to continuously interact with the surrounding vacuum space, including the motion of the vacuum space's superfluid itself, causing the particle to vibrate rather than spin in place. This vibrational motion also leads to more complex interactions and mechanical effects between particles.

(5) Other Motions

In addition to the aforementioned motion patterns, particles are also influenced by their own initial excitation-induced linear motion, or by the oscillations, flows, and other movements of the vacuum space itself, resulting in different additional motions. When a large number of particles gather or merge, they can form special vibrational motion patterns, and the frequencies of these vibrational motions can lead to more complex effects on the interactions between particles.

The various Kranz patterns formed by the vibrational frequencies on a plane are a simple representation of the influence of this vibration on the interaction of microscopic particles, which can be simply understood as a special mechanical effect resulting from the vibration of a large number of particles under their own vibrations and common vibrations, as shown in Figure 8-11.

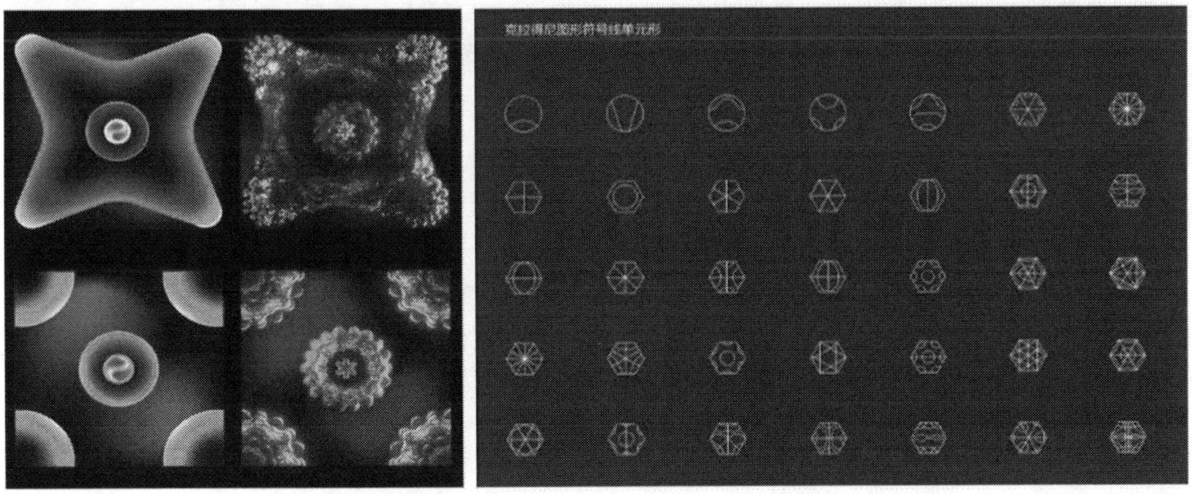

Figure 32: Figure 8-11 Clérambault Figure

However, the current Kranz graphs are merely simple simulations on a plane. In reality, particles form composite particles, interact, and create various macroscopic materials under the influence of their own vibrations and the vibrational modes and frequencies of all particles as a whole, including the empty space

Krahn figures are more complex. Experiments can be conducted in a vacuum and weightless environment to excite dust particles with electromagnetic field vibration frequencies, observing three-dimensional Krahn figures and simulating and understanding the specific mechanisms and processes of microscopic particle formation in the creation of matter.

(4) Particle Energy Levels and Flow Field

Vortex balls, as fundamental particles, possess a closed circulation of internal and external flow fields and have a high-speed spin, which provides the basic conditions for a rigid particle. In this case, the external flow field forms the shell of the rigid particle, giving it closure.

For composite particles, their internal and external enclosed flow fields are formed by the flow fields of their internal nuclear particles. When a composite particle has multiple nuclear particles, its enclosed flow field is formed by merging the flow fields of these nuclear particles into a unified internal and external flow field, which is also referred to as the shell of the composite particle. As the number of nuclear particles increases, the shell of the composite particle formed together also increases, resulting in different energy levels of the composite particle. Atoms are a type of composite particle, and their outer energy levels such as K、 L、M、 N、O、P are precisely these composite flow field shells.

At the same time, the increase in nuclear particles will lead to different structural distributions among them, forming a nuclear particle array. The existence of this array causes the shell of composite particles to differentiate into energy levels, creating distinctions between main energy levels and sub-energy levels. For atoms, the main energy level corresponds to the aforementioned atomic K、 L、M、 N、O、P energy levels, as shown in Figure 8-12, while the sub-energy level refers to the s、p、 d、f、 g orbitals within each energy level.

Figure 33: Figure 8-12 Different flow field layers of particles constitute different energy levels

These orbitals and energy levels together determine the various reactive properties of atoms and influence the distribution of electrons among different energy levels and orbitals.

(5) Particle's flow field and internal elements

For the basic quantum soliton vortex, its flow field components are the fundamental solitons. As the particle's energy increases, these components are excited into vortices, vortex rings, vortex balls, or even composite particles. In addition to the fundamental solitons, the flow field components also include particles with lower energy, such as vortices, vortex rings, vortex balls, or even smaller vortex balls.

It can be said that the greater the difference in energy between the components of the particle's field and the particle itself, the more stable the particle becomes. When the energy of the elements in the field approaches that of the particle itself, its stability decreases. Once the external compressive forces are removed, such less stable particles will quickly collapse into independent nuclear particles and other particles with higher energy in the field. This is the process of particle decay.

The decay time T of a particle is proportional to the ratio of the mass of the particle nucleus M_0 to the mass of the large particle in the flow field M_1, and also proportional to the ratio of the spin energy of the particle nucleus E_0 to the spin energy of the large particle in the flow field E_1, that is:

$$T \propto \frac{M_0}{M_1} \propto \frac{E_0}{E_1}$$

Or in other words, the decay time T of a particle is proportional to the ratio of its nuclear mass M_0 and energy E_0 to its total mass M and energy E

$$T \propto \frac{M_0}{M} \propto \frac{E_0}{E}$$

The total mass of a particle includes the mass of the particle's core (nuclear particle), all individual particles in the flow field shell, all individual particles in the space within the particle body, and all particles between the particle and its flow field shell (such as electrons, etc.), that is:

$M = M_0 + M_1 + M_2 + M_3$

Equation 8.1 is the equation for the total mass of a particle, where:
M: The total mass of the particle.
M_0: The mass of the particle's core (nuclear particle).
M_1: The total mass of all solitary particles in the flow field shell.
M_2: The total mass of all isolated particles within the particle's internal space.
M_3: The total mass of all particles (such as electrons, etc.) between the particle's flow field shells.
Similarly, the total energy of a particle includes the energy of the particle's core (nuclear particles), all isolated particles in the particle's internal space, all isolated particles in the particle's internal space, and the energy of all particles (such as electrons, etc.) between the particle's flow field shells. That is:

$E = E_0 + E_1 + E_2 + E_3$

Equation 8.2 is the total energy equation for the particle, where:
E: The total energy of the particle
E_0: Energy of the particle core (core particle)
E_1: The energy of all solitary particles in the flow field shell.
E_2: The energy of all solitary particles in the space within the particle.
E_3: The energy of all particles (such as electrons) between particle flow field shells.
Although electrons and similar particles exist between particle shell flow fields, they do not participate in the circulation of the internal and external flow fields of the particle. They can be considered as peripheral particles corresponding to the core particles of the particle.

From the above particle decay formula, it is known that all particles will decay. The stronger the spin energy of the particle, and the greater the mass-energy ratio of the core particles compared to the elements in the flow field, the more stable the particle is; otherwise, it is more likely to decay.

In addition, the decay of particles is also influenced by factors such as the composition of internal nuclear particles and the arrangement of the flow field (i.e., the energy levels and orbital arrangements of the particles).

8.3 Formation of Electric and Magnetic Fields [35][36]

In the previous overview of spatial wave dynamics, we discussed the origin of all forces and provided principles of electric fields, magnetic fields, and electromagnetic forces, along with corresponding models, including matrix models and breathing models for electric fields and charges. However, according to the basic principles of spatial particle dynamics discussed earlier, the stacking and evolutionary development of the most fundamental base particles and composite particles in the universe should all be simple. The models discussed above clearly do not meet this requirement and appear to be overly complex, so they are at least incomplete.

Next, we will conduct a more in-depth analysis of electric fields, magnetic fields, electric forces, magnetic forces, including electromagnetic waves and fundamental particles, from the perspective of the principles of the generation and evolution of vacuum space superfluid and the most basic particles, providing more reasonable models and evolutionary paths.
(1) Charge
(1) Particle Spin
In the previous discussion on the formation and evolution of particles, we pointed out that all particles have spin motion. Spin is the reason for the particle's closure and rigidity, and it also gives the particle stronger stability.

Corresponding to vortex rings, vortex spheres, and higher-level composite particles, they also have circulating inner and outer flow fields in addition to spin. This circulation creates an entrance (S pole) for the outer flow field to enter the inner flow field and an exit (N pole) for the inner flow field to flow out to the outer flow field. The existence of the S pole and N pole allows the spin of the particle to have directional distinctions. When observed from the N pole direction, the spin of the particle can be classified as positive spin (clockwise) or negative spin (counterclockwise), as shown in Figure 8-13.

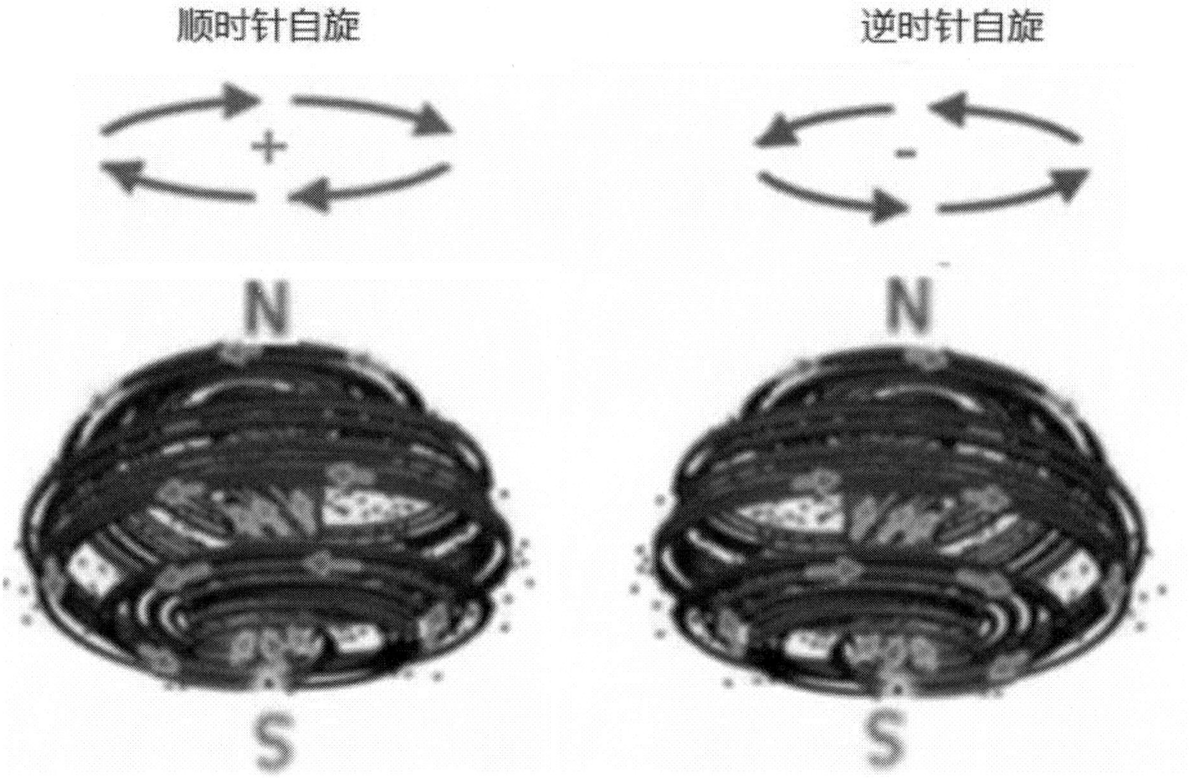

Figure 34: Figure 8-13 Spin Direction of Particles

Due to the different spin directions of particles, they produce different mechanical interactions with each other in the superfluid flow field of the vacuum space, and also have different interactions among themselves. The result of these interactions is that particles acquire differences in their charge properties and exhibit electric and magnetic forces.

(2) Positive and negative charges

We consider two particles, one spinning clockwise and the other counterclockwise. When the two particles approach each other, the direction of the fluid flow in the empty space between them is the same, and the flow speed increases. Meanwhile, the flow fields on their outer sides are opposite in direction, and under mutual collision and compression, the flow speed decreases. According to Bernoulli's equation in fluid mechanics, this results in reduced fluid pressure between them, forming a negative pressure zone, while the fluid pressure on their outer sides increases, forming a positive pressure zone. Consequently, a spatial pressure that causes them to move closer together is generated, manifesting as an attractive force, as shown in Figure 8-14.

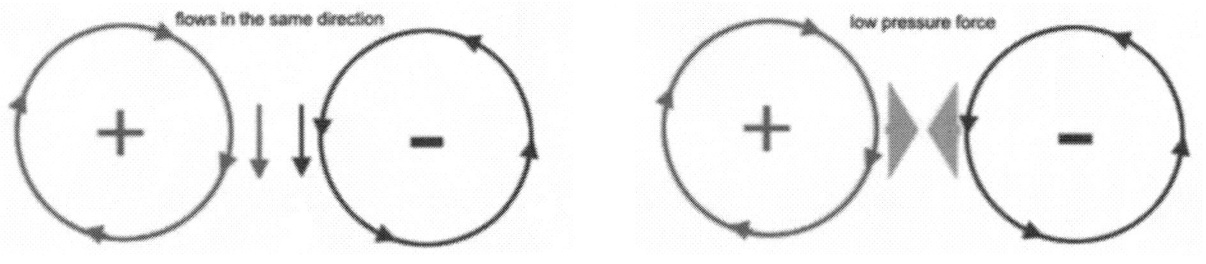

Figure 35: Figure 8-14 Attraction between particles with different spin directions

If two particles both spin clockwise or both spin counterclockwise, when they approach each other, the direction of the flow field in the empty space between them is opposite, causing a collision and compression that slows down the flow speed. Meanwhile, the flow fields on their outer sides have the same direction and increased speed. According to Bernoulli's equation in fluid mechanics, this results in increased fluid pressure between them, forming a positive pressure area, and decreased fluid pressure on their outer sides, forming a negative pressure area. Consequently, a mutual repelling spatial positive pressure is generated between them, manifesting as a repulsive force, as shown in Figure 8-15.

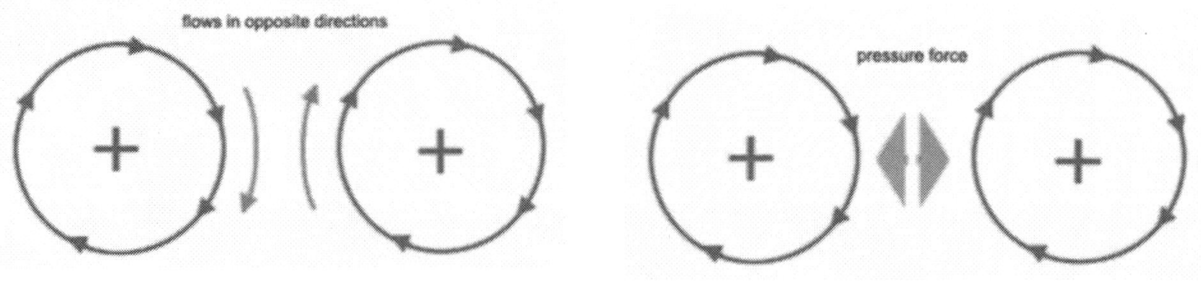

Figure 36: Particles with the same spin direction repel each other

The force of attraction or repulsion between particles caused by their spin is the electric field force we are familiar with. Particles with different spin directions thus acquire the property of electric charge, similar to the mechanical effects between two vortex flow fields, as shown in Figure 8-16.

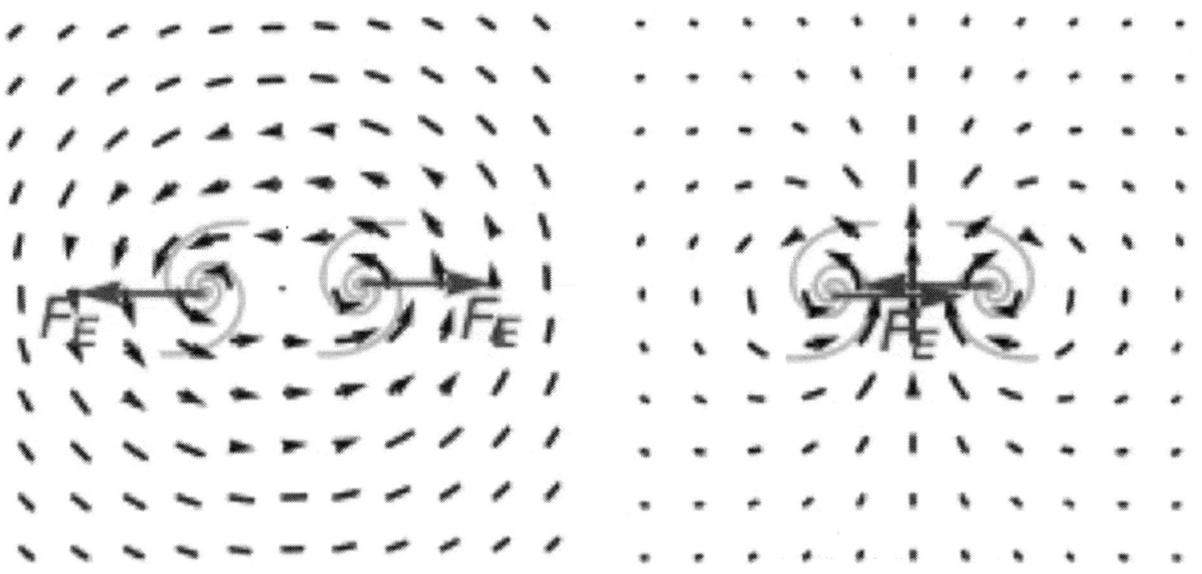

Figure 37: Figure 8-16 Mechanical Effects Between Two Vortex Flow Fields

We set the observation from the north pole in the flow field direction, with positively spinning (clockwise) particles as positive charge particles and negatively spinning (counterclockwise) particles as negative charge particles, then the above phenomenon manifests as mutual interaction between like charges

Repulsion, like charges attract each other.
(3) Charge quantity
The magnitude of the particle's spin energy E_s determines the strength of the electric field force between particles, thus also determining the amount of charge the particle carries, i.e., the charge magnitude. For primary solitary particles or composite particles with regular nuclei, their spin energy E_s is proportional to the charge they carry Q, that is:

$$E_s \propto Q$$

At the same time, the different charges, spins, etc., of nucleons within a particle will affect the overall charge expression of the particle, which is manifested as the particle's overall shell flow field will have different layers and directions, flow speeds, etc. Therefore, the charge of irregular multi-nucleon particles or complex inner-nucleon particles is not only influenced by the particle's spin energy, but also by the differences in the direction of flow fields in different shells (layered spin directions) within the particle's flow field, and it is the sum of the energies of all different flow directions in the particle's shell flow field, that is:

$$Q \propto E_s = \sum E_+ - \sum E_-$$

Where E_+ represents the magnitude of the positive charge field spin energy, E_- represents the magnitude of the negative charge field spin energy, the difference E_s determines the overall charge of the particle, and the sign of E_s determines the sign of the charge. If E_s is zero, the particle appears neutral with no charge.

Therefore, the charge of a particle can be summarized as its ability to cause a unified rotational flow field in the surrounding vacuum space. The larger the range of the rotating flow field and the higher the flow velocity, the greater the charge.
(2) Electric Field and Electric Force
(1) Electric Field
Different spins and vortices of particles form different interactions, causing the particles to appear to have different charges.

Since particles and all matter are immersed in the pervasive superfluid vacuum, the spin of particles also interacts with the surrounding vacuum space, causing it to be dragged along and generate corresponding rotation. The rotational flow field of this vacuum space forms a flow intensity gradient field centered on the particle, which is the electric field we are familiar with. The strength of the electric field is represented by the rotational intensity of the flow field in different regions, as shown in Figure $8-17$.

When a particle spins in the superfluid vacuum space, the rotational strength of the surrounding superfluid vacuum space's rotational flow field it generates decreases with increasing distance from the particle, and its magnitude is inversely proportional to the square of the distance from the particle, that is:

$$E \propto \frac{1}{r^2}$$

Or

$$E = K\frac{Q}{r^2}$$

Where $ E $ is the electric field at a point around a particle or charge, $ r $ is the distance from that point to the particle or charge, $ Q $ is the charge of the particle, and $ \epsilon_o $ is the proportionality constant, also known as the Coulomb's constant. This aligns with current physical theory predictions and experimental verification results.

The electric field strength at a point around a charge is a manifestation of the rotational intensity of the vacuum field flow at that point, so it can be further inferred that the rotational intensity of the flow is related to the density and rotational speed of the vacuum space at that point, so we have:

$$K \propto \rho v$$

ρ is the density of the vacuum space in this region, and v is the rotational flow velocity of the vacuum space flow field at this point.
At the same time, according to the formula for the electrostatic constant:

$$K = \frac{1}{4\pi\varepsilon_0}$$

we can see that the vacuum permittivity ε_0 is related to the density and flow velocity of the surrounding empty space. When the effects of the flow velocity v of the empty space are not considered, the vacuum permittivity ε_0 is inversely proportional to the density of the empty space ρ (Equation 3.9):

$$\varepsilon_0 \propto \frac{1}{\rho}$$

According to the speed of light formula:

$$C = \frac{1}{\sqrt{\varepsilon_0 \mu_0}}$$

It is known that the speed of light C and the vacuum permeability μ_0 are also related to the density of the void space ρ .

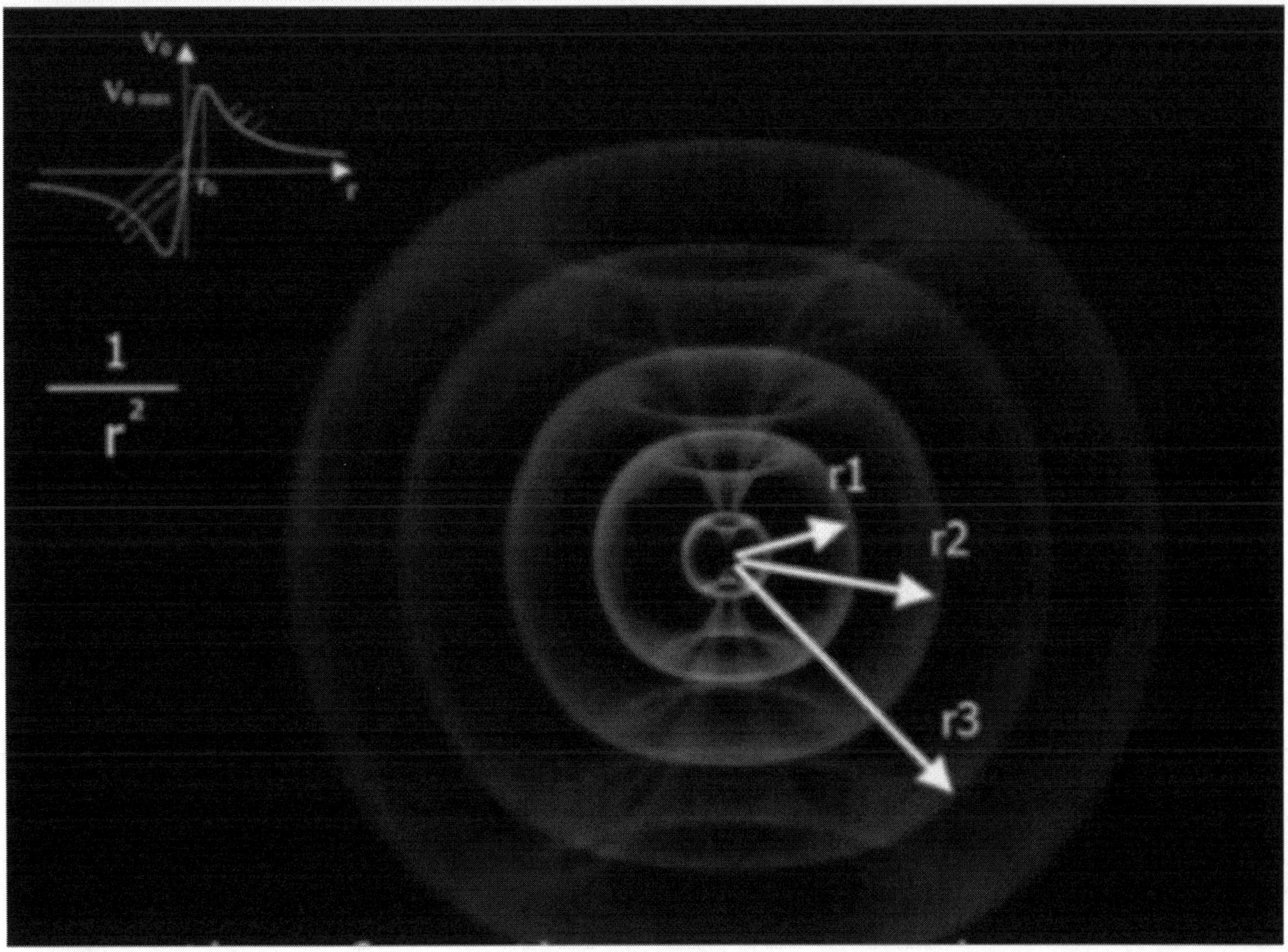

Figure 8-17 Electric Field: Field of Electric Field Strength Gradient around Charges

(2) Equipotential Surfaces and Equipotential Lines

According to our previous analysis, the electric field strength around a point charge is a manifestation of the rotational strength of the surrounding vacuum space caused by its spin. Since the vacuum superfluid is quantized, this rotational field is also quantized, meaning it is composed of individual quantum rotating spherical surfaces.

Each quantum rotating sphere is an independent flow field, where each point is at the same distance from the charge, or has the same rotational speed, thus possessing equal electric field strength and appearing as an equipotential surface.

The electric field around a point charge is composed of an overall spatial flow field made up of countless such independent equipotential surfaces, as shown in Figure 8-17.

If the electric field is sliced along any axis with the charge at the center, the resulting cross-section will show a series of equipotential lines centered on the charge. Observing the equipotential lines around the center of the charge on a plane, we get such a result, as shown in Figure 8-18.

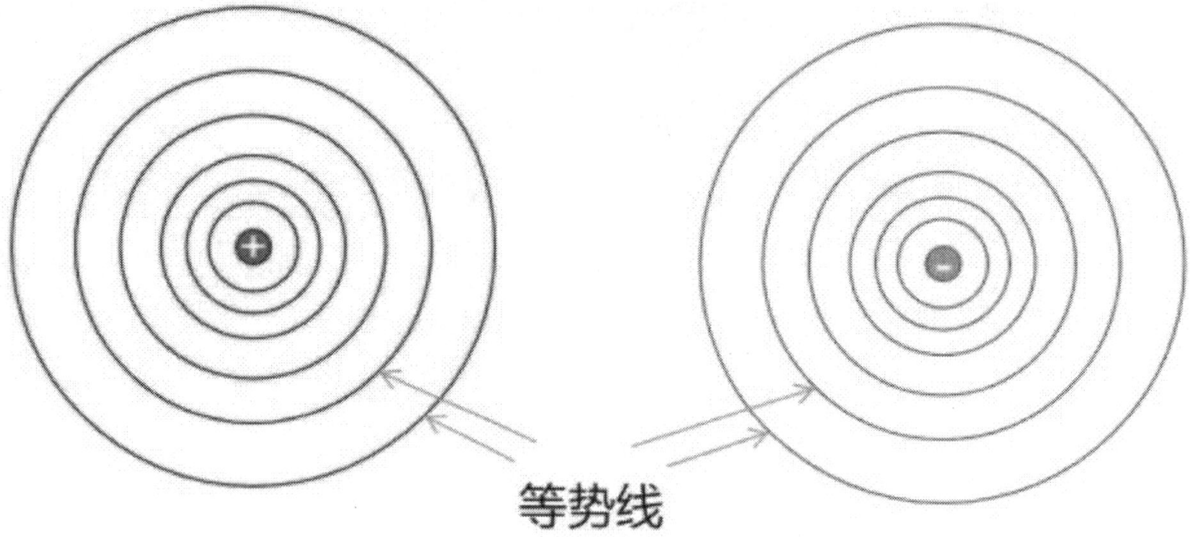

Figure 38: Figure 8-18 Electric Field Equipotential Lines

(3) Electric Field Force

The electric field of a particle is a rotating flow field around its center, generated by its spin in the surrounding empty space. Due to the circulation of the internal and external flow fields of the particle itself, this spatial rotating flow field not only has a horizontal direction but also a vertical direction. Therefore, when two free charged particles meet, the force caused by this spatial flow field will cause changes in their spatial states, aligning their spin axes parallel and their S and N poles in the same direction, as shown in Figure 8-19

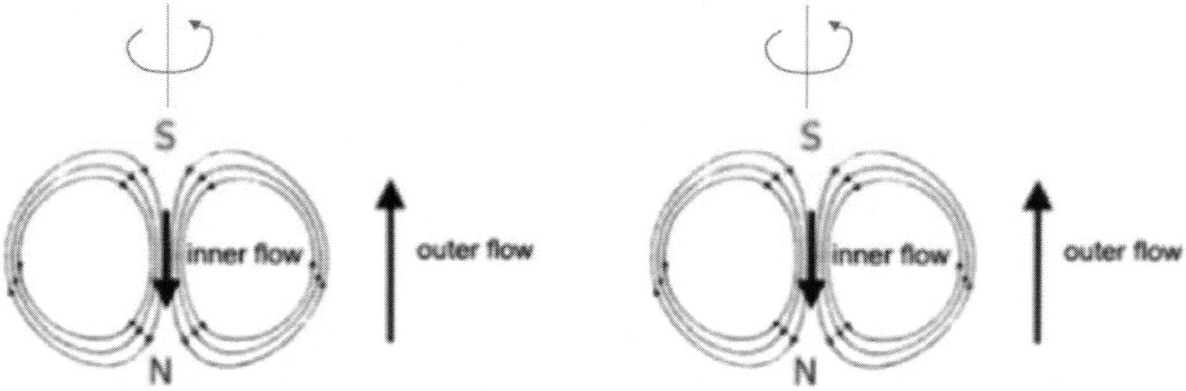

Figure 39: Space State Synchronization (Fig. 8-19)

This is the spatial state synchronization between charged particles.
In this case, when a point charge Q1 enters the electric field of another opposite charge Q2, since the spin directions of the two charges are opposite, the direction of the spatial flow field is the same between them and opposite on their back sides. This will create a spatial negative pressure in the void space between the two charges, resulting in an attractive force. Conversely, if it enters a like charge particle, a repulsive force will be generated, as shown in Figure 8-20:

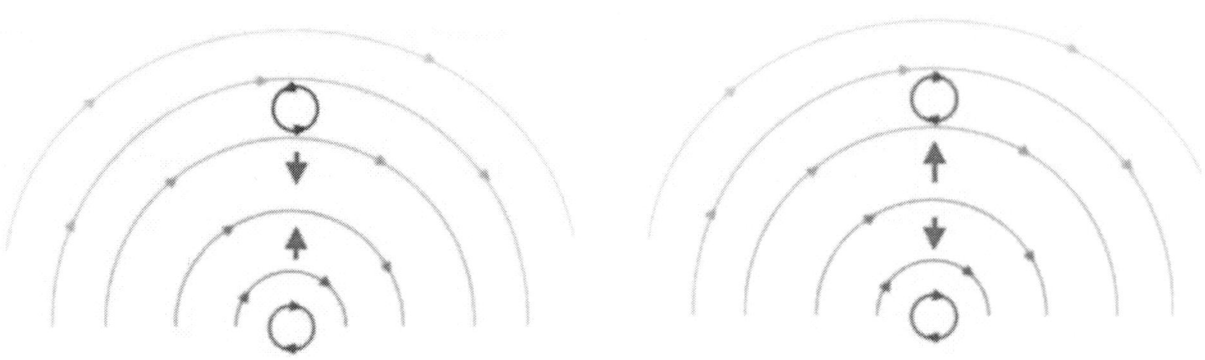

Figure 40: Figure 8-20 Attraction and Repulsion Forces of Particles Entering an Electric Field

Let the distance between two charges be r, then according to the previous analysis, the electric field strength of Q_1 at that point is:

$$E = K\frac{Q_1}{r^2}$$

The electric field force on charge Q2 is:

$$F = K\frac{Q_1 Q_2}{r^2}$$

Therefore, according to the previous analysis, the magnitude of the electric field force between two charges is related to the difference in velocity v between them and the space behind them, and also depends on the density ρ of the superfluid in the vacuum of the electric field.

(3) Magnetic Field and Lorentz Force

(1) The relationship between magnetic and electric fields

As analyzed above, charge, due to its own high-speed rotation, creates a high-speed rotational flow field in the surrounding vacuum space superfluid, and the intensity gradient of this flow field is the electric field.

when charges begin to move, an infinite number of vortex rings are formed around them, which are also

There are high-speed rotating flow fields and high-speed vortices along the flow direction. A large number of such vortex rings form the flow field, which is the magnetic field. The direction of the flow field's rotation is the direction of the magnetic field, so the magnetic field is a closed flow field, as shown in Figure 8-21.

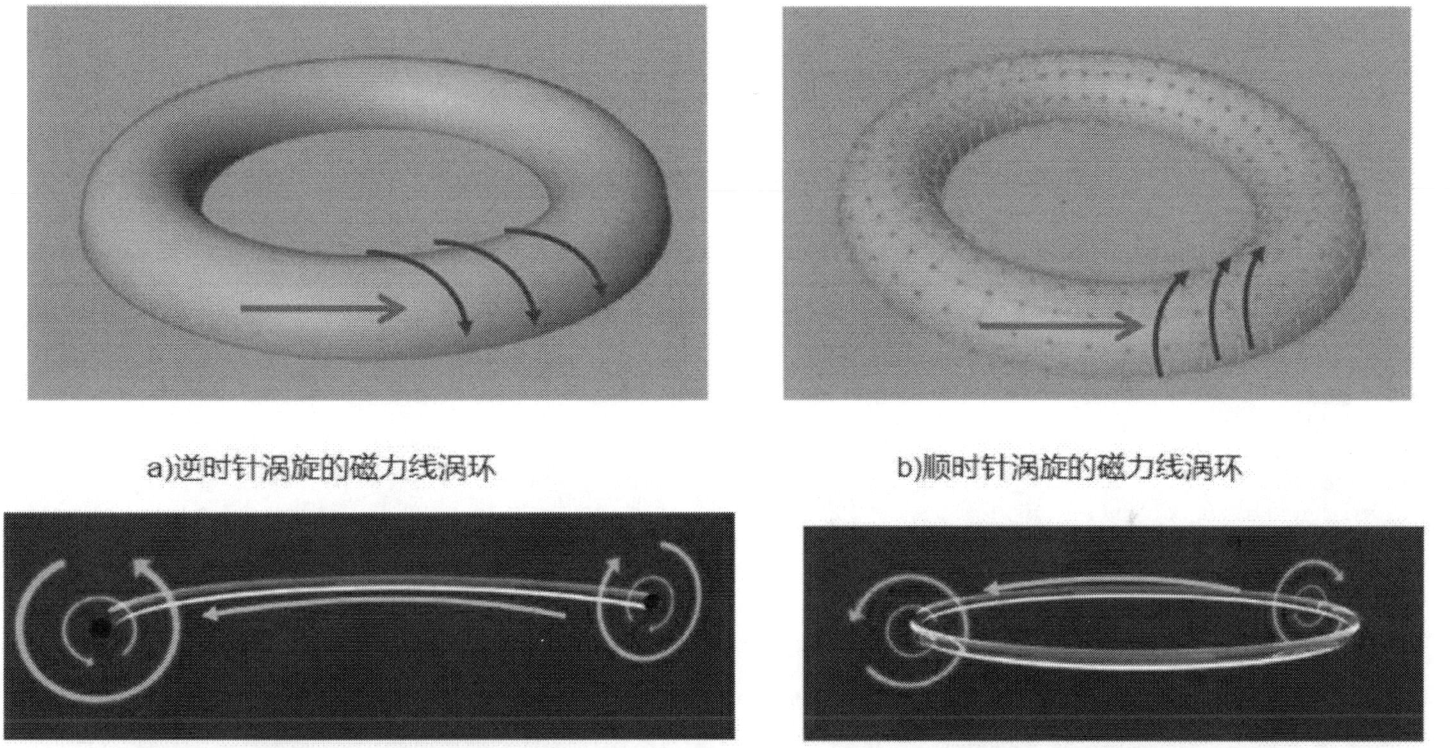

a)逆时针涡旋的磁力线涡环

b)顺时针涡旋的磁力线涡环

c)逆时针涡环磁力线的磁场方向和截面涡旋流场方向

c)逆时针涡环磁力线的方向和涡环流场

Figure 8-21: The formation of a magnetic field by a spatial closed-loop vortex flow field

The high-speed rotating superfluid in the surrounding vacuum space, due to its quantum nature, consists of countless field lines. Each field line contains wave particles or other solitary particles, micro-particles, etc., which themselves possess spin and orbital motion. Therefore, each field line can be considered as a magnetic field line. Hence, the spatial flow field that forms an electric field can also be regarded as a magnetic field. However, this magnetic field is source-based, and its effect is entirely the same as that of an electric field, so it can be considered equivalent to an electric field.

Therefore, the carrier of electric and magnetic fields is the same, namely the vortex flow field of the superfluid in vacuum space. The difference lies in that the electric field is a gradient distribution field of the vortex flow field strength, while the magnetic field is a vortex direction field of the vortex flow field. The two directions are perpendicular to each other.

In other words, when the vortex flow field of the superfluid in space is continuously excited by the spin of electric charges, the gradient effect of the flow field strength is very strong, mainly manifesting as an electric field effect. When the vortex flow field of the superfluid in space is an inertial maintenance vortex flow field without electric charges, the gradient effect of the flow field strength is very weak, meaning the electric field effect is also very weak.

Is mainly manifested as a magnetic field effect.

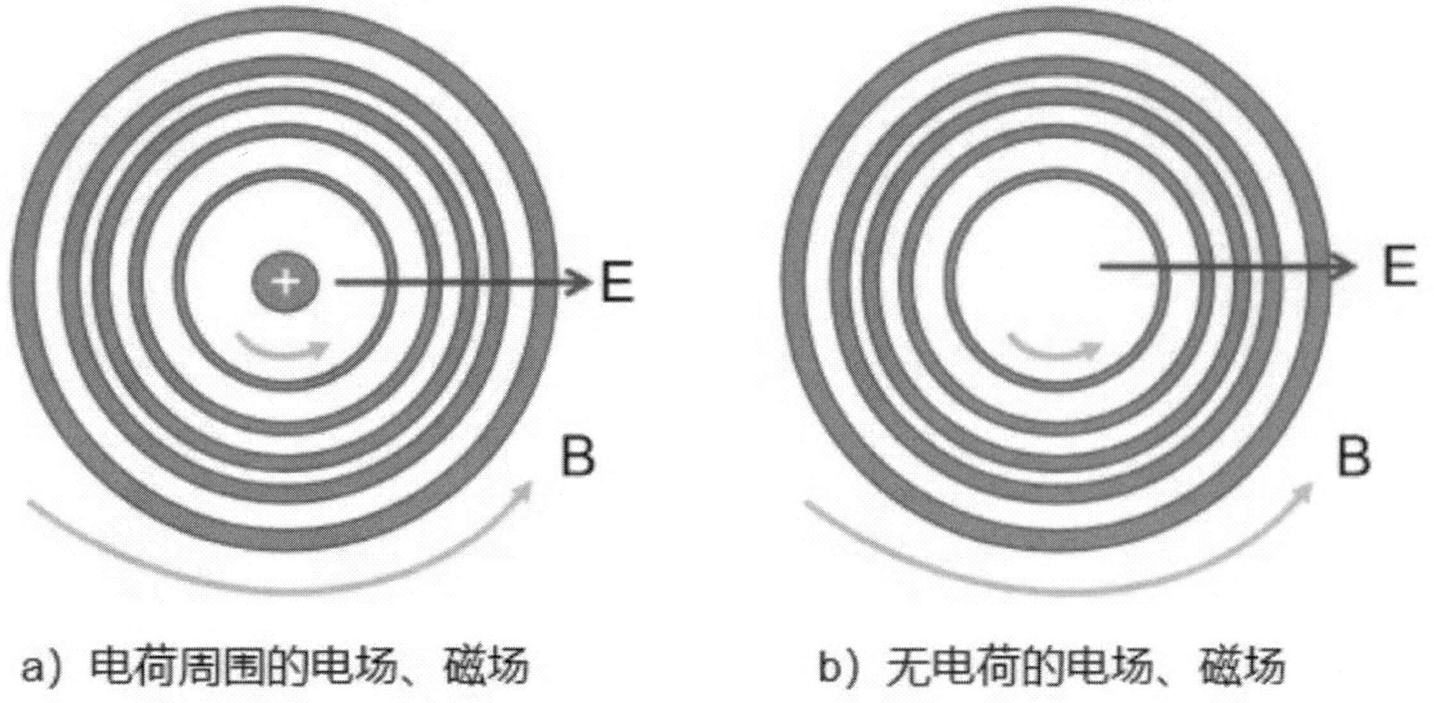

Figure 8-22 Active electric field, magnetic field and passive electric field, magnetic field

From this perspective, it can be considered that the electric field is an active magnetic field, and the magnetic field is a passive electric field. Both are two different descriptions of the same spatial superfluid vortex field phenomenon, as shown in Figure 8-22.

(2) Principle of Magnetic Field Produced by Moving Charges

When a charge is stationary, it can be considered to produce only an electric field. When the charge is in motion, the electric field it generates transforms into a vortex field that persists behind the moving charge in the direction of motion, and this vortex field is what we traditionally refer to as a magnetic field.

There will always be a gradient of field strength around a charge, so there will also be an electric field effect. However, because this field conversion is real-time, the electric field effect of the charge's movement is almost entirely converted into a magnetic field effect. In other words, when a charge moves, it mainly manifests as a magnetic field effect, as shown in Figure 8-23.

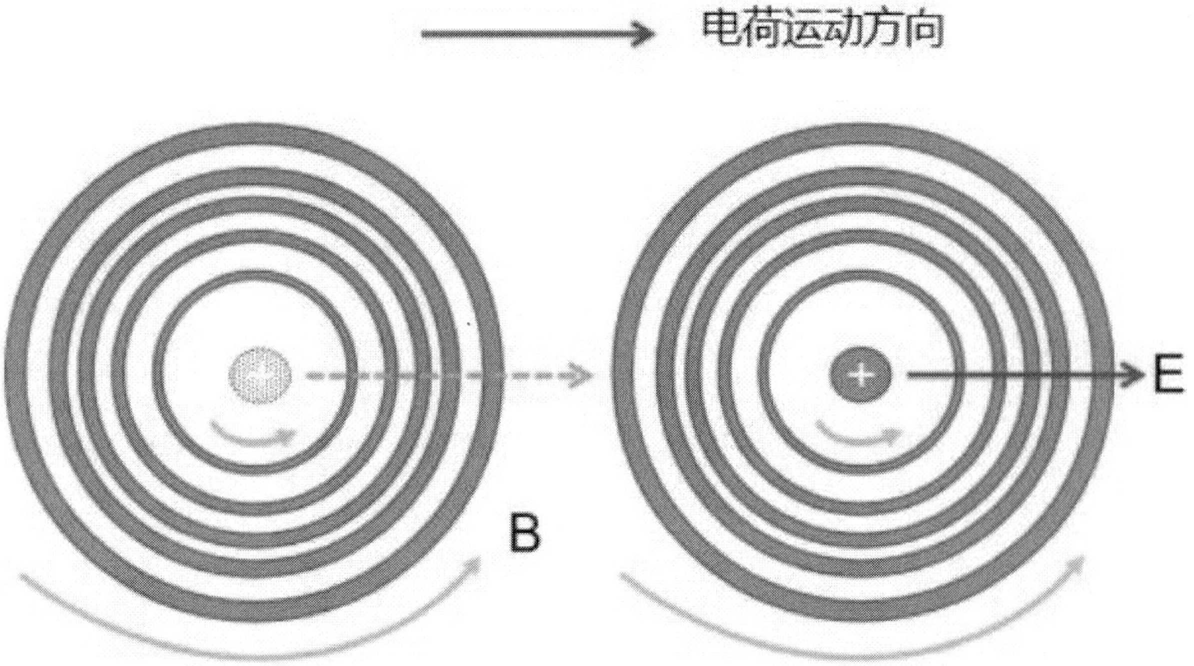

Figure 41: a) The spatial vortex field left by a moving charge around and behind it: magnetic field

b) The continuous excitation produced around a charge generates a spatial vortex field.
Spatial Vortex Field: Electric Field

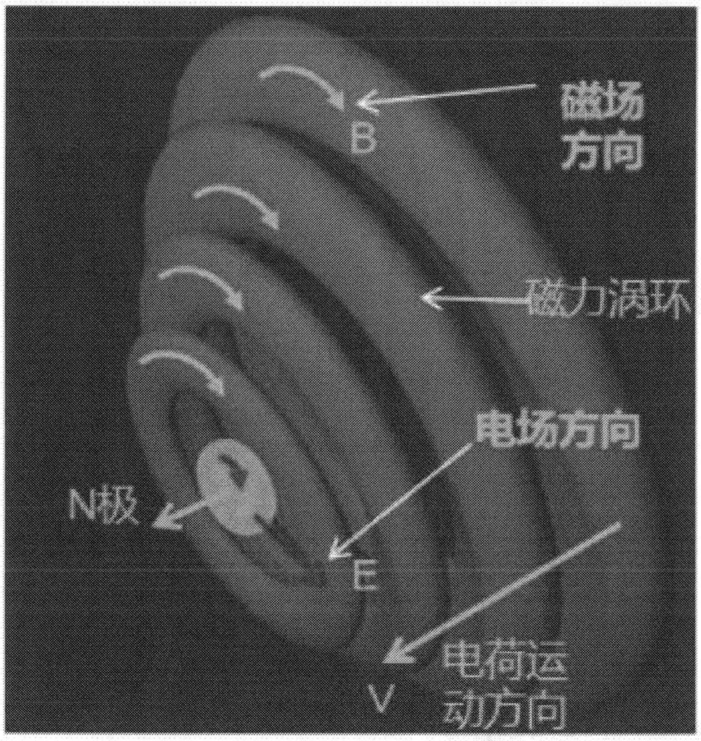

Figure 42: c) Schematic of charge motion producing a magnetic field

Figure 8-23 The electric field and magnetic field produced by moving charges

The static field of charges disappears quickly because it lacks continuous excitation and maintenance from charges, unless there is continuous movement of charges to excite and maintain it, forming a stable magnetic field.

It can be considered that the vortex flow field around moving charges manifests as a magnetic field. As the distance from the charge decreases, the gradient of the flow field becomes stronger, showing more pronounced electric field effects. At the point where the charge is located, the electric field effect is strongest.

When the charge moves away from its current position or returns to rest, the magnetic field it produced at that location does not immediately disappear but requires a certain amount of time, although this time may be very short. This process conforms to the new localization principle.

There have been many theoretical studies in academia regarding the nature and relationship of electric and magnetic fields, proving that a magnetic field is the motion effect of an electric field, and that they are unified. However, due to the lack of a fundamental physical explanation for their nature, there are many ambiguous areas, which can lead to misunderstandings or paradoxes. For example, suppose there are two observers, A、B, and A, with A moving at a constant velocity in sync with the charge Q, while B is stationary. When A and B coincide in position, B can observe the magnetic field produced by the moving charge Q, whereas A, being relatively stationary with respect to Q, cannot observe the magnetic field. This leads to contradictions or ambiguities, and so on.

Here, according to our explanation of the nature of electric and magnetic fields, we can understand that both individuals will observe a magnetic field, because here the magnetic field traditionally considered is an electric field that is source-free and produced by moving charges, while the traditional electric field is a source field, or a field with a source, produced by charges. The electric and magnetic fields observed here are two different phenomena of flow fields.

(3) Principle of Current Producing a Magnetic Field

Moving charges can produce a magnetic field, but this field is difficult to maintain for a long time and quickly disappears. When there are a large number of charges in a medium, the disorder in the spin directions of the charges leads to disorder in the N-S pole directions, causing the generated magnetic fields to not remain consistent and thus cancel each other out. As a result, the overall magnetic field effect is not uniform.

When charges move under the pressure of an electric potential difference, they adjust their direction, that is, the N end of the $N-S$ axis points in the direction of motion, making their outflow field direction parallel to their relative motion direction with respect to the external vacuum space superfluid, thus achieving the principle of minimum force or minimum energy consumption.

Therefore, when a large number of charges form a unified motion under voltage drive, since the direction of the $N-S$ axis of the charges aligns with the direction of motion, the magnetic field vortex flow field they form—i.e., the direction of the magnetic field—also becomes unified, resulting in the merging and enhancement of the magnetic field flow intensity. At the same time, due to the continuity of the current, these magnetic fields are continuously excited by the flow of countless charges behind them, forming a stable existence.

Thus, a continuous magnetic field will appear around the energized wire, meaning that the current produces a stable magnetic field, as shown in Figure 8-24.

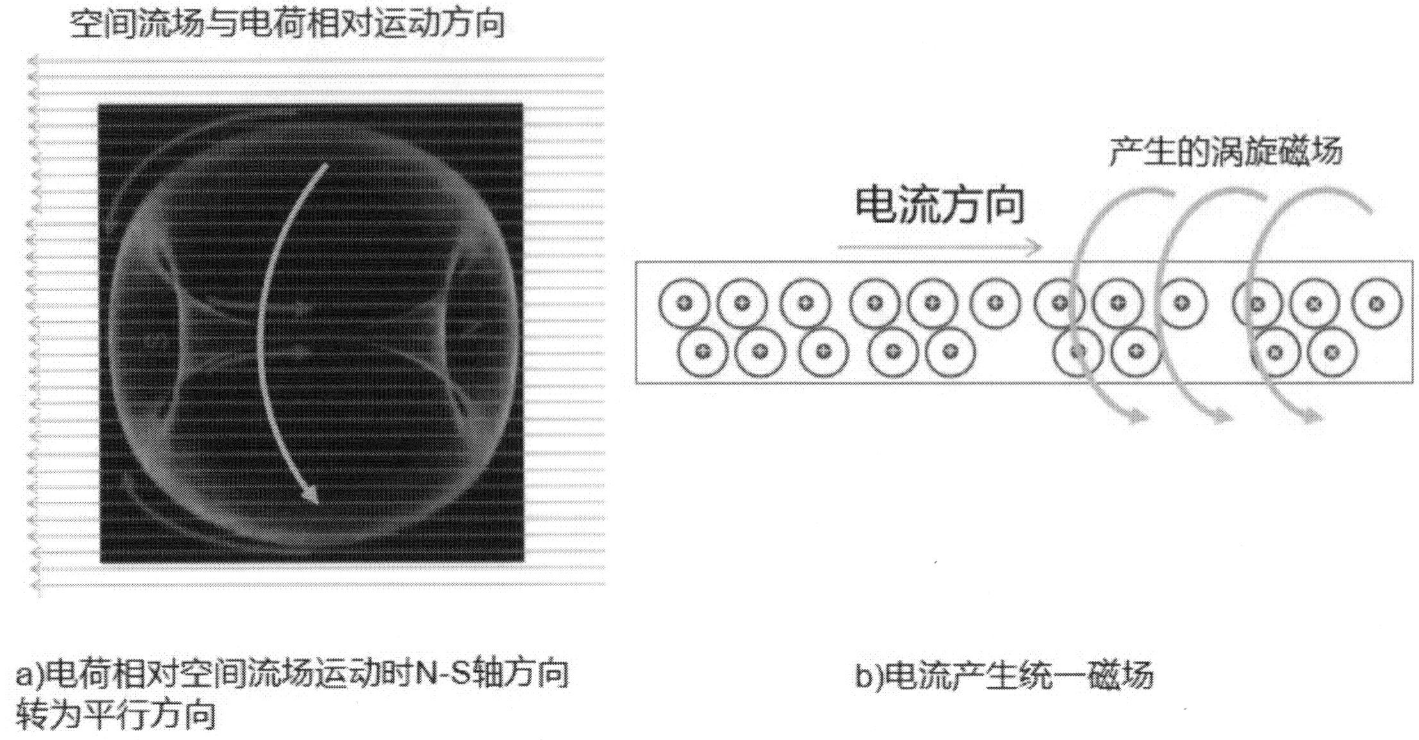

Figure 8-24 Current motion produces a magnetic field

(4) Force on a charge in a magnetic field

When a charge is stationary in a vacuum space without a current field, its N − S axis points in a random direction. However, when it is in a spatial flow field, such as a magnetic field, it adjusts its direction so that the $N-S$ axis is parallel to the flow field direction. Moreover, the flow direction of the charge's external field is the same as that of the magnetic field flow, achieving the minimum force condition.

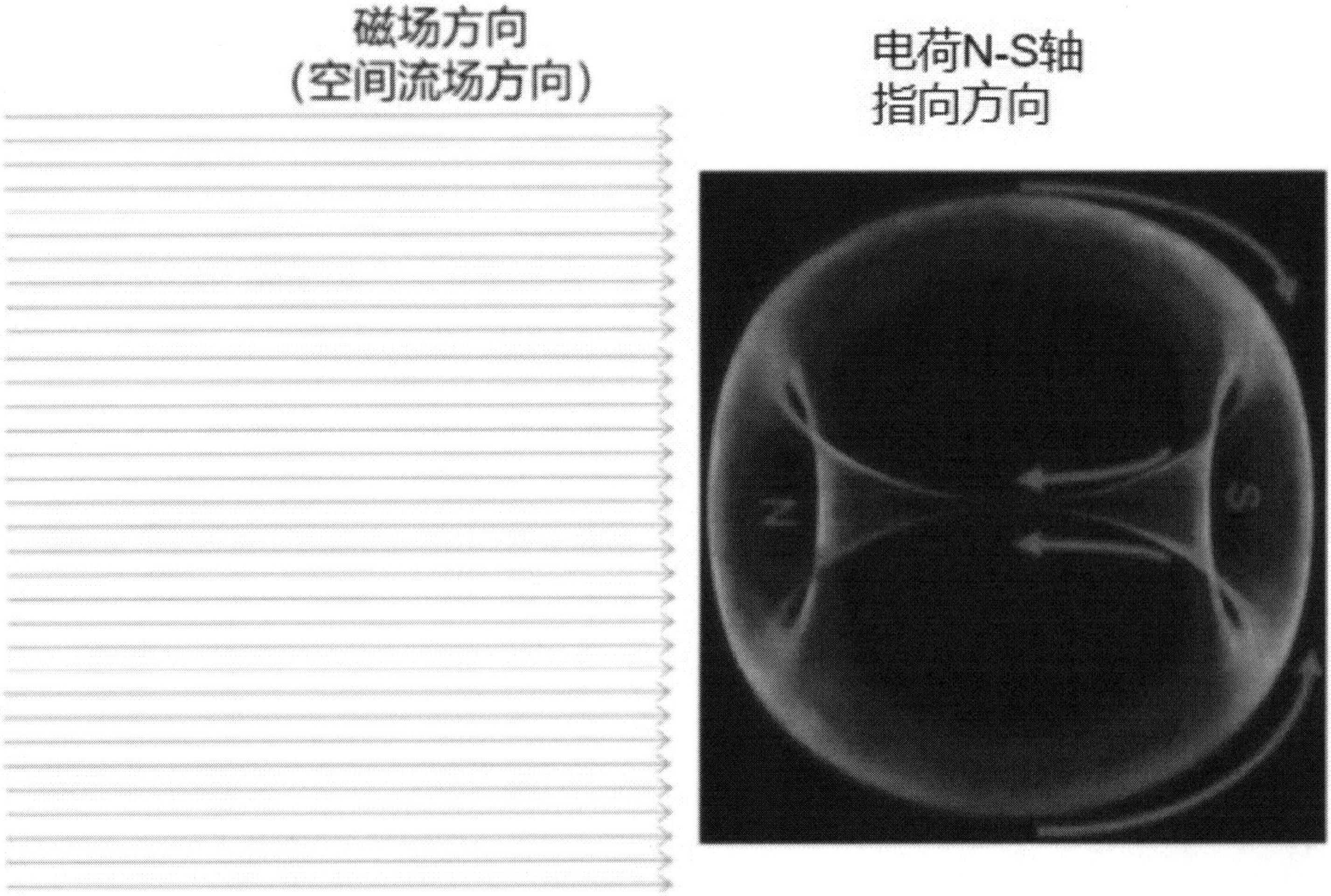

Spatial State Adjustment of a Charge at Rest in a Magnetic Field

In this case, the forces exerted by the field on the charge in a uniform magnetic field are almost equal in all directions, so the charge is in an equilibrium state. The axial impact force from the magnetic field is relatively small and does not produce a noticeable effect. Therefore, when a charge is at rest in a uniform magnetic field, it experiences no net force.

When the magnetic field is non-uniform, the forces exerted by the field on the charge from different directions differ. This causes the charge to continue altering its axial direction, allowing the overall force to remain approximately zero and enabling it to continue being at rest in the magnetic field, as shown in Figure 8 - 25.

When charged particles move in a spatial flow field, they adjust their state, with the N - S axis aligning to be parallel to the relative motion direction with respect to the spatial superfluid, the N pole deflects in the direction of motion, making the external flow field direction consistent with the relative motion direction of the spatial flow field, thereby reducing the force exerted by the void space during movement. Here, the relative motion direction of the charge flow field and the spatial magnetic flow field is the combined direction of the charge motion and the magnetic field flow direction.

At the same time, the motion of charges generates a magnetic field, which is the excitation of a large number of helical flow field vortex rings around it. These vortex rings interact with the magnetic field flow field, creating a flow field pressure difference due to the difference in relative flow velocities on both sides. This pressure difference acts on the particle, causing it to experience a force from the background magnetic field and deflect from its forward direction.

The force is called the Lorentz force.
According to the Biot-Savart law, the greater the charge Q of an electric charge particle, and the faster it moves with velocity V, the stronger the magnetic field B_1 it generates, resulting in a larger Lorentz force interacting with the background magnetic field B. At the same time, the stronger the background magnetic field B, the greater the Lorentz force. Therefore, the Lorentz force formula can be expressed as:

$$F = QVB \sin \theta$$

F is the Lorentz force, Q is the particle's charge, B is the strength of the background magnetic field, V is the particle's velocity, and θ is the angle between the particle's velocity and the direction of the background magnetic field B.

Because the direction of the vortex field formed by the opposite spin directions of positive and negative charges is also opposite, the final force exerted by the background magnetic field acts in opposite directions, resulting in a deflection opposite to the direction of motion. Positive charges follow the left-hand rule, while negative charges follow the right-hand rule, as shown in Figure 8-26.

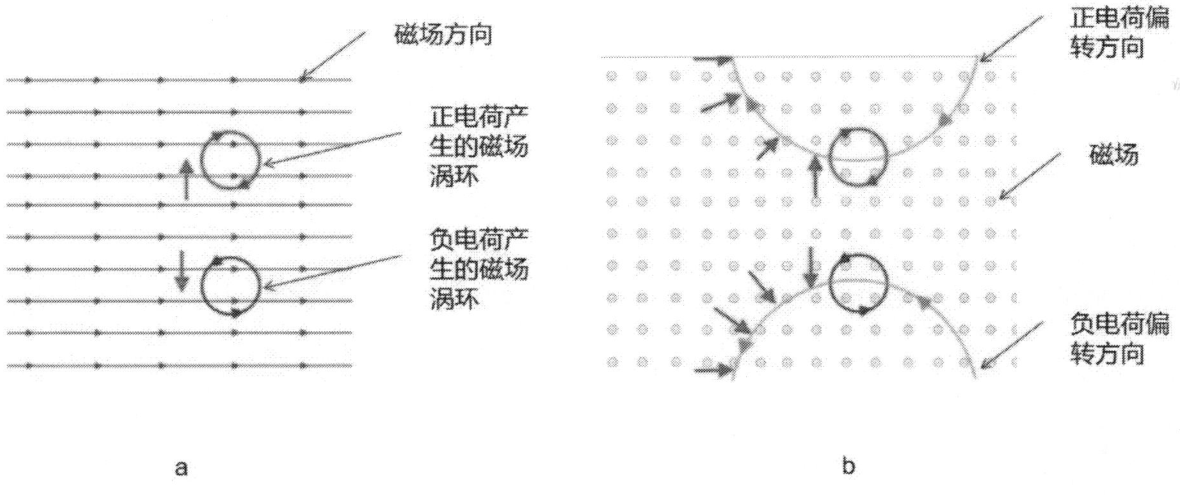

Figure 43: Figure 8-26 Force on a charge moving in a magnetic field

(5) The Relationship Between Charge Velocity Direction and Lorentz Force
In the previous section, we analyzed the mechanism of Lorentz force generation, which is the result of the interaction between the magnetic field produced by the moving charge and the background magnetic field, leading to a flow field's mechanical effect. However, why does the magnitude of the Lorentz force depend on the direction of the charge's motion relative to the magnetic field?

We set the flow field velocity of the magnetic field as C (assuming the flow field velocity equals the speed of light), and the charge's motion relative to the magnetic field

The velocity of the charge is V, as shown in Figure 8-27(a).
In this case, the angle between the charge's velocity and the magnetic field B is θ , and the direction of the N pole of the charge also points in the direction of the velocity, i.e., forming an angle θ with the magnetic field B, as shown in Figure 8-27(b).

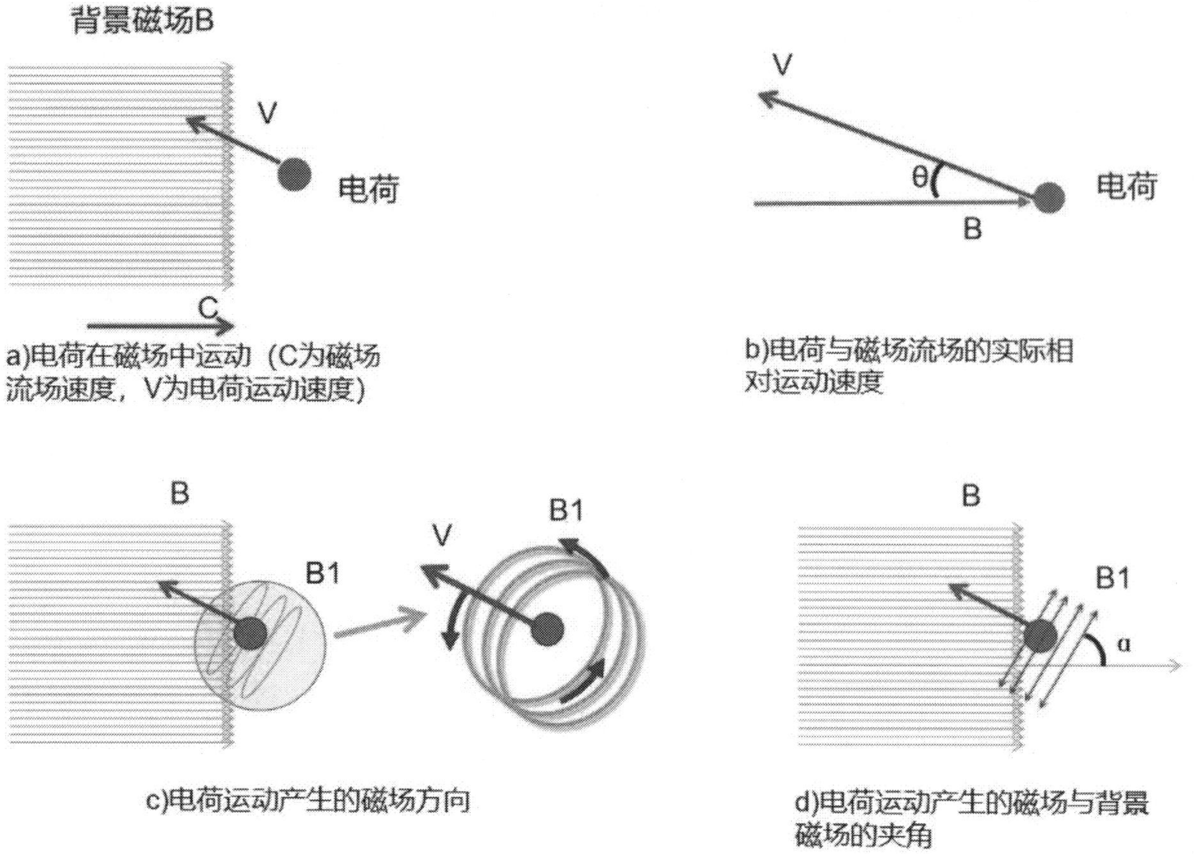

Figure 44: Figure 8-27 Charge motion direction and the direction of the generated magnetic field

Since the charge's motion generates a magnetic field B1, according to our previous discussion, B1 is a vortex field centered on the charge, with its plane direction perpendicular to the charge's motion direction, as shown in Figure 8-27(c). Let the angle between the magnetic field B1 and the background magnetic field B be α , as shown in Figure 8-27(d). Then:

$\alpha = 90^\circ - \theta$

Because the magnetic field B1 surrounds the direction of charge motion, the magnetic field directions on both sides perpendicular to the motion direction are opposite. Under the interaction with the background magnetic field B, the flow velocities on both sides are inconsistent, leading to a hydrodynamic difference according to Bernoulli's principle, which forms a total force. This total force is the Lorentz force, directed perpendicular to both the magnetic field B direction and the charge motion direction V.

Because the component of magnetic field B1 perpendicular to the background magnetic field B does not produce a field flow with the magnetic field B

Mechanical effects, so the magnitude of the Lorentz force is only related to the parallel component of the magnetic field B1 with respect to the background magnetic field B, that is:

$\mathrm{F} \propto \mathrm{B}_1 \cos\alpha$

Or:

$\mathrm{F} \propto \mathrm{B}_1 \sin\theta$

As shown in Figure 8-27, when the charge velocity V increases, the strength of the generated magnetic field B1 increases; as the velocity angle θ increases, the angle α between the magnetic field B1 and the background magnetic field B decreases, resulting in a larger component of B1 in the direction parallel to B. This leads to a stronger hydrodynamic effect. Therefore, the greater the velocity of the charge perpendicular to the background magnetic field, the greater the Lorentz force it experiences, that is:

$\mathrm{F} \propto \mathrm{V} \sin\theta$

Since the magnitude of the Lorentz force is proportional to the charge quantity and the strength of the background magnetic field, we have the Lorentz force equation:

$\mathrm{F} = \mathrm{QVB} \sin\theta$

Where: F is the magnitude of the Lorentz force, Q is the charge of the particle, B is the strength of the background magnetic field, V is the velocity of the particle, and θ is the angle between the particle's velocity and the direction of the background magnetic field B.

Therefore, the magnetic field B1 produced by the motion of charge only has components parallel to the background magnetic field B, because the opposite velocities of the flow fields on both sides generate a net force with the magnetic field B, i.e., the Lorentz force; the components of B1 perpendicular to B do not produce a net force effect. This is the physical reason why the magnitude of the Lorentz force is proportional to the component of the charge's velocity that is perpendicular to the magnetic field, and why the direction of the Lorentz force is always perpendicular to the plane formed by the direction of the magnetic field and the direction of the charge's motion.

We can further consider: suppose the charge velocity is perpendicular to the direction of the background magnetic field B, i.e., $\theta = 90^\circ$, then the Lorentz force equation becomes:

$$\mathrm{F = QVB}$$

When the charge velocity reaches the speed of light, the Lorentz force is:

$$\mathrm{F = QCB}$$

Where $ C $ is the speed of light.
Also because of the relationship between the magnetic field and the electric field:

$$B = \frac{E}{C}$$

Where E is the electric field strength, substituting into the Lorentz force equation gives:

$$\mathrm{F = QE}$$

When the speed of a charge in a magnetic field reaches the speed of light perpendicular to the magnetic field, the magnetic force acting on it will be equal in magnitude to the electric force.

In the previous text, we assumed that the speed of the magnetic field flow is equal to the speed of light, but in reality, the speed of the field in a fluid can be greater than, or even much greater than, the speed of longitudinal waves in the field. Therefore, the speed of the magnetic field vortex can be much greater than the speed of light. In this case, when the particle's speed exceeds the speed of light and continues to increase, the magnetic force it experiences will also continue to increase.

(6) The reason electrons in an atom do not emit electromagnetic waves and magnetic fields

According to the basic principles of spatial relativity 6: the principle of spatial relativity, charges exist in relative motion with the superfluid vacuum, and can only generate magnetic fields and electromagnetic waves when reaching the corresponding threshold.

Additionally, since electrons in atoms form $\mathrm{N-N}$ pole pairs or $\mathrm{S-S}$ pole pairs with protons and are not orbiting the atom, there is no rotational acceleration, and thus no electromagnetic waves are produced.

Another important reason is that electrons in atoms are enclosed by the atom's energy levels within its flow field boundary. Therefore, even if electrons generate magnetic fields and electromagnetic waves, they will be sealed and converted into internal flow field energy by the atom's flow field boundary, and will not be released. This internal flow field energy balances with protons, electrons, and internal flow fields, allowing the electron-proton pairs to exist stably.

8.4 Analysis of the Principle of Electromagnetic Induction [37][38][39][40]

(1) The Essence of Electromagnetic Induction

According to our previous discussion on the formation principles of electric and magnetic fields, we can conclude that electric and magnetic fields are two different descriptions of the same vortex flow field in the vacuum space superfluid. The electric field is the gradient field of the vortex flow field strength, while the magnetic field is the vortex strength field, and they are perpendicular to each other.

In terms of dimensions, the vortex flow field strength described by the two is not the same type of strength. Just as momentum and kinetic energy are both used to describe the intensity of motion, they are not the same kind of energy.

In principle, both exist simultaneously, but at different times, they are described in different directions: when charges are stationary, the electric field is used for description; when charges are in motion, the magnetic field is used. The electric field has sources and is irrotational, while the magnetic field is source-free and has vorticity.

Therefore, there is no mutual induction between electric and magnetic fields; they are two aspects of the same entity, existing simultaneously. The essence of electromagnetic induction is the related effects caused by the movement of charges resulting from changing electric or magnetic fields.

However, the current electromagnetic theory, including the most fundamental Maxwell's equations, has provided an incorrect interpretation of this.

(2) Principle of electric field induction by a changing magnetic field

We look at the description of electromagnetic induction in Maxwell's equations in vacuum, which is in the form of:

$$\nabla \times \mathrm{E} = -\frac{\partial \mathrm{B}}{\partial \mathrm{t}} \quad (8.1)$$

Faraday's Law of Electromagnetic Induction is given by equation 8.1, which means: the rate of change of magnetic flux through a surface is equal to the circulation of the induced electric field.

We have analyzed before that a change in magnetic flux or a change in the magnetic field does not induce an electric field, and the electric field is the gradient field of the vorticity field strength of the vacuum space superfluid, so it is a solenoidal field with no curl, and its curl is

zero, the change in magnetic flux does not produce the circulation of an electric field.

We know that the electromotive force of electromagnetic induction can be divided into motional electromotive force and induced electromotive force, among which the motional electromotive force is:

$$\mathrm{E} = \mathrm{BL}V \sin\theta$$

In the formula, B is the magnetic field, L is the length of the moving conductor, V is the velocity of the conductor, and θ is the angle between the velocity V and the magnetic field.

The motional electromotive force arises from the Lorentz force exerted by the magnetic field on charged particles in the moving conductor. According to the Lorentz force formula $\mathrm{F} = \mathrm{qvB}$, when charged particles in a conductor move with velocity v in a constant magnetic field B, they experience the Lorentz force, which causes them to move along the conductor under its constraints, forming an approximate electric field force, and thus manifesting as "an electric field E is generated."

That is, the source of motional electromotive force is the relative motion between charges in the moving conductor and the magnetic field, which generates the Lorentz force, and forms within the conductor under the action of internal constraining forces.

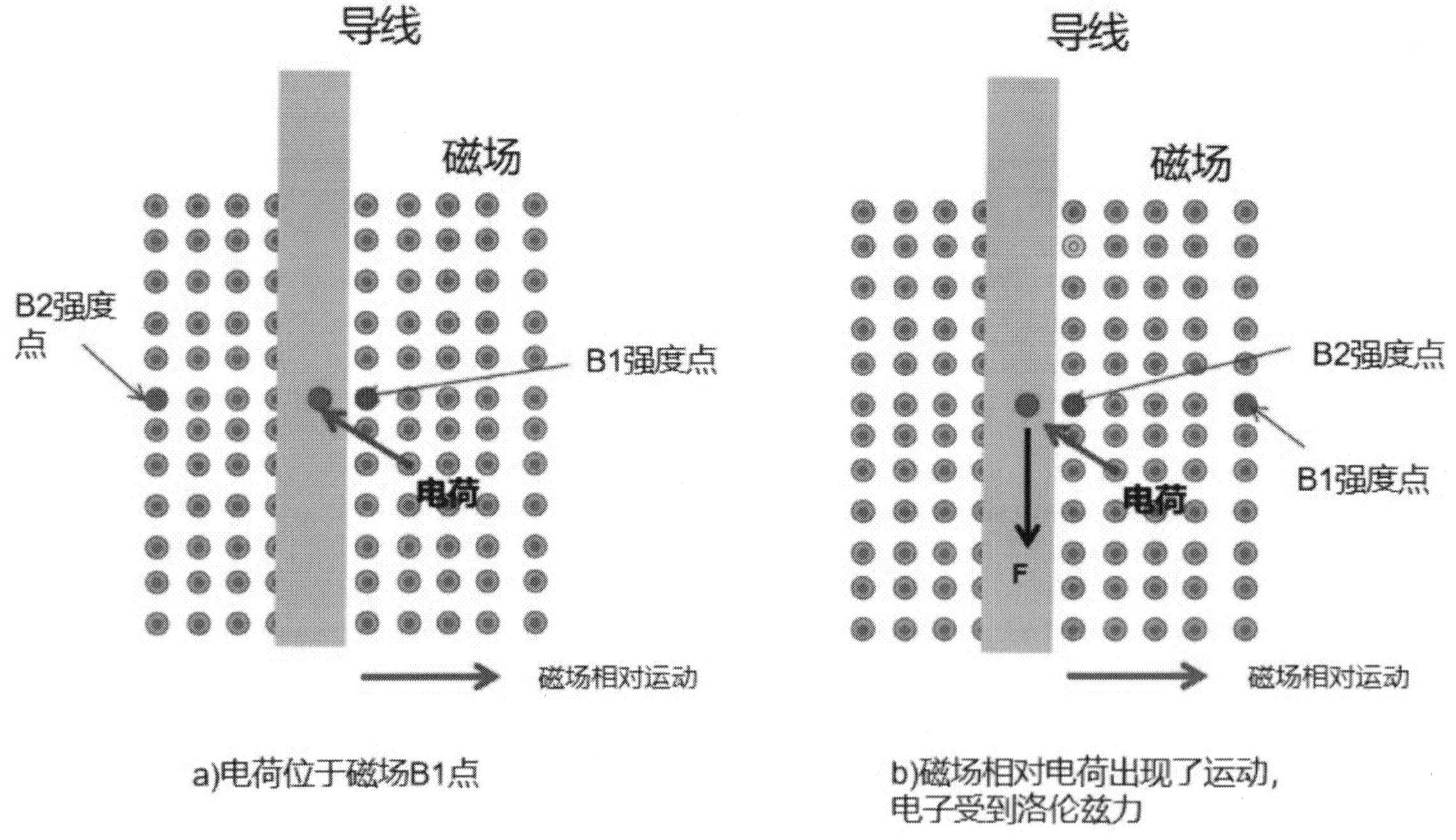

Figure 8-28 Magnetic field changes cause relative motion between the magnetic field and the charge

Similarly, we can think of the formation principle of induced electromotive force, which should also be similar, according to induced electric

Power formula (for single-loop circuits):

$$E = \frac{\Delta\Phi}{\Delta t}$$

Here, the magnetic flux changes, which is equivalent to a change in the magnetic field. In this case, it is as if the charges in the circuit are moving relative to the magnetic field. However, the relative motion here is initiated by the magnetic field itself: the magnetic field strength at the location of the charge changes from B1 to B2, meaning the magnetic field appears to move relative to the charge. The magnetic field lines at B2 move to the location of the charge, as shown in Figure 8-28.

This change causes relative motion between charges and the magnetic field, resulting in the Lorentz force that makes the charge move, and under the constraint of the conductor, forms motion along the conductor's direction, which is the current, showing the electric field effect.

The rate of change of magnetic flux is equivalent to the rate of change of position between charges and the magnetic field, meaning that a relative velocity is generated between them:

$$\frac{\Delta\Phi}{\Delta t} \equiv \frac{\Delta l}{\Delta t} = v'$$

Equation 8.2 is the equivalent equation of a changing magnetic field and the relative motion of charges, where Δl is the distance of relative movement between the magnetic field and the charge, and v ' is the relative velocity between the magnetic field and the charge.

$$E = \frac{J}{\sigma} = \frac{nqv}{\sigma} = \frac{\Delta\Phi}{\Delta t}$$

or

$$\nabla \times \left(\frac{nqv}{\sigma}\right) = -\frac{\partial B}{\partial t}$$

Where σ is the conductivity, n is the number of free charges moving in the conductor, q is the charge quantity, and v is the overall velocity of the charges in the conductor.

(3) Principle of Induced Magnetic Field by Changing Electric Field

is equal to the circulation of the magnetic induction B around any closed path in a steady magnetic field

The algebraic sum of the currents enclosed multiplied by the permeability. This conclusion is called Ampère's circuital law (Am−).

Ampère's Circuital Law can be derived from the Biot-Savart Law, and it reflects the property of steady magnetic fields where magnetic field lines are linked with current-carrying conductors.

Its differential form is:

$$\nabla \times B = \mu_0 j$$

where E is the electric field strength, B is the magnetic induction, j is the current density, and μ_0 is the vacuum permeability.
However, Maxwell used a circuit connected to a time-varying voltage source with a capacitor to illustrate the contradiction in this situation, as shown in Figure 8-29. At this time, there is a time-varying conduction current μ_0 in the circuit, which establishes a time-varying magnetic field. Choosing an open surface S_1 bounded by the closed path C, Ampère's Circuital Law gives $\oint H \cdot dl = \mu_0 I\_enc$. But when choosing another open surface S_2 bounded by the same closed path C, since the conduction current through S_2 is zero, it gives $\oint H \cdot dl = 0$. The same magnetic field strength vector H yields contradictory results when integrated around the same closed path C.

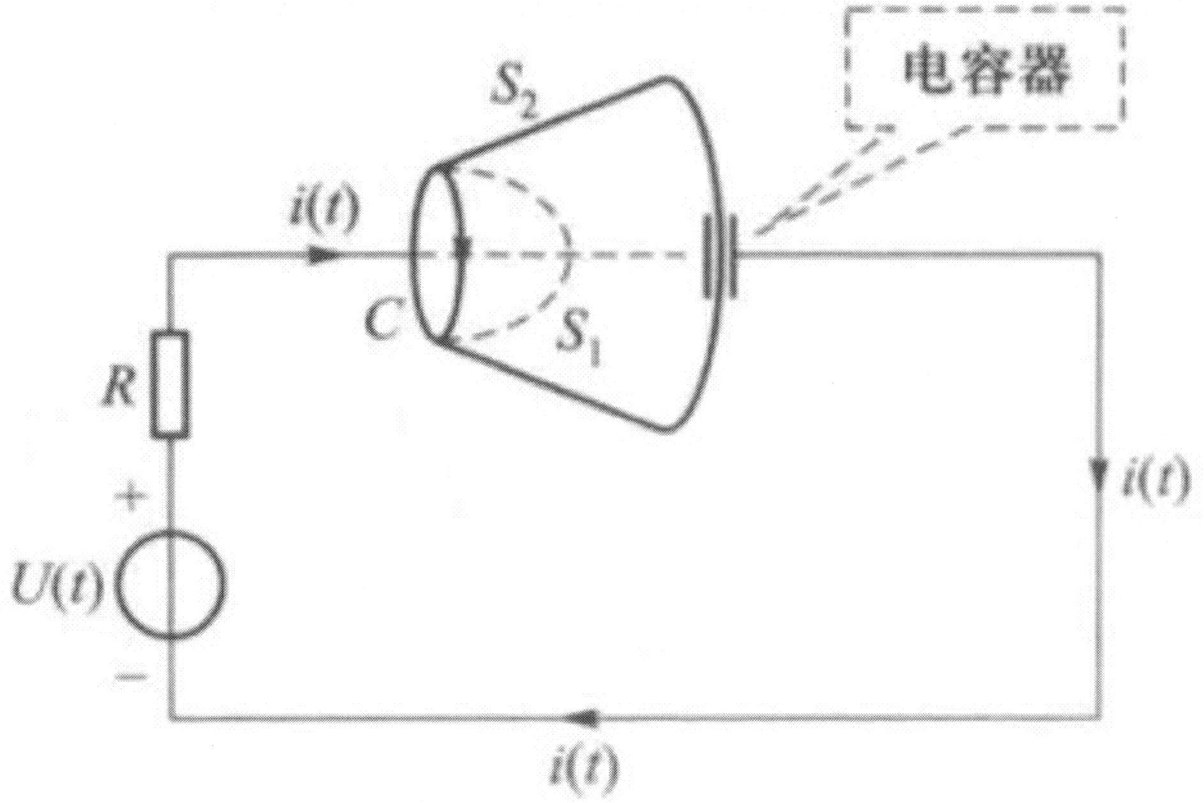

Figure 45: Capacitor Connected to a Time-Varying Voltage Source

Maxwell believed that this indicated the Ampère's circuital law derived from a steady magnetic field was not applicable to time-varying fields, and that there must be another form of current between the two plates of the capacitor, known as displacement current:

$$J_d = \frac{\partial D}{\partial t}$$

Where $ J_{\text{displacement}} $ is the displacement current density.
Therefore, he added the displacement current term to the Ampère's circuital law, forming the Ampère-Maxwell law equation, whose differential form is

$$\nabla \times \mathrm{B} = \mu_0 \varepsilon_0 \frac{\partial \mathrm{E}}{\partial \mathrm{t}} + \mu_0 \mathrm{j}$$

Comparing it with Ampère's circuital law, it simply adds the term for the changing electric flux on the right side, referencing Faraday's law of electromagnetic induction. Maxwell believed that a changing electric field would also produce a magnetic field.

However, as we analyzed earlier, electric field and magnetic field are two attributes of the vortex field in the vacuum space superfluid, they are two sides of the same coin, not mutually induced.

The "contradiction" shown in Figure 8-29 actually arises from confusing the mechanism of magnetic field generation. We all know that the relative motion between charges and the vacuum space superfluid produces a magnetic field. The Ampère's circuital law corrected this by stating that a steady current can produce a magnetic field, which actually broadens the conditions and scope for magnetic field generation. Therefore, although it is generally correct in most cases, it begins to fail when facing the situation in Figure 8-29.

In reality, a magnetic field is produced as long as charges move within a conductor, it is not necessary to form a complete current loop.

The induced magnetic field is independent of the displacement current, which has been discussed in many scholars' papers (such as references [36], [37], [38], [39], etc.), and this is consistent with our theoretical analysis.

Therefore, Maxwell deliberately added a displacement current to solve this problem, which was entirely unnecessary.

In fact, what we need to do is correct Ampère's circuital law by simply accounting for the movement of charges in equation 8.3.

Since the current density J can be expressed as:

$$\mathrm{J} = \mathrm{nqv} (8.4)$$

where n is the number of free charges in the conductor, q is the charge quantity, and v is the charge moving speed. Therefore, equation 8.3 can be corrected to:

$$\nabla \times \mathrm{B} = \mu_0 \mathrm{nqv} (8.5)$$

When electric charges in a conductor move collectively in the same direction, a circulating magnetic field can be generated. In this case of a circuit being truncated by a capacitor as shown in Figure 8-29, the same principle applies: although the circuit is truncated and no overall current is produced, there is still overall charge movement in the circuit conductors, thus generating a magnetic field.

(4) Modification of Maxwell's Equations

Maxwell's equations are very important in the field of electromagnetism and even in physics, mainly consisting of four equations in the field of electromagnetism, whose differential form is as follows:

$$\nabla \cdot \mathrm{E} = \frac{\rho_e}{\varepsilon_0} (8.6)$$
$$\nabla \times \mathrm{E} = -\frac{\partial \mathrm{B}}{\partial \mathrm{t}} (8.7)$$
$$\nabla \cdot \mathrm{B} = 0 (8.8)$$
$$\nabla \times \mathrm{B} = \mu_0 \varepsilon_0 \frac{\partial \mathrm{E}}{\partial \mathrm{t}} + \mu_0 \mathrm{j}$$

Where E is the electric field strength, B is the magnetic induction, j is the current density, ε_0 is the vacuum permittivity, and μ_0 is the vacuum permeability.

Gauss's electric field law, equation 8.6, states that the electric flux through a closed surface is proportional to the enclosed charge.

Faraday's law, equation 8.7, indicates that the rate of change of magnetic flux through a surface is equal to the circulation of the induced electric field.
Gauss's magnetic field law, equation 8.8, means that the magnetic flux through a closed surface is always 0.

Equation 8.9 is the Ampère-Maxwell equation, which states that the rate of change of electric flux through a surface is equal to the circulation of the induced magnetic field around the surface.

According to our previous analysis, electric fields and magnetic fields are actually two aspects of the spatial superfluid vortical flow field, the two sides of the same coin, so there is no mutual induction phenomenon between them.

Electric fields are sources and irrotational, while magnetic fields are source-free and rotational.
At the same time, a changing electric field "induces" a magnetic field, and a changing magnetic field "induces" an electric field. In reality, this is not true induction, but rather the changing electric or magnetic field exerts a mechanical force on the free charges in a conductor, causing them to move accordingly and thereby producing the corresponding magnetic or electric field effects.

Therefore, we make the corresponding corrections to Maxwell's equations based on the above points.
Because the electric field is solenoidal and has sources, we have:

$$\nabla \cdot \mathrm{E} = \frac{\rho_e}{\varepsilon_0}$$
$$\nabla \times \mathrm{E} = 0$$

When a magnetic field changes, it exerts a force on charges within a conductor, causing them to move, that is:

$$\nabla \times \left(\frac{\mathrm{nqv}}{\sigma}\right) = -\frac{\partial \mathrm{B}}{\partial \mathrm{t}}$$

Where σ is the conductivity, n is the number of freely moving charges in the conductor, q is the charge quantity, and v is the overall velocity of the charges in the conductor.

This movement of charges in the conductor exhibits an apparent phenomenon of an induced electric field, that is:

$$\mathrm{E}' = \frac{\mathrm{nqv}}{\sigma}$$

Where E' is the equivalent induced electric field. This apparent electric field effect is actually a magnetic effect. Since the magnetic field is source-free and has vorticity, we have:

$$\nabla \cdot B = 0$$
$$\nabla \times B = \mu_0 j$$

Where μ_0 is the vacuum permeability, and j is the current density in the conductor of a closed circuit.
For conductors in non-closed circuits, there will not be a complete continuous current, so at this moment we have:

$$\nabla \times \mathrm{B} = \mu_0 \mathrm{nqv}$$

Among them, n is the number of free charges that can move in a conductor, q is the charge quantity, and v is the overall movement velocity of the charges along the conductor. That is, there can also be overall charge movement in a disconnected circuit, forming a break current (as shown in Figure 8-29):

$$\mu_0 \mathrm{j}' = \mu_0 \mathrm{nqv} = \nabla \times \mathrm{B}$$

Where $ J_{\text{breaker}} $ is the breaker current density, which is the current density formed by the movement of charges in the non-closed current. Therefore, the corrected Maxwell's equations can be expressed as follows:

$$\nabla \cdot \mathrm{E} = \frac{\rho_e}{\varepsilon_0} (8.10)$$

$$\nabla \times \mathrm{E} = 0 (8.11-1)$$

$$\nabla \times \left(\frac{\mathrm{j}}{\sigma}\right) = -\frac{\partial \mathrm{B}}{\partial \mathrm{t}} (8.11-2)$$

$$\nabla \times \left(\frac{\mathrm{j}'}{\sigma}\right) = \nabla \times \left(\frac{\mathrm{nqv}}{\sigma}\right) = -\frac{\partial \mathrm{B}}{\partial \mathrm{t}} (8.11-3)$$

$$\mathrm{E}' = \frac{\mathrm{nqv}}{\sigma} (8.11-4)$$

$$\nabla \cdot \mathrm{B} = 0 (8.12)$$

$$\nabla \times \mathrm{B} = \mu_0 \mathrm{j} (8.13-1)$$

$$\nabla \times \mathrm{B} = \mu_0 \mathrm{j}' = \mu_0 \mathrm{nqv} (8.13-2)$$

Among them,
E : Electric field strength.
E$'$: Equivalent induced electric field strength (electric field strength equivalent to magnetic field effect).
ρ_e : Charge density.

B: Magnetic induction intensity.
J: Closed circuit current density.
J': Open circuit current density.
Vacuum permittivity.
Vacuum permeability.
Conductivity.
n: The number of free charges that can move in a conductor.
q: Charge quantity.
v : The overall velocity of charge movement within a conductor.
Equation 8.10 is Gauss's electric field law, stating that the electric flux through a closed surface is proportional to the enclosed charge.

The four formulas of equation 8.11 are corrected forms of Faraday's law of electromagnetic induction, relating the rate of change of magnetic flux through a surface to the induced current.

Equation 8.11-1 indicates that the electric field is irrotational, and equation 8.11-2 shows the relationship between changing electromagnetic fields and induced currents in a closed circuit; equation 8.11-3 illustrates the relationship between changing electromagnetic fields and induced currents in an open circuit, as shown in the situation of Figure 8-29; equation 8.11-4 represents the equivalent induced electric field produced by changing electromagnetic fields.

Equation 8.12 is the Gauss's law for magnetism, stating that the magnetic flux through a closed surface is always zero.
The two formulas of equation 8.13 are the corrected forms of the Ampère-Maxwell equation, relating the current enclosed by a surface to the circulation of the induced magnetic field. Among them, equation 8.13-1 applies to a closed circuit, while equation 8.13-2 applies to an open circuit, as shown in Figure 8-29.

Therefore, the above correction to Maxwell's equations mainly involves Faraday's law of electromagnetic induction and the Ampère-Maxwell law, because the essence of electromagnetic induction is formed by the relative motion between charges and magnetic fields.

It can be the active motion of charge, or the active motion of the magnetic field, which are equivalent, not a direct mutual induction between the magnetic field and the electric field.

(5) The relationship between the magnetic field and the vortex flow field

According to Ampère's circuital law, the circulation of the magnetic induction B equals the enclosed current multiplied by μ_0

$$\oint_C B \cdot dl = \mu_0 I$$

Where μ_0 is the vacuum permeability, and I is the current in a closed loop.
The theorem shows that magnetic fields are produced by currents, and the vorticity of the magnetic field is directly related to the distribution of the current, that is, the circulation of the magnetic field around an infinite straight conductor depends only on the current in the conductor and is independent of the radius of the closed path.

That is, the magnetic field circulation around any closed path with any radius centered on this wire is equal, making it a conserved quantity.

Similarly, in our fluid mechanics, there is also the same vorticity conservation quantity: the circulation of velocity, as described by Kelvin's theorem [30] : In the ideal fluid flow under the conditions of potential mass forces and a barotropic flow field, the circulation of velocity along any arbitrary closed fluid line does not change with time:

$$\frac{d\Gamma}{dt} = 0$$

Where Γ is the velocity circulation:

$$\Gamma = \oint_L v \cdot dl$$

where $ v $ is the fluid velocity.
The circulation of a vortex ring in an ideal fluid is conserved and does not change with the expansion of the vortex ring. When the vortex ring expands, the vortex tube is stretched, leading to a decrease in cross-sectional area and an increase in vorticity density, but the total vorticity remains constant.

The theorem of vortex tube strength conservation (if the fluid is ideal, the flow field is barotropic, and the body forces are potential, then the strength of any vortex tube remains constant throughout the entire motion) and others also describe conserved quantities in a vortex flow field.

This is consistent with the conservation of magnetic field circulation in the Ampère's circulation theorem, indicating that the magnetic field should also be a closed vortex flow field formed in the vacuum space superfluid, which aligns with our previous analysis on the nature of the magnetic field.

According to our previous analysis, both the magnetic field and the electric field are two different directional descriptions of the same phenomenon—the vortex flow field of the vacuum space superfluid. The magnetic field describes the strength field of the vortex flow, while the electric field represents the strength gradient field of the vortex flow. This is similar to the theoretical analysis in classical fluid dynamics regarding the vortex circulation field and the strength gradient field. Combining classical fluid dynamics with electromagnetic field theory can better analyze and understand the essence of the electromagnetic field, leading to a more complete and accurate mathematical description.

The parameters such as ε_0 in electromagnetic field theory are related to the density and other properties of the superfluid vacuum space, which is also one of the reasons for the difference between electromagnetic field theory and classical ideal vortex hydrodynamics in mathematical description.

(6) Principles of Electromagnetic Wave Generation and Propagation

As we analyzed above, electric fields and magnetic fields do not induce each other. The apparent phenomenon of "mutual induction" we observe is actually the mechanical effect produced by changing electric or magnetic fields on charges within a conductor, not a true "mutual induction."

Electric fields and magnetic fields are two directional descriptions of the same vortex field entity; they are two aspects of one thing, existing simultaneously and naturally, so there can be no "mutual induction."

However, in a vacuum, there are no charges, so electric and magnetic fields cannot induce each other. The theory that we currently believe electromagnetic waves propagate in a vacuum through continuous mutual induction is actually a misunderstanding of the image of electromagnetic wave propagation. The principles of electromagnetic wave generation and propagation will be analyzed in detail in Chapter 8.10.

8.5 Electronic Energy Level Arrangement of an Atom

(1) Atomic Energy Levels and Orbits
Protons are particles with positive charge and are nuclear vortices with closed internal and external flow fields, having an internal flow field outlet (N pole) and an external flow field inlet (S pole). When protons are part of the nucleus in an atom, the atom also possesses these characteristics.

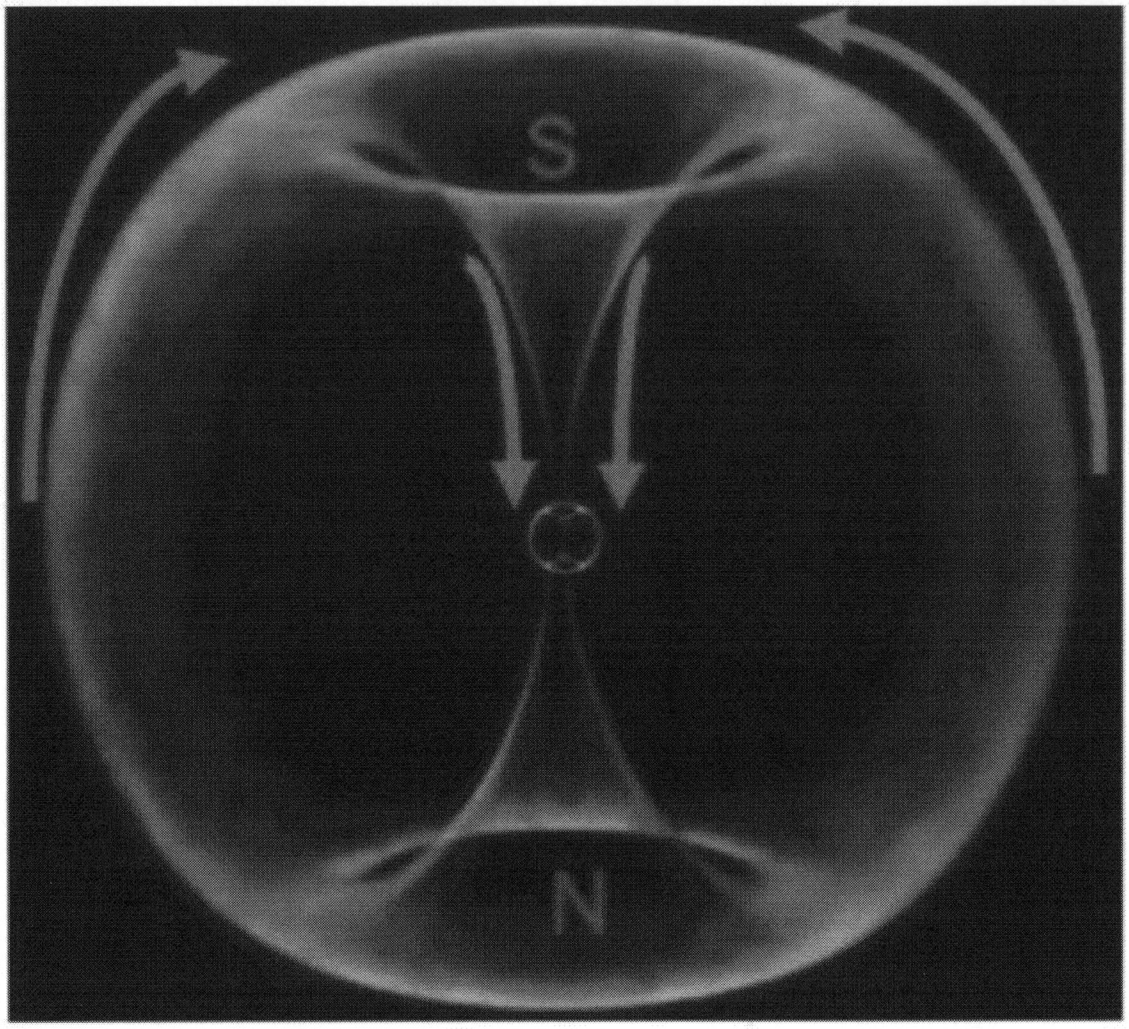

Figure 46: Figure 8-30 Atom - Forms an orbit with N and S poles

An atom with one proton will have one N pole and one S pole. This pair of corresponding N and S poles constitutes an orbit of the atom, with the N pole and S pole being the N pole end and S pole end of the orbit, respectively, as shown in Figure 8-30.

A single or several such orbits constitute a shell in the atomic shell field, which is also an energy level.

As the number of protons in the atomic nucleus increases, these protons form various three-dimensional structures, allowing the atom to have more energy levels and orbits [41] .
(2) Atomic structure
Similar to the most basic and smallest unit of a living organism—the cell—its interior also has a clear structure, which can be divided into the cell nucleus, cell wall, cytoplasm, cell sap, and more specific functional parts, just as an atom.

As the smallest unit of macroscopic matter, it also has a clear internal structure. By distinguishing and naming these structures, we can better understand the internal structure and composition of atoms.

As shown in Figure 8-31, atoms and other fundamental particles contain the following structure:
(1) Particle wall: It includes the inner wall and outer wall of the particle, corresponding generally to the boundaries formed by the internal and external flow fields respectively. This boundary, especially the outer wall boundary, gives the particle its enclosure and rigidity, ensuring its stability.

For an atom, the outer wall of the particle corresponds to its energy level. When multiple protons form the atomic nucleus inside, the outer wall expands into multiple layers, forming multiple energy levels. Each energy level is further divided into multiple sub-layers, which are the sub-orbitals of each energy level.
(2) Particle orbit: That is, the entrance of the outer flow field into the inner flow field, known as the S pole mouth, and the exit of the internal flow field to the external flow field, known as the N pole mouth.
(3) Particle body: That is, the spatial part between the inner wall and outer wall of the particle.
(4) Particle mass: The interior of the particle is not a vacuum, but rather contains a dense particle liquid composed of a

large number of void space superfluid wave particles. In this particle liquid, there are also a large number of excited solitary particles and micro-particles. These particles, together with the particle liquid, undergo dense interactions, forming complex internal flow fields and mechanical effects.
(5) Particle nucleus: The inner core of the particle, such as the atomic nucleus, is the core that forms and maintains the circulation of the internal and external flow fields. It can be a single nucleus or a nuclear matrix formed by multiple nuclei. The stronger the nucleus or the more powerful the nuclear matrix, the stronger the internal and external flow fields of the particle, and the more pronounced its mass, rigidity, and stability.

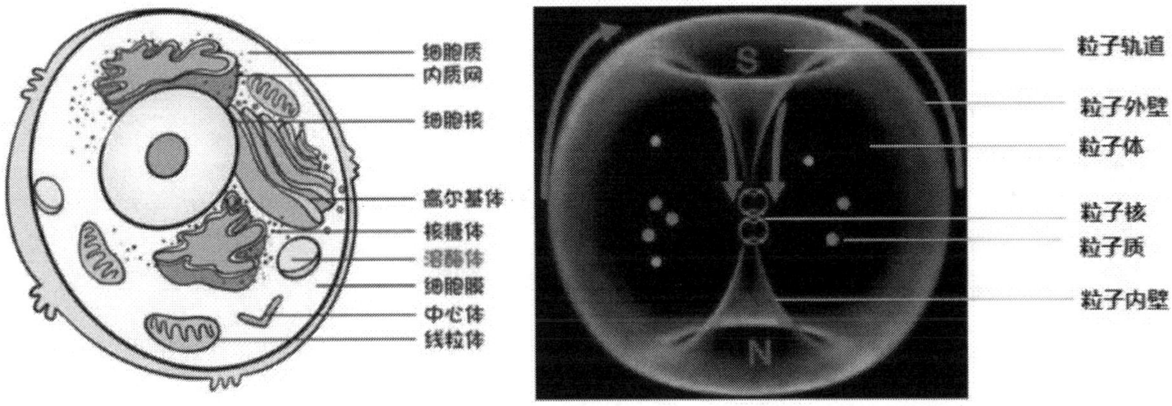

Figure 47: Figure 8-31 Particle Structure

(3) The electron arrangement rules of atoms

Because the presence of protons gives atoms a positive charge, they attract and come close to negatively charged electrons. When electrons approach and come into contact with the atomic shell, the attraction between the electron and the atom weakens due to the different directions of the atomic shell's field and the complexity and variation of the medium in the atomic field. Without higher energy input, the electron cannot continue to enter the atom. During the interaction with the atomic shell field, the electron enters the N pole or S pole orbit of the atom, forming a stable polar pair and thus remaining on the atom's energy level and orbit, as shown in Figure 8-32.

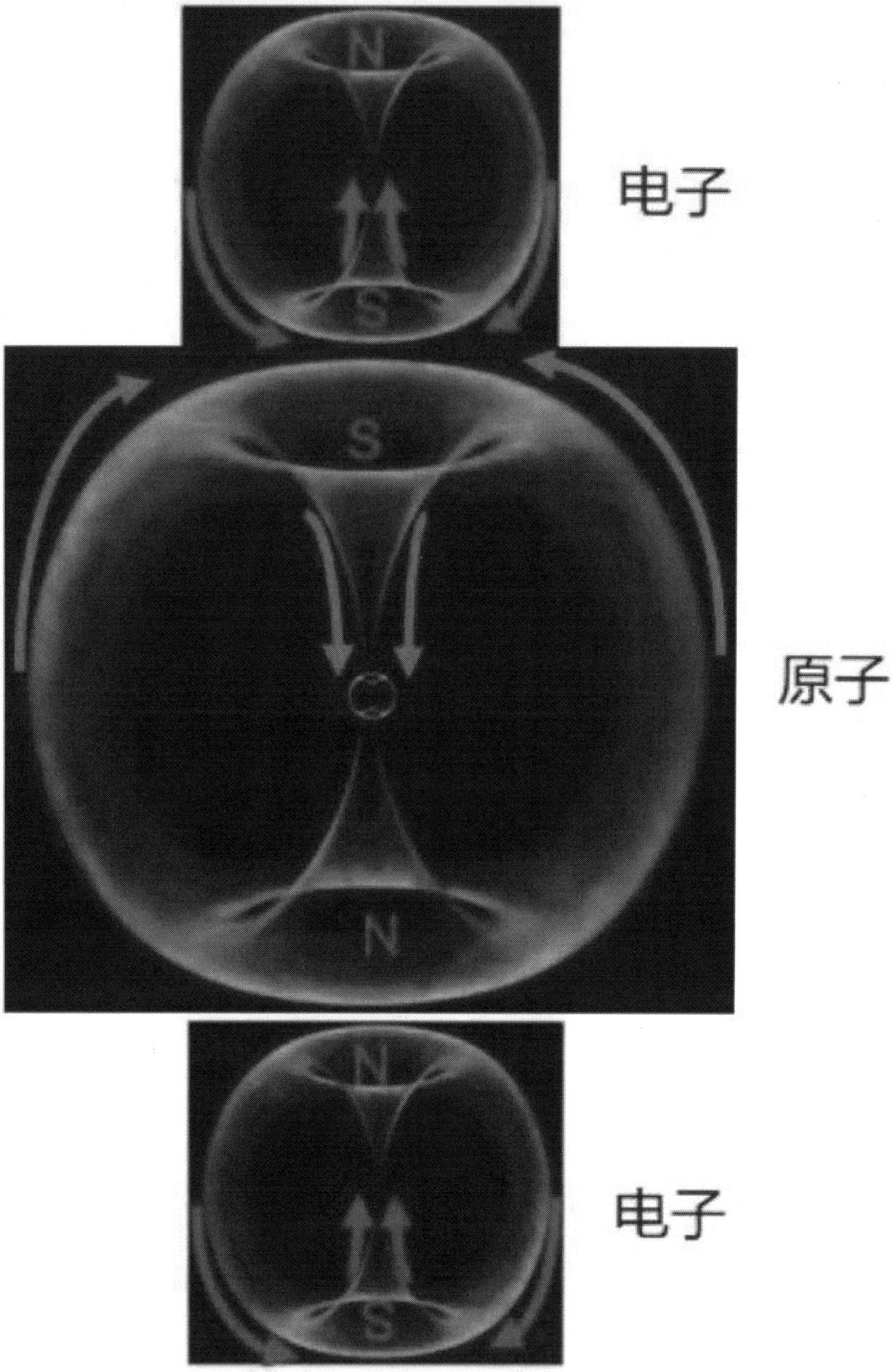

Figure 48: Figure 8-32 An atomic orbital pairs two electrons

Thus, this electron becomes an orbital electron within the atom.
The electron residing at the N pole orbital mouth is also the N pole corresponding to the atomic orbital, because in this case, the direction of the electron's N pole outflow field aligns with that of the atom's N pole outflow field, forming a spatial flow field negative pressure between them, which allows the electron to more stably reside at the orbital mouth.

Similarly, the electron residing at the S pole mouth is also the S pole corresponding orbit's S pole mouth.
So, the two orbitals of an atom's orbital can accommodate two electrons with opposite spin directions, which is the Pauli exclusion principle: an atomic orbital can hold only two electrons, and they must have opposite spins

Electron spin must be opposite.
At the same time, because the electrons in the $S-S$ pole pairing and orbital mouth are both flowing from the outer field to the inner field, similar to an inhalation effect, forming better negative pressure stability, this pairing is relatively more stable than the $N-N$ pairing. When an electron enters a new orbital, it will preferentially enter the S pole orbital mouth to form $S-S$ pole pairing. When an energy level has multiple orbitals, electrons will follow the order of first completing all

$S-S$ pole pairings, and then proceeding to $N-N$ pole pairings. This is Hund's rule: when an atom has multiple empty orbitals, electrons prefer to occupy different orbitals rather than pair up in the same one, and the initial electrons occupying different empty orbitals have the same spin direction.

For example, the $2p$ orbital of a nitrogen atom has three electrons, which will preferentially occupy three different p orbitals, and the spin directions of these three electrons will be the same. Here, each orbital is a $N-S$ orbital, and each electron will preferentially perform $S-S$ pole pairing to occupy separate orbitals, so the initial spin directions of the three electrons (relative to the direction facing the nucleus) are the same, while the subsequent electrons that occupy other orbitals will have opposite spin directions. This aligns with our theoretical prediction, as shown in Figure 8-33.

Atom	$1s$	2s		$2p$		Electronic configuration
Li	↑↓	↑	□	□	□	$1s^2 2s^1$
Be	↑↓	↑↓	□	□	□	$1s^2 2s^2$
B	↑↓	↑↓	↑	□	□	$1s^2 2s^2 2p^1$
C	↑↓	↑↓	↑	↑	□	$1s^2 2s^2 2p^2$
N	↑↓	↑↓	↑	↑	□	$1s^2 2s^2 2p^3$
O	↑↓	↑↓	↑↓	↑	↑	$1s^2 2s^2 2p^4$
F	↑↓	↑↓	↑↓	↑↓	↑	$1s^2 2s^2 2p^5$
Ne	↑↓	↑↓	↑↓	11	11	$1s^2 2s^2 2p^6$

Serway, Physics for Scientists and Engineers, 5 / e
Figure 42.14 Harcourt, Inc.

Figure 8-33 Pauli Exclusion Principle and Hund's Rule for Electron Orbital Filling

It can be predicted that when two electrons are subjected to sufficient pressure, such as extremely low temperatures or very slow superfluid flow in the surrounding vacuum, the S poles of the two electrons will be more easily combined due to greater negative pressure, forming $S - S$ pole pairs, which are commonly known as Cooper pairs in superconductivity.

(4) Neutron Production and Decay

Electrons and protons attract each other due to electric field forces, but because of differences in various flow field walls, they cannot merge. Therefore, electrons and protons eventually form a force field balance existing as $S - S$ polar particle pairs. At this moment, a new vortex flow field will also be formed between the S poles of the two, which we can call the K flow field.

When electrons and protons are subjected to a strong external force, the K field between them is intensely excited, forming two new particle pairs: electron neutrinos and anti-electron neutrinos. Eventually, the electron neutrino is emitted, while the anti-electron neutrino acts as a buffer between the electron and proton, creating a more tightly bound composite particle: the neutron.

This is the typical weak interaction process in which a proton combines with an electron to form a neutron, as shown in Figure 8-34.

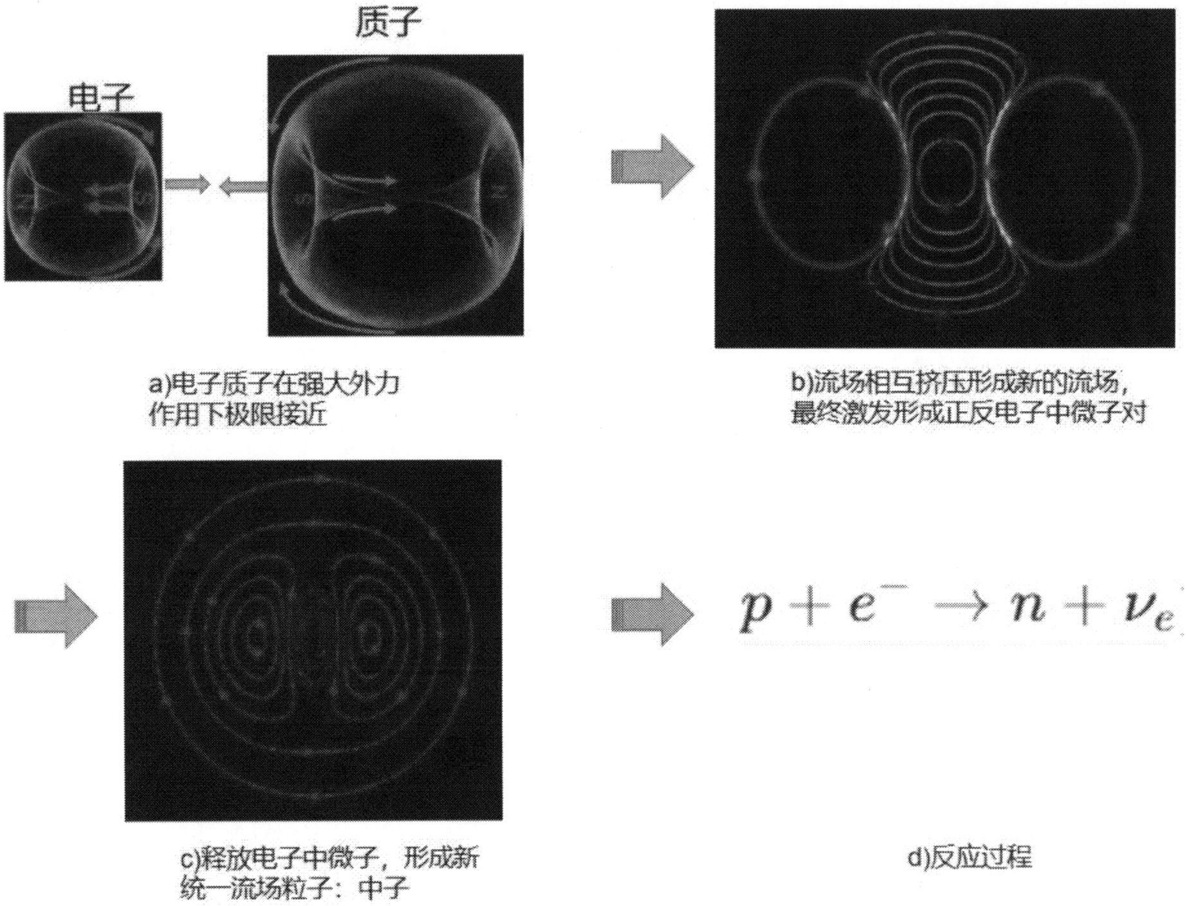

Figure 49: Figure 8-34 Proton and Electron Combining to Form a Neutron

Due to the large amount of energy involved in the combination of electrons, antielectron neutrinos, and protons, once the neutron leaves the atomic nucleus and is no longer suppressed by the outer nuclear field, it immediately becomes unstable. Under the influence of the antielectron neutrino energy in the middle, the electron and proton gradually destroy the polar field between them, eventually completely breaking it apart. The antielectron neutrino gains freedom and is released, while the electron and proton are separated, and the neutron decays and disappears.

This is the classical process of neutron decay.

(5) Extension of atomic energy levels and orbitals

(1) Atomic K-shell energy levels

A proton can allow an atom to form a $N-S$ orbital, and two protons can also form a proton pair corresponding to such a $N-S$ orbital, so an atom's orbital can correspond to a maximum of two protons.

When an atom has one proton, although it has two orbital ports N and S that can hold two electrons, this is the first energy level of the atom: the K layer.

Because the energy level of a single proton is insufficient, the second pair of electrons will be very unstable and tend to leave easily, while the empty N orbital ports will also make the atom highly reactive and unstable. Only when two protons work together to form an orbital $N-S$ can a stable pairing of two electrons be achieved.

Therefore, a proton forming a hydrogen atom with one $N-S$ orbital has relatively low stability, while two protons forming a helium atom with one $N-S$ orbital has very high stability, which aligns with the observed cosmic elemental data and chemical reaction experimental results, as shown in Figure 8-35.

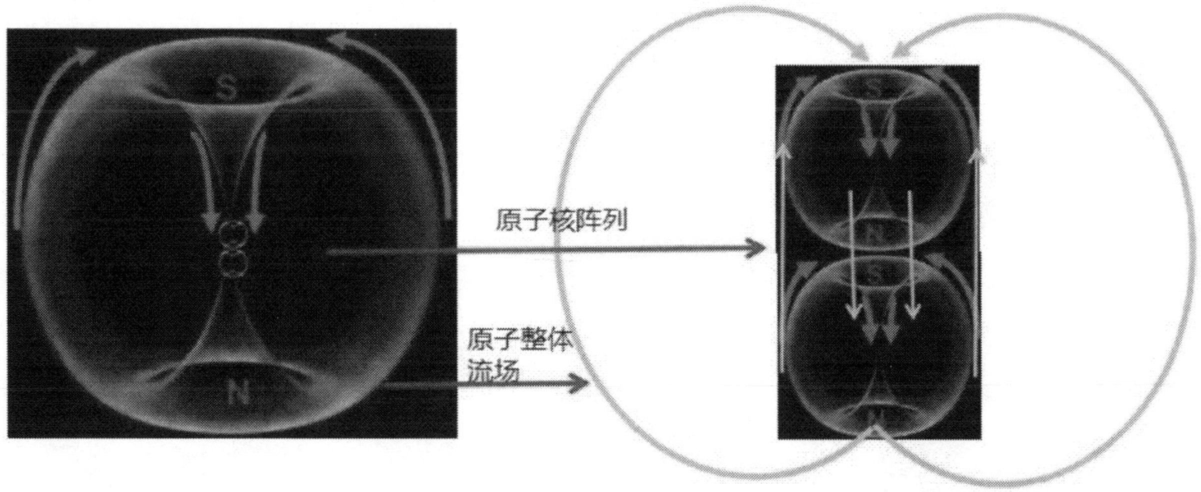

Figure 50: Figure 8-35 Proton pairs form a nuclear matrix, collectively creating an N-S orbital

The two protons forming the atomic nucleus have partially connected inner fields, and partially merged outer fields into the atom's outer field, enhancing the particle outer wall of the atom. In this case, the atom's outer field will possess a stronger $N-S$ orbital, capable of holding two electrons, which form $S-S$ and $N-N$ polar pairs respectively.

The K shell energy level of an atom contains only one orbital: the S orbital. The S orbital has one N-S orbital that can accommodate a pair of electrons with opposite spins.

(2) Atomic L shell energy levels

When the atomic nucleus increases by adding another proton, three protons exceed the energy limit of one S orbital in the K shell. The newly added proton will merge into the flow field nuclear matrix of two protons, allowing the atom to expand and add a new flow field wall, which is equivalent to adding a new energy level: the L shell.

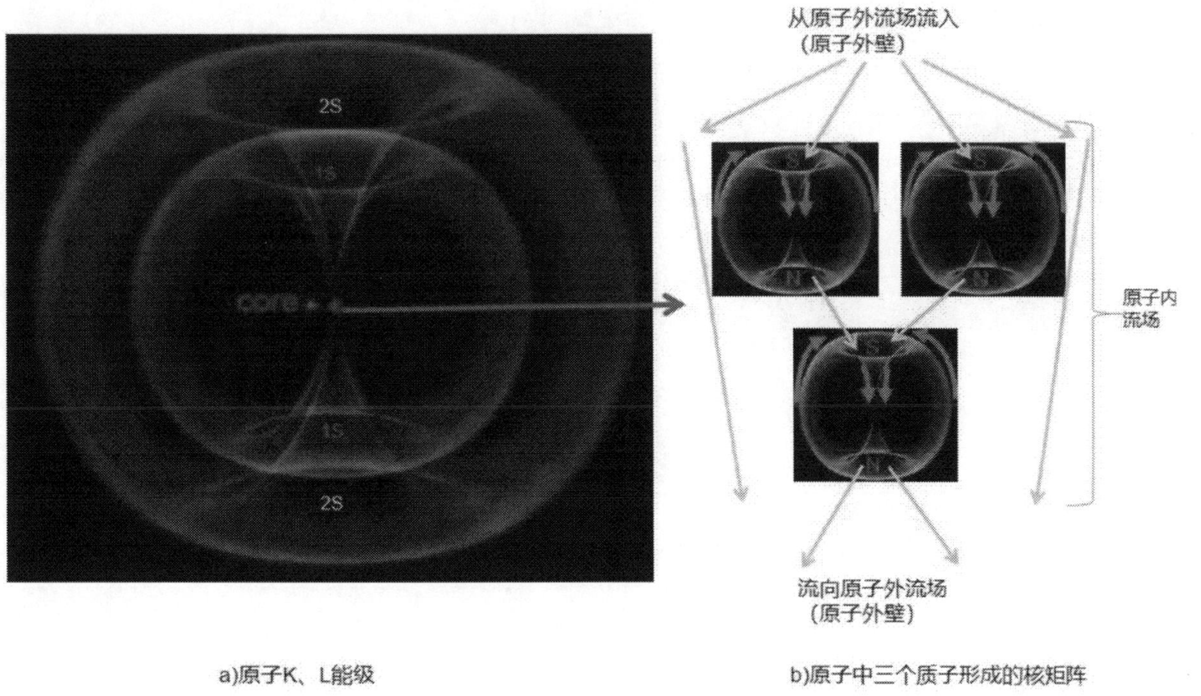

Figure 51: Figure 8-36 Nuclear matrix and flow field of three protons

The expansion of each energy level in an atom involves the flow field emerging and recirculating from the N-S orbitals of the previous energy level, forming a closed loop. Therefore, the first orbital added to the L energy level is identical to the K energy level, also being an S orbital containing one N-S orbital, capable of holding two electrons, as shown in Figure 8-36(a).

The three protons now form a new nuclear matrix, jointly supporting the atomic flow field wall: K energy level and L energy level, as shown in Figure 8-36(b).

When reaching 4 protons, the energy of the S orbital in the L energy level also reaches its limit. Adding more protons will then introduce a new orbital, the P orbital, in the L energy level. Because of the energy foundation provided by the two protons in the K shell, the L energy level can support a total of 8 additional protons in the nuclear matrix, including one $\mathrm{N}-\mathrm{S}$ orbital from the S orbital and three N - S orbitals from the P orbital.

The three $\mathrm{N}-\mathrm{S}$ of the newly added P orbital are distributed symmetrically along the X、Y、Z three axes on the atomic energy sphere, ensuring minimal interaction forces between them, as shown in Figure 8-37(a).

At this moment, the nuclear matrix composed of 8 protons will also form a stable structure similar to a "unit cell," which we can call the proton unit cell. Meanwhile, different numbers of neutrons will be distributed within these nuclear matrices to ensure the unit cell achieves strong stability, as shown in Figure 8-37(b).

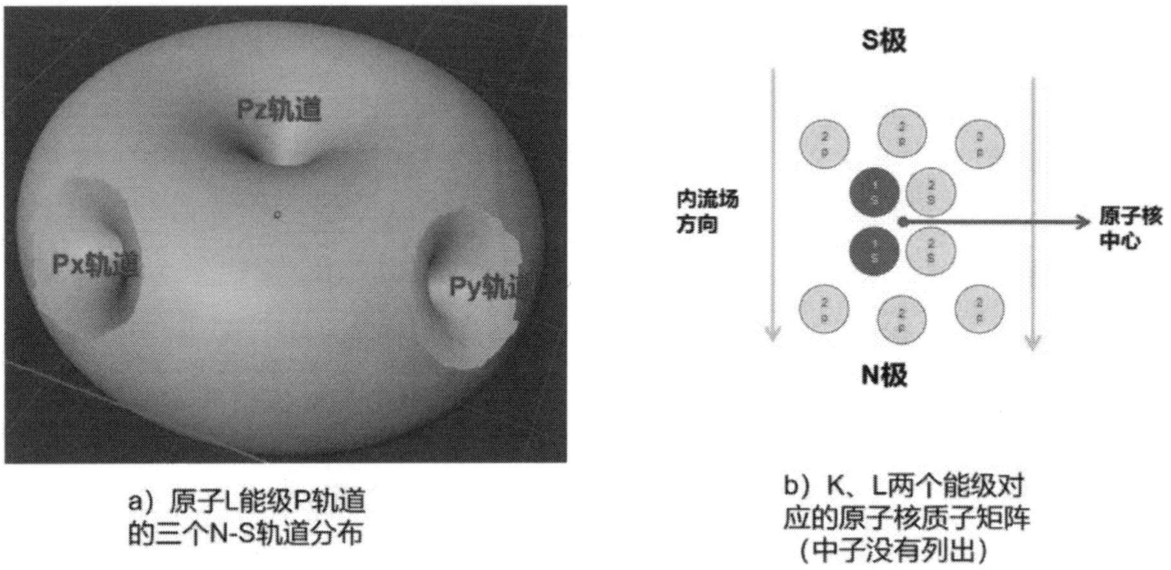

Figure 52: Figure 8-37 Atomic Orbit and Nucleon Matrix

(3) Atomic M Level and Higher Energy Levels
After the L level reaches the upper limit of 8 protons, when adding more protons to the atomic nucleus, a new layer will be added to the proton matrix, resulting in a new layer for the atom's outer field wall, which means the addition of a new energy level: the M energy level.

Due to the structure of the nuclear proton matrix, like the L level energy, all new energy levels are added on top of the previous one.

The N - S poles form a closed field with inflow and outflow, so each new energy level will include all the orbits from the previous energy level, adding a new orbit on top of that.

Therefore, the M energy level will have the S and P orbits that the L energy level has, and will also add a new orbit: the d orbit. At this moment, the proton matrix inside the atomic nucleus is shown in Figure 8-38:

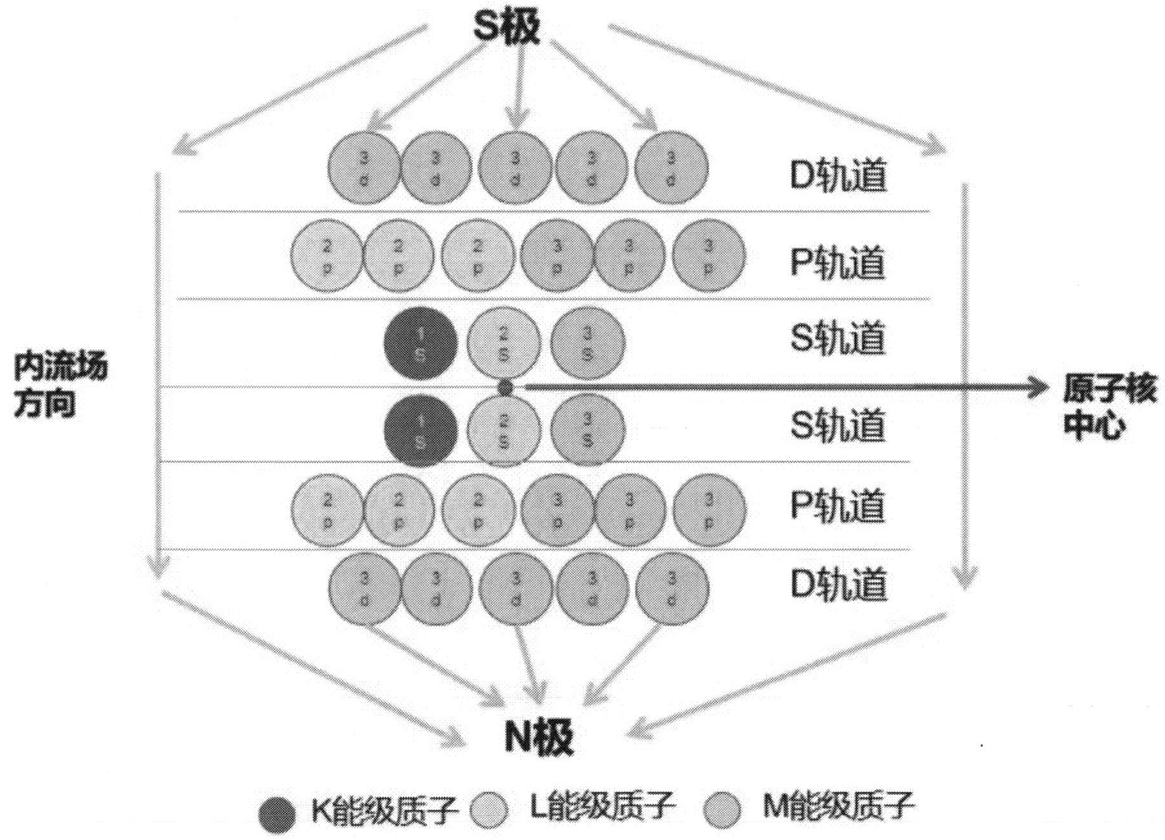

Figure 53: K、 L、M Proton matrix corresponding to three energy levels (neutrons are not listed)

Proton matrix in the atomic nucleus with three energy levels as shown in Figure 8-38

Each time an atom adds an energy level, the proton matrix within the atomic nucleus also adds a layer. Protons in each layer form a three-dimensional polygonal structure, and together with neutrons, they create a dense polyhedral proton-neutron cell structure. Due to the combination of their individual field interactions and the shared internal field of the atom, a stable structure is formed, collectively supporting the powerful energy of the atom.

This atomic nucleus cell structure exists in a dynamic equilibrium state within the atom's internal field. It undergoes changes such as compression, rotation, and displacement of outer protons due to the atom's movement and interactions with its surrounding environment, leading to different structural model deformations, such as the shell model, liquid drop model, and collective model.

All other new energy levels also function similarly to the M energy level, adding a new orbit based on all the orbits of the previous energy level. Through this stacking method, all atoms in the entire periodic table are generated.

8.6 Antimatter Principle

(1) Production of Antimatter

We have analyzed before that when the superfluid vacuum space is excited, it produces high-speed spinning solitons, or solitary waves. These solitons, due to their high-speed spinning, possess good closure, rigidity, and stability, which provide them with the fundamental conditions to become particles.

As the excitation energy increases, the generated solitons become stronger, forming solitons that can further merge and stack, leading to the creation of more powerful and complex composite particles, which in turn form the numerous fundamental particles we observe.

Therefore, the source of particles lies in spin, with high-speed spinning solitary particles, micro-particles, and even the particles themselves existing within each composite particle. Spin is the fundamental basis for the generation of particles.

We consider that when the flowing field of empty space impacts a strong particle, it generates vortex pairs behind the particle due to negative pressure, similar to how fluid in our macroscopic world forms vortices behind obstacles. These vortex pairs consist of two solitons with opposite spins. Due to the quantum nature of empty space, these solitons have almost identical mass, spin velocity, or energy, as well as almost identical structures, including composite particle structures.

Such pairs of particles are the positive and negative particles we are familiar with, as shown in Figure 8-39.
Therefore, positive and negative particles are often produced in pairs due to the interaction between the superfluid vacuum space and particles.

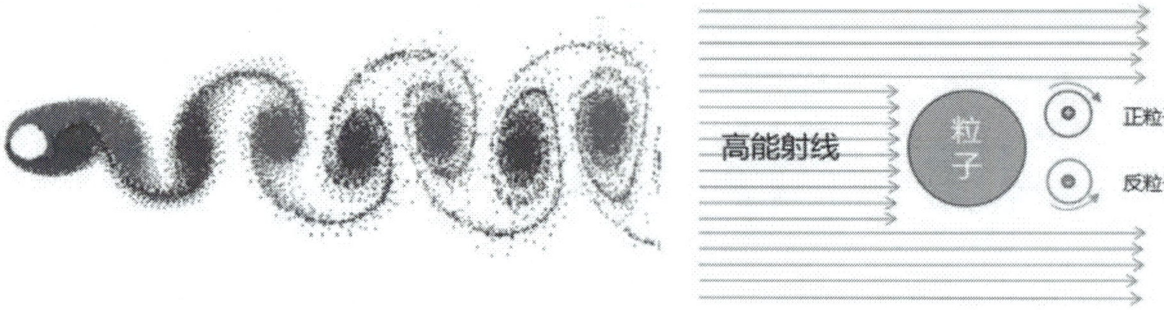

a)流体在阻挡物后形成的正反涡旋（涡街）

b)高能射线冲击粒子后形成的正反自旋的正反粒子

Figure 54: Formation of Matter and Antimatter Particles (Figure 8-39)

Of course, the superfluid vacuum space will also generate individual positive particles or antiparticles due to its own vortices. The solitons and other fundamental micro-particles it initially produces are examples of this phenomenon.

Other times, when the interaction energy between the superfluid vacuum space and particles is insufficient, or due to their spin states, individual positive or negative particles can be produced.

Because their spin directions are opposite, positive and negative particles exhibit opposite charge effects, resulting in them having opposite charges. Since their volume and spin speed are the same, the magnitude of their charges is equal.

Now, in experimental observations, high-energy gamma rays interacting with particles are an effective method for producing matter-antimatter pairs. For example, the Shanghai Institute of Optics and Fine Mechanics uses a femtosecond petawatt laser facility and a high-pressure gas target to generate a large number of high-energy electrons. These high-energy electrons then interact with high-Z material targets, producing intense gamma rays through bremsstrahlung radiation. The gamma rays subsequently interact with high-Z atomic nuclei to create electron-positron pairs.

This aligns with our predictions.

(2) Annihilation of Matter and Antimatter

Since particles and antiparticles have opposite spins and opposite charges, they attract and move closer to each other. When the outer flow fields (i.e., the particle's outer boundary) of two particles meet, if the flow field contains not only the basic solitons but also more other solitons, solitary particles, etc., the spins of these solitons and solitary particles are also completely opposite, making them each other's antiparticles. Therefore, they will continue to attract and approach each other until their outer flows meet

Field fusion begins, and the particles start to merge with each other.
At this time, because the particles in their respective fields and particle bodies follow the spin motion of the particles, their directions of movement begin to reverse relative to each other, leading to mutual collisions and interference, which disrupt their internal motion and cause collapse, ultimately resulting in the complete destruction of the external fields and mutual annihilation of the particles.

If the two particles contain nuclear particles or nuclear matrices inside, these nuclear particles and matrices are also completely opposite spins and each other's antiparticles, so they will continue the process of mutual attraction, approaching, fusion, and collapse as described above, until all particles are completely destroyed and disappear.

This is the principle and process of particle-antiparticle annihilation, as shown in Figure 8-40.

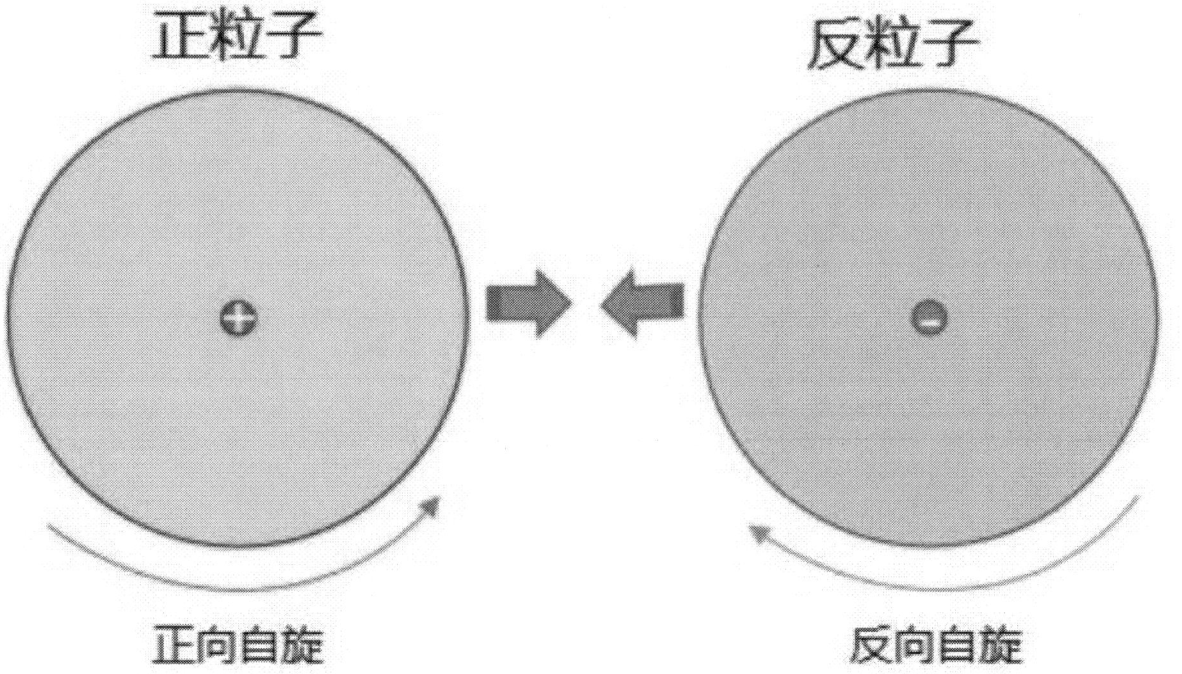

Figure 55: a) Attraction and approach of particle and antiparticle

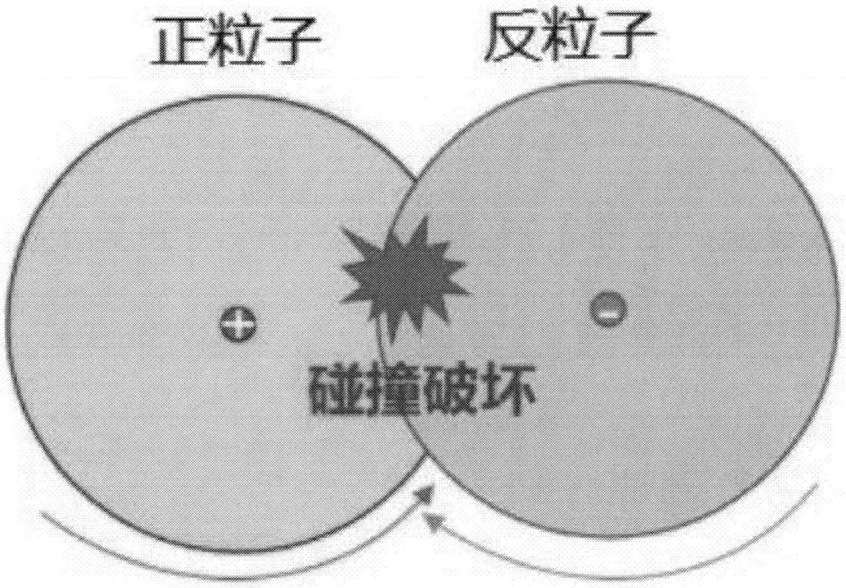

Figure 56: b) Fusion and collision destruction of particle and antiparticle

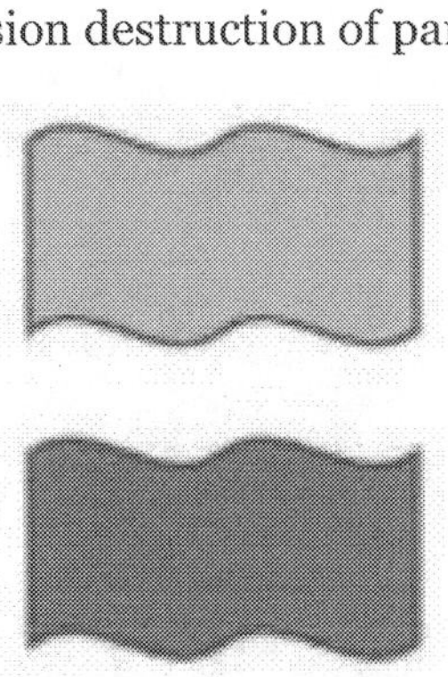
Figure 57: c) Annihilation of matter and antimatter into electromagnetic radiation
Space Wave

Figure 8-40 Annihilation of particles and antiparticles

The final products of the annihilation of particles and antiparticles are the void space wave, i.e., the wave superfluid fluctuation, and electromagnetic wave rays. The photons in the electromagnetic wave rays are themselves antiparticles; if their spin directions are also opposite, they will continue to annihilate into the most fundamental wave particles, converting all the energy of the two particles into the fundamental space wave fluctuation energy.

Because antiparticles or their core particles spin at high speeds, even exceeding the speed of light, they possess strong spin kinetic energy. When they annihilate and release gamma rays or wave particle oscillations, these energy changes convert into ray energy or spatial wave kinetic energy.

Since the ray propagates at the speed of light, assuming all the particle's kinetic energy is converted into electromagnetic radiation, the energy of the electromagnetic ray is:

$$E = mC^2$$

E is the energy of the emitted rays, m is the total mass of the particle and antiparticle, i.e., the total mass of the emitted photons and bosons, and C is the speed of light.

This equation is the mass-energy equation. Here, after the annihilation of matter and antimatter particles, their mass does not disappear but is entirely converted into the mass of photons and wave particles. This process conforms to the basic principle of our space relativity 7: the principle of mass-energy conservation, which means both mass and energy are conserved.

A large amount of energy is released during nuclear fission or fusion reactions. The same principle applies, and according to equations 8.14 (particle total mass equation) and 8.15 (particle total energy equation), there is no mass loss in this process. Instead, the observable mass of the particle is converted into the mass of unobservable photons and wavicles, releasing internal energies such as the particle's spin energy, nuclear particle spin energy, particle mass motion energy, and spin energy, which become the energy of electromagnetic radiation.

(3) The difference between matter and antimatter particles and particles with positive and negative charges

Matter and antimatter particles carry opposite charges and attract each other, eventually annihilating, whereas ordinary charged particles like electrons and protons, although they also attract and come close to each other, do not annihilate. This is mainly influenced by the following factors:

(1) Different particle compositions

Although electrons and protons have opposite spins, their compositions are different, including the solitons, types of micro-particles, and spin structures in their respective flow fields. Therefore, when electrons and protons approach each other closely, due to differences in the spin and hierarchical structure of the flow field itself, they cannot continue to approach or merge, resulting in the disappearance of gravitational attraction or the cancellation of gravitational and repulsive forces.

(2) Different particle volume shapes

The flow field motion of positive and negative charge particles includes not only spin flow fields but also internal and external circulation flow fields. The direction of all external flow fields is from the N pole vortex upward toward the S pole. Especially when particles have multiple $N - S$ orbits, the flow field becomes more complex. Therefore, when positive and negative charge particles approach each other, the originally attractive flow field will change

The field direction will change, leading to attraction compatibility or even repulsion, so it will not continue to approach infinitely.

(3) Differences in the elements of particle fields

Positive and negative charges, such as electrons and protons, have different masses and energy levels, leading to differences in the types of micro-particles included in their respective particle wall flow fields. For example, the particle wall flow field of an electron may include bosons, photons, and other elements, while a proton's flow field may contain other micro-particles with greater energy. When these two particles approach each other closely enough, the sub-flow fields within their respective particle wall flow fields will show significant differences, which can weaken or cancel out the attraction between them, and even create a short-range repulsion, preventing them from getting closer or merging.

(4) Characteristics of Quantum Fluid Field

Due to the superfluid properties of the vacuum space, the fluid field around the particle wall is filled with a large number of complex vortices and turbulent flows, rather than smooth laminar flows. This makes the fluid field effects around the particle wall very complicated. Since positive and negative particles are basically consistent in terms of mass, volume, and composition, the difference in turbulent forces within the细分 fluid fields of the particle wall is not particularly significant. However, for particles with opposite charges such as electrons and protons, there are very large differences in terms of mass, volume, and composition, so the turbulent flow field conditions in their respective particle walls will also differ greatly, forming significant differences in the fluid field.

It is precisely due to the above factors that protons and electrons do not merge, but instead interact under the influence of the flow field to form stable $S - S$ or $N - N$ poles, creating new composite particles: atoms.

Other charged particles with positive and negative charges behave in the same way.

This is also the reason why particles and antiparticles annihilate each other, while positive and negative charges do not.

Of course, when positive and negative charges are subjected to sufficiently strong external driving forces, they will overcome the barrier of the outflow field shell and form a stronger mutual interaction.

(4) Principle of Asymmetry Between Matter and Antimatter

(1) Initial Asymmetry

The superfluid vacuum space in the universe is disturbed to produce vortical flow fields, leading to the formation of the first quantum vortices, which create the initial solitons. These solitons are also the original matter particles.

As the energy of the vortex field in the void space increases, or as these initial quantum solitons further merge, more stable micro-particles are formed, which have stronger energy and rigidity, as well as being closed systems, becoming the fundamental particles that exist long-term.

Because the vortical flow field of the superfluid vacuum space has direction, different directions produce different vortex solitons, which in turn generate matter particles with different spins, corresponding to either positive or negative matter particles.

Due to the asymmetry of the energy in the cosmic vacuum superfluid vortical flow field, the number and energy of the produced positive and negative particles will also be asymmetric. This initial asymmetry between positive and negative matter particles directly influences and even determines whether there is more positive or negative matter in the subsequent universe.

These initial matter particles, which we call primordial matter particles.

From the current distribution of matter in our universe, the initial matter particles should be more positive than negative particles. In other words, the positive vortical energy of the superfluid vacuum space was greater than the negative vortical energy at the beginning, leading to the final remaining of matter particles in the universe, which are the most widespread and form the basis of the material world, being positive particles.

(2) Secondary Asymmetry

According to the principle of the generation of positive and negative matter particles discussed earlier, after the appearance of the first set of initial matter particles, the huge disturbance energy of the superfluid would interact with these initial matter particles to produce a large number of positive and negative particle pairs. However, since the initial matter particles are positive matter particles with the spin direction of positive particles, this would result in a greater number of positive particles being produced, with higher energy, greater stability, and the ability to generate more composite positive particles. In contrast, the number of negative particles with opposite spin would be relatively smaller, with less energy, less stability, and more prone to decay. The composite negative particles that can be formed through fusion would also be fewer, as shown in Figure 8-41.

This belongs to the Magnus effect in fluid mechanics, when the angular velocity vector of a rotating object

When the object's flight velocity vector does not coincide with it, a lateral force is generated in the direction perpendicular to the plane formed by the rotational angular velocity vector and the translational velocity vector. This lateral force arises because the object's rotation can cause the surrounding fluid to rotate, increasing the fluid velocity on one side and decreasing it on the other.

As the mutual annihilation of positive and negative particles occurs, the advantage of positive particles continues to grow. Through countless repetitions of the process of producing positive and negative particles, more positive particles are generated while the antimatter is widely annihilated and disappears, ultimately forming the foundation of the material universe and evolving into the cosmic world we see today.

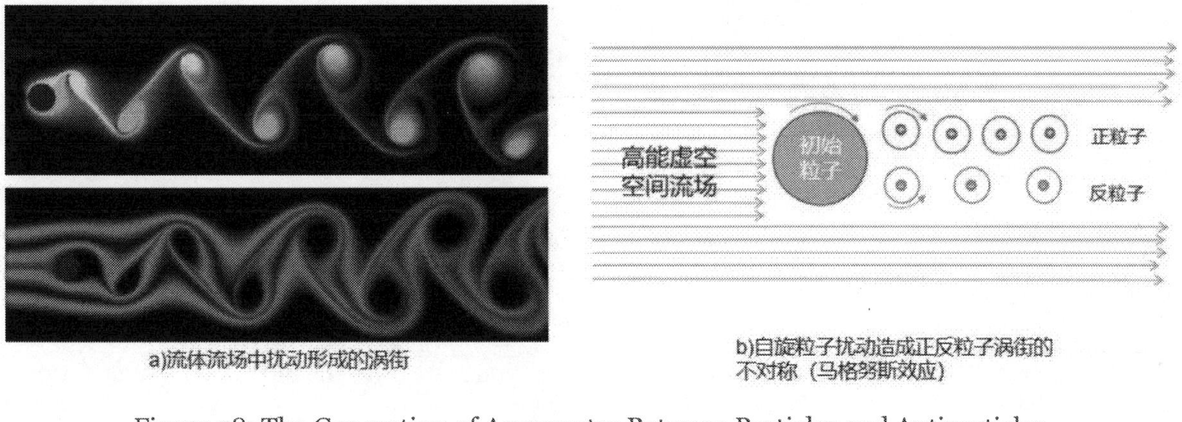

Figure 58: The Generation of Asymmetry Between Particles and Antiparticles

At the same time, the interaction between a particle matrix composed of a large number of particles and the spatial superfluid becomes more complex. For example, a large number of positive particles as a source of matrix disturbance will make the generated particle and antiparticle fields from the vacuum superfluid more complex, producing more high-energy composite particles and antiparticles. It will also significantly amplify the asymmetry in the number and energy between particles and antiparticles, as shown in Figure 8-42.

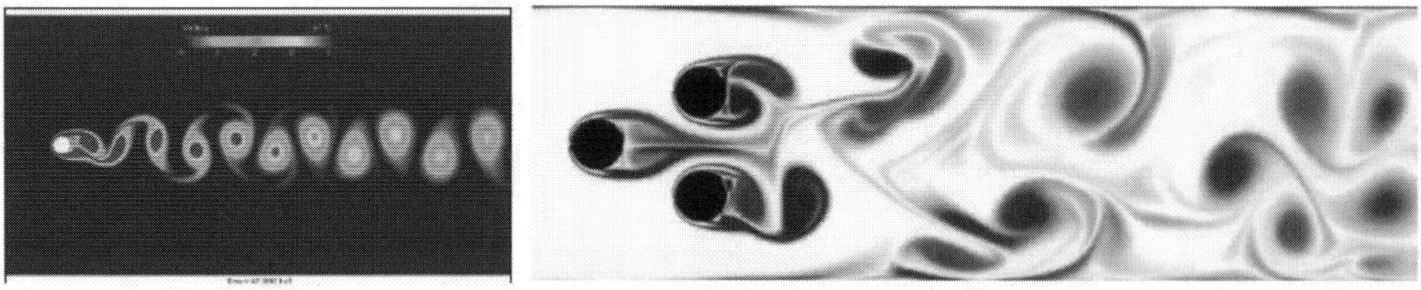

a)单粒子扰动　　b)多粒子矩阵扰动

Figure 8-42 The Interaction of Spatial Superfluid with Matrix Perturbation Sources

In the case of ultrastrong rays interacting with particles to produce particle-antiparticle pairs that we now understand,

Since the particles themselves are positive particles with positive spin, the energy or the number of positive and negative particle pairs produced will differ.

In contrast, if the acting particle itself is an antiparticle, then because it has antiparticle spin, the energy or the number of produced particle-antiparticle pairs will also differ, but in this case, the antiparticle will have greater energy or be more numerous.

(3) Fusion of Matter and Antimatter Particles

After the initial production of matter and antimatter particles, some particles did not annihilate or disappear, but instead, due to complex flow fields and particle interactions, merged with other particles to form new composite particles.

Therefore, these antiparticles are not actually disappearing, but are included within other particles. When a large amount of antimatter particles are contained within fundamental particles, it manifests the asymmetry between matter particles and antimatter particles.

(4) Differentiation between matter and antimatter particles

Antimatter particles and matter particles have completely opposite spin directions, which also result in opposite charge characteristics, while other fundamental properties remain consistent. Therefore, we can consider that matter and antimatter particles should be distinguished and defined by their spin directions.

Matter particles have positive spin, meaning they spin clockwise when observed from the S pole, while antimatter particles have opposite spin, meaning they spin counterclockwise when observed from the S pole. This is consistent with our understanding of charge spin, as charge itself is the spin-generated effect of a void space flow field.

In this sense, protons belong to the category of positive matter particles, while electrons belong to the category of antimatter particles. Therefore, our current material world actually contains a large amount of both positive and negative particles. What we are now puzzled about is why the corresponding positive particles of antimatter particles such as electrons, and the corresponding antimatter particles of positive particles such as protons, have disappeared.

This question can be understood from three aspects: the initial energy asymmetry of the vacuum space, the asymmetry in the secondary production of matter and antimatter, and the asymmetry in the fusion of matter and antimatter. It is precisely these three reasons that lead to the current situation

There exist antimatter particles such as electrons in the universe, but no corresponding positrons as matter particles, and there exist matter particles such as protons, but no corresponding antiprotons as antimatter particles.

The asymmetry of the initial perturbation energy in the cosmic vacuum superfluid leads to the asymmetry in the generation of initial particles and the selectivity in the subsequent particle production, resulting in the selectivity of particle evolution paths.

Based on the above four points, the emergence of matter in the universe is due to the presence of energy in the vacuum superfluid. The asymmetry between matter and antimatter in the universe arises from the asymmetry of the vortex energy in the vacuum superfluid.

Similar to the chaotic systems we are familiar with, the initial slight energy asymmetry in the cosmic vacuum superfluid undergoes multiple long-term iterations, leading to significant differences and resulting in the current distinction between antimatter and matter. This is an emergent outcome of the chaotic system of the cosmic vacuum superfluid. Our current material world is also an emergent result of the vacuum superfluid system.

At present, this property of the spatial flow field of the universe may still be influencing the difference between matter and antimatter, such as the vortex characteristics of the universe's spatial flow field itself, the particle characteristics of the universe, etc.

Modern astronomical observations have shown that there is an asymmetry in the spin direction of galaxies in the universe, clockwise and counterclockwise, which can be considered as a manifestation of the asymmetry in the vortex energy of the universe's vacuum space superfluid. Recently, István Szapudi from the University of Hawaii at Manoa and his collaborators proposed in a paper published in the Monthly Notices of the Royal Astronomical Society that adding a small amount of rotation to the cosmological model can perfectly resolve the anomaly in the measurement of the Hubble constant, all of which are consistent with our theoretical analysis.

8.7 The Principle of Charge Conservation

The law of charge conservation is a fundamental principle in physics, which describes that the total amount of electric charge in an isolated system remains constant. Charge cannot be created or destroyed; it can only be transferred from one object to another or from one part of the system to another, but the total charge of the entire system remains unchanged.

In most cases, we can consider the law of charge conservation to be universally applicable and an important foundation for our understanding and interpretation of electromagnetic phenomena. However, in certain special situations, such as some processes in particle physics, charge non-conservation may occur, for example, in the annihilation of positrons and electrons into photons, where the charge is not conserved. These phenomena appear to go beyond the scope of the charge conservation law.

In fact, according to our previous analysis, the charge of a particle is a mechanical effect caused by the vortex flow field of the superfluid in the surrounding vacuum space due to its spin. The magnitude of the charge depends on the energy of the particle's spin. Therefore, from this perspective, the total charge in a system, including the entire universe, may not be conserved.

The non-conservation of charge mainly has two reasons:
(1) Spin Energy Change
Because a particle's charge is the mechanical effect of the vortex flow field caused by the particle's spin in the surrounding vacuum space, the magnitude of the charge depends on the size of the particle's spin energy. In fact, spin energy is a form of kinetic energy from the particle's motion, and this energy can be converted into other types of kinetic energy.

For example, when a particle is subjected to an external force, its spin energy is reduced. Once this spin energy is reduced sufficiently, the charge effect it exhibits will also weaken, showing signs of charge non-conservation.

Even when particles break apart due to reasons such as charge bombardment, transforming into the most basic wave particles that move in the form of spatial waves, that is, when spin energy converts into wave kinetic energy, the charge they carry, or the charge effect they manifest, will also disappear in this scenario.

Protons and electrons, although appearing very stable, are constantly interacting with the surrounding vacuum space. If there is no energy replenishment, their energy will continuously decrease, eventually leading to decay. In such cases, the amount of electric charge they carry may "disappear." Moreover, due to this energy reduction, the electric charge carried by electrons or protons may be constantly decreasing, although the decrease is far too small to be observable and does not affect the stability of matter.
(2) Spin without charge effect

In complex physical interactions, chemical reactions, or particle reactions, the total charge of the system is conserved. However, some particle spins produce charge effects, while others do not. For example, electrons and protons, with sufficient mass and spin energy, can trigger a noticeable vortex field effect in the surrounding vacuum space, thereby generating charge. On the other hand, particles such as photons and waves, due to their small mass and spin energy, cannot cause sufficient vortex field effects in the vacuum space, and thus do not form observable or noticeable charges.

When some charged particles in a physical reaction transform into photons, waves, or other microscopic particles, even if the total spin energy and mass of all particles remain conserved and unchanged, the spin of photons and waves does not produce a charge effect. Therefore, in these physical reactions, the total charge may not be conserved.

For example, the annihilation of an electron-positron pair, which we mentioned earlier, is one manifestation of this phenomenon, even though the total electric charge remains conserved, the individual total positive and negative charges are no longer conserved.
(3) The sealed spin effect
In some particle reactions, charged particles may combine with other particles or even be encapsulated within larger particles, thereby sealing their own charge effects inside the new particle. As a result, the overall charge may not be conserved.

For example, in the proton-proton chain reaction in the Sun, two protons convert into a deuterium nucleus (D) through the weak interaction, releasing positrons and electron neutrinos, etc. In such reactions, processes like exciting new charges (electron-positron pairs) and encapsulating charges (new electrons being sealed within proton-electron pairs) occur. Although most of the time the overall charge remains conserved, there may be asymmetric reaction processes that lead to the system's total charge being conserved, but the overall charge exhibited by the system may not be conserved due to asymmetric encapsulation of charged particles.

It can be expected that, for the above reasons, the total amount of positive and negative charges will not be conserved in other physical reactions

The phenomenon of constant difference is only noticeable when the difference is large enough to be observed.
The cosmic vacuum space superfluid constantly generates new electrons, protons, etc., due to energy excitation, so the total

charge in the universe is actually always changing. However, this does not affect the general conservation of charge in our macroscopic experiments and material systems.

In summary, charge conservation essentially reflects the conservation of particle spin energy. However, in cases where some particles' spin does not produce a charge effect, or when spin energy is converted into other forms of kinetic energy, charge conservation may not hold.

8.8 The Composition and Evolution of Particles and the Universe [42]

(1) Solitons

(1) Formation and Classification of Solitons

Solitons are the most fundamental solitary waves formed by the excitation of a superfluid flow field in the vacuum space, also known as solitary waves or solitons. The most basic form of soliton is a vortex, which is also the foundation of all particles. The mass and energy accumulated within the vortex are the sources of mass and energy for all particles.

Solitons include several types such as vortices, vortex pairs, vortex matrices, vortex particles, and vortex rings.
A single vortex, due to its weak energy, will quickly dissolve through interactions with other surrounding vortices, and has poor stability. It only possesses an initial spin and partial stability, but not a fully stable particle form.

When two vortices form a closed vortex pair, stronger stability can be achieved.
When multiple vortices form closed loops in the flow field, creating a vortex matrix, it can be referred to as a vortexon. The overall energy and stability of the vortexon increase significantly, adding extra closure and rigidity, allowing it to exist for a long time.

When a large number of vortices form a unified closed flow field, a new three-dimensional stable structure called a vortex ring can be formed.
A vortex ring can be considered a type of vortex matrix, with each cross-section regarded as a vortex

A closed flow field formed by the sequential connection of numerous vortices, which includes not only the vortex flow in the cross-section but also a high-speed flow around the entire ring, thus forming a relatively stable closed body.

Vortex rings have high mass and energy, and possess characteristics of closure, stability, and basic rigidity similar to particles, and can be considered the most fundamental form of a particle-like structure.

The number of wavelets included in the overall structure varies depending on the size of the vortex ring. We can infer that the number of wavelets in a relatively stable, particle-like vortex is on the order of 10^{20} , while the number of wavelets in a vortex ring is on the order of 10^{30} .
(2) Possible soliton examples: photons
Photons are the most numerous and have the smallest mass among the macroscopic particles we can observe. Almost all matter emits light, continuously releasing a large number of photons. Based on this phenomenon, photons should be the most fundamental particle level with simple structure and low energy, such as vortices, vortex matrices, and vortex particles, etc., which are solitons.

At the same time, photons have high stability, long survival periods, and are very easy to interact with matter, such as being absorbed, so it can be inferred that photons are vortices produced by the excitation of the superfluid in empty space, forming closed flow fields.

The mechanism of photon production, as discussed earlier, is the continuous vortex street generated by the superfluid flow field of the vacuum space interacting with particles and other obstacles. Under slightly higher energy conditions, the vortices produced in the vortex street often form closed three-dimensional structures, which are referred to as vortex clusters composed of multiple vortices.

Because the superfluid flow field of the vacuum space interacts continuously with various particles, generating numerous vortex streets and composite vortex fields, it can produce a large number of vortices, which is consistent with the vast number of photons.

The high-speed vortex energy also ensures good stability and closure, forming the basic particle structure. The energy and mass of these vortices are much greater than those of the most fundamental vortices, allowing them to transmit over long distances and for extended periods.

Because of its simple structure, relatively small volume, and lack of complex internal structure and high rigidity,

it is very easily absorbed by the flow fields of other high-energy particles and also very easily excited by the flow fields of other particles.

(3) Possible soliton instance: neutrino

The number of neutrinos is also extremely large, with a large amount being produced every moment, and their mass is very small, so neutrinos should also be one of the most fundamental solitons.

At the same time, neutrinos have a subtle mass that we can observe, indicating their energy is greater than that of photons. Neutrinos rarely interact with matter and have extremely strong penetrating power, so they are also solitons with very small volume and mass, and without the magnetic and electric field effects produced by vortex flow fields.

Based on these points, we speculate that neutrinos are vortex rings within solitons, composed of numerous vortices, with greater energy and mass than photons. However, as vortex rings, they do not form sufficient vacuum space flow fields around them, so they do not produce corresponding electric and magnetic field effects. Their much larger volume and mass also make it difficult for them to be easily absorbed by matter, resulting in difficulty in interacting with substances.

Based on our previous analysis of solitons in the composition and evolution of particles, it can be inferred that the number of bosons contained in neutrinos is on the order of 10^{30} , and the number of bosons contained in photons increases with their energy. Their mass is at least 10^{10} times that of the photon's mass, as shown in Figure 8-3.

The upper limit of neutrino mass observed in modern physics experiments is less than 0.45 eV, and the upper limit of photon mass is less than $8.4 \times 10^{-17}\mathrm{eV}$, which is basically consistent with the prediction.

(4) Proper Mass and Apparent Mass

It is important to note that the mass of the particle we are discussing is the intrinsic mass, which is the total mass consisting of the number of bosons multiplied by the mass of each boson, that is:

$$\mathrm{M} = \mathrm{nm_0}$$

Equation 8.14 is the eigenmass formula of a particle, where M is the eigenmass of the particle, n is the number of wave particles contained in the particle, and $\mathrm{m_0}$ is the mass of the wave particle.

When macroscopic particles experience resistance due to their interaction with the superfluid vacuum space during motion, if we do not consider this effect, we would also attribute this resistance to the particle's mass, along with other similar effects. This mass is referred to as the apparent mass of the particle

$$\mathrm{M_c} = \mathrm{nm_0} + \sum \mathrm{m_f}$$

Equation 8.15 is the apparent mass formula of a particle, where M_c is the apparent mass of the particle, n is the number of wave particles contained in the particle, $\mathrm{m_0}$ is the mass of a wave particle, and $\mathrm{m_f}$ is the additional mass exhibited by the particle due to various neglected and unconsidered interaction scenarios.

The mass we observed in our experiment is often apparent mass.

(2) Initial particles

When the superfluid vacuum space is further energized, the energy of the vortex ring soliton increases, forming a higher energy vortex ball.

The vortex ball possesses an internal and external circulation flow field. The closed ring-shaped flow field in the vortex ring evolves into the high-speed self-rotation of the vortex ball's flow field wall, while the vortex flow fields in each cross-section of the vortex ring evolve into the internal and external circulation flow fields of the vortex ball's flow field wall. This internal and external circulation flow field allows the vortex ball to have a flow field wall similar to a particle wall. The increase in flow field walls enhances the vortex ball's closure, rigidity, and the stability of particles, making it the initial particle.

The number of wave particles included in a vortex ball varies depending on its energy level, and we can estimate that the number is on the order of 10^{35} .

(3) Microparticles

(1) Formation and Classification of Micro Particles

The initial particles, vortex balls, have no core, so their mass and energy are limited. When the superfluid vacuum space is further excited with energy, during the process of complex and dense turbulent quantum vortices gathering to form vortex balls, some vortex rings or small vortex balls will be incorporated, becoming the core of the vortex ball. At this point, the

The vortex ball becomes a composite particle: the vortex ball with a nucleus.
The addition of a nuclear particle to the vortex ball enhances its flow field, making it more powerful and stable. Its mass and energy increase significantly, thereby achieving greater stability and becoming a rigid particle capable of producing strong interactions.

Relative vortex rings as the nucleus of the vortex ball, the vortex ball with a nucleus has greater energy, mass, spin, etc., and can form more distinct flow field mechanical effects such as electromagnetic fields. It can also possess charge properties.

At this point, the nuclear vortex sphere officially becomes a stable micro-particle.
Because of the quantum properties of quasiparticles in the superfluid vacuum space, all the flow fields and within the particles of micro-particles exhibit dense and complex turbulent vortical motion, forming complex fluid effects. In the flow field of the particle wall and within the particle body of micro-particles, not only do the basic quasiparticles exist, but also elements such as vortices formed by large numbers of quasiparticles, vortex matrices, and even initial particles, which further increase the complexity of micro-particle structures and enhance their energy strength.

The number of wave particles included in the overall composition varies depending on the energy of the micro-particles, and we can estimate that the number of included wave particles is on the order of 10^{40} .
(2) Possible micro-particle example: electron
The mass of an electron reaches 511 keV, which is a million times that of a neutrino, so it has already surpassed the soliton level and reached the level of initial particles or even above that of micro-particles.

At the same time, since the electron's spin is $1/2$, a half-integer, the electron should have an internal structure, that is, a nuclear particle. This is because the spin speed of the nuclear particle and the overall flow field of the particle differ, leading to a half-integer spin. The initial particle's vortex ball particle is non-nuclear, and its spin should be an integer. Therefore, the electron should at least be considered a micro-particle: that is, a nuclear vortex ball particle.

Electrons are the lightest particles among fermions, widely existing in nature, and are the basic units that form the material world. They have very high stability, and their structure should not be overly complex. We can speculate that

Electrons are a type of micro-particle: the single-nucleus vortex sphere particle.
(4) Macroscopic particles
(1) Formation of macroscopic particles
When the excitation energy of the superfluid vacuum space is further increased, or when micro-particles interact and further merge to form new, more powerful particles, these particles will have more nuclear particles, forming a nuclear particle matrix. This will further significantly enhance the properties of the particles such as mass, energy, spin, and volume, making them the majority of macroscopic particles we can observe.

Most of the particles observed in our particle physics, such as hadrons and quarks, belong to this category of macroscopic particles.
(2) Classification of Macroscopic Particles
It is worth noting that in the nucleus of a composite particle, the matrix formed by multiple particles within the nucleus merely undergoes a change in form, which can cause changes in the flow field and thus affect the force effects of the flow field. This is similar to the principle of the Halbach magnet array, where merely adjusting the direction of the magnets can lead to significant differences in the overall magnetic force of the array. In such cases, the composition of the composite particle remains unchanged, but its displayed properties such as mass, energy, isospin, and strangeness will all change, leading to more forms or types of composite particles. This is also one of the reasons for the observed octet classification of baryons, mesons, etc., in particle physics, as shown in Figure 8-43.

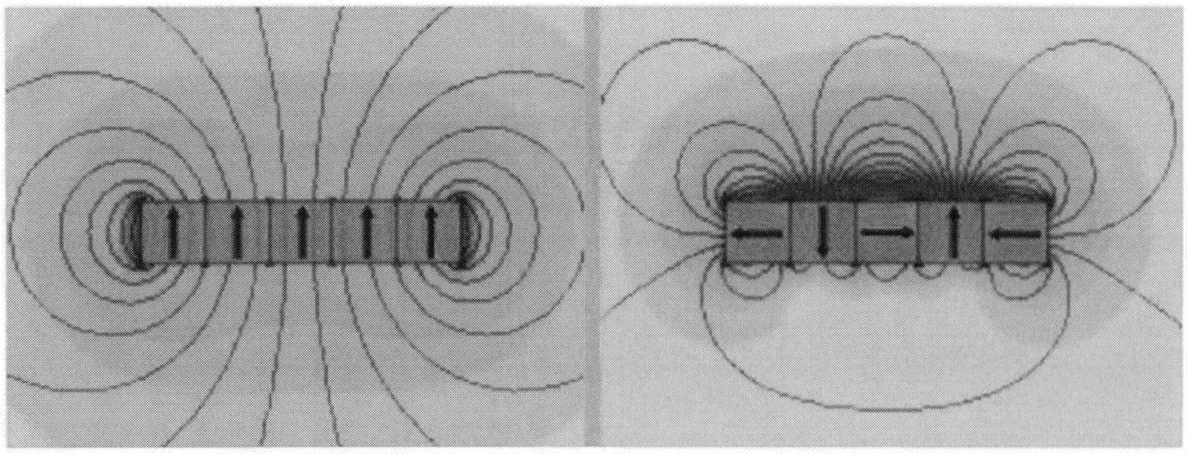

Figure 59: a) Magnetic field distribution of magnets in sequential arrangement

b) Magnetic field distribution of magnets in Halbach array

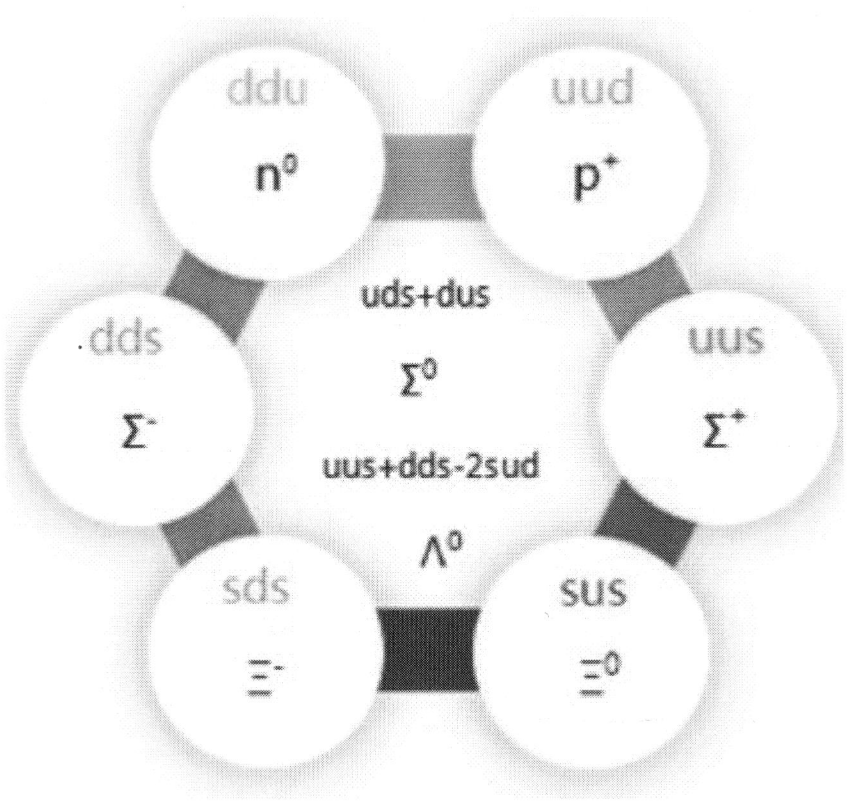

Figure 60: c) Octet classification of particles

Figure 8-43 Magnetic field configurations of different magnet arrays and octet classification of particles

(3) Spin Quantum Numbers of Macroscopic Particles
In modern physics, we can observe that macroscopic particles all have the characteristics of spin quantum numbers. In fact, according to our previous discussions on particles, macroscopic particles as well as earlier microscopic particles, initial particles, and even solitons, all possess quantum properties of spin.

It is noted that for relatively simple particles without internal composite particle structure, such as vortices and nucleus-free vortex particles, their spin returns to the initial state after one full rotation, so their spin quantum number is generally an integer like 1.

However, for nuclear vortex ball particles, even those with nuclear matrices, which are composite particles, the whole consists of multiple parts formed by the fusion of multiple particles. Each particle has its own independent spin, which may cause the spin of the overall particle's flow field and the nuclear particle, or the nuclear particle matrix flow field, to be out of sync. For example, if the nuclear particle rotates once, but the overall particle only rotates half a turn, in such a case, the overall particle needs to rotate twice to return to its initial state. Therefore, the spin quantum number of this type of particle is $1/2$.

Other composite particles that contain nuclear particles or nuclear particle matrices often also exhibit this situation, so their spin quantum numbers will be half-integers.

The synchronization of the spin of the particle field and the spin period of the nuclear particle field is the reason why we observe different spin quantum numbers for macroscopic particles, such as integers and half-integers.

The number of wave particles included in a macroscopic particle varies depending on its energy level. We can infer that the number of included wave particles is on the order of 10^{45} .
(4) Macroscopic particle examples: protons, neutrons, etc.
The mass of a proton is approximately 1836 times that of an electron, making it a macroscopic particle far more complex than an electron. It is a particle with a nuclear matrix, and even the nuclear particles themselves should be composite particles to form a sufficiently large mass and stability.

Neutrons are similar to protons, both belonging to macroscopic particles with particle matrices. Most other particles with mass greater than an electron are generally considered macroscopic particles.

(5) Matter Particles

(1) Spin and Charge Effects

Macroscopic particles can form particle complexes due to electromagnetic force, strong force, and weak force, and further form complex material structures, resulting in familiar material particles such as atoms and molecules. At this moment, due to the complexity of their internal structure, the overall spin of these particles will be greatly reduced, no longer exhibiting significant charge effects caused by their own spin.

(2) Gravitational Effects

The interactions and stacking between material particles form various different material configurations, generating diverse overall properties, which serve as the foundation for the formation of the real material world.

In this process, gravity will exhibit an increasingly significant effect.
(3) New Viscous Fluid Effects
A large number of material particles form new fluids such as liquids and gases, which also exhibit complex vortices and turbulence, similar to the superfluid in the void space. At this moment, these material particles behave like new entities on a macroscopic level, but unlike individual wave particles that are smooth on the surface and only subject to inertia, the fluid in the void space has a superfluid effect with extremely low viscosity. Therefore, when these massive material particles form a fluid, their viscosity increases, leading to various phenomena typically associated with viscous fluids, such as laminar flow.

The number of wave particles included in the overall composition of material particles varies depending on their mass. We can estimate that the number of included wave particles is on the order of $10^{45} - 10^{50}$.
(4) Matter particle examples: atoms, molecules
All larger-scale particles composed of fundamental particles such as macroscopic particles, microscopic particles, initial particles, solitons, etc., such as atoms and molecules,

(6) Planet particles

(1) Spin and charge effects

A large amount of matter particles accumulate to form macroscopic matter. As the mass of matter increases and its spin energy decreases, the forces between matters gradually become dominated by gravity. When matter accumulates to the maximum, it forms the planets we are familiar with in the macroscopic universe.

As a complete entity, a planet also has spin motion, so it also possesses "charge." Its spin can create a drag effect on the surrounding vacuum space superfluid, thus forming electric and magnetic fields.

The current planet is like a giant particle. In macroscopic conditions, we tend to ignore and find it difficult to observe the electric and magnetic field effects, mainly considering the gravitational interaction between the planet and other surrounding "planet particles."

The magnetic fields we observe for planets are mostly the result of magnetic effects generated by the movement or rotation of the planet's core, rather than the planet's own rotation. For example, Earth's magnetic field is caused by the motion of its core. Since the rotational speed of such macroscopic planets is very slow, they are unable to form sufficiently strong vortex field effects in the surrounding empty space, resulting in observable macroscopic electric and magnetic field effects.

(2) Gravitational Effects

The force between stars mainly manifests as gravitational effects, as discussed in the introduction to spatial hydrodynamics earlier. Gravitational effects are longitudinal waves formed by the thermal motion of matter exciting the surrounding vacuum space superfluid, leading to changes in the velocity of the superfluid flow field in the vacuum space, and thus exhibiting a macroscopic flow field mechanical effect, similar to electric and magnetic fields, which all conform to Bernoulli's principle of fluid dynamics.

(3) Stellar Fluid

At the macrocosmic level, stars are like "drops" on a new level, forming a kind of "fluid" on the new cosmic level under the influence of gravity and range gravitational fields. Under the common action of the superfluid in the void space, they are excited by energy to produce vortex effects similar to fluid flow fields, resulting in countless vortex galaxy systems.

Therefore, galaxies can be considered as vortex solitons formed by a large number of stars as a fluid, excited by energy, similar to the basic vortex particles in the void superfluid, as shown in Figure 8-44.

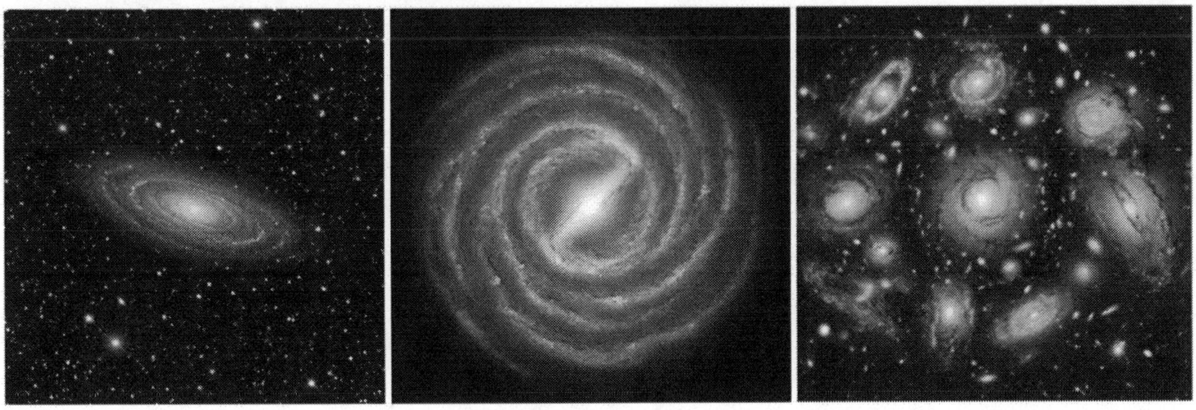

Figure 61: Figure 8-44 Macroscopic "fluid" formed by countless stars

The number of material particles included in a celestial body varies depending on the mass of the star particle, and we can estimate that the number of included material particles is on the order of $10^{40} - 10^{60}$.

(7) Galaxy Particle

As mentioned above, galaxies such as the Milky Way can be considered as fluid formed by countless stellar particles, which are coherent structures of waves excited by energy. These waves have a certain degree of closure and stability, and even rigidity, similar to a special kind of particle.

Therefore, galaxies can be considered as macroscopic entities in the universe, formed by stellar fluids as new "solitary waves" or "solitons," which are essentially the most basic solitons: vortices, similar to the fundamental solitons formed when a superfluid vacuum space is excited, producing a large number of wave particles.

This is a new level of super-macro galactic vortex particle, as shown in Figure 8-45(a).
And galaxies such as the Milky Way, as super-macroscopic galaxy particles, can form galaxy fluids when a large number of galaxy particles come together. Furthermore, galaxy fluids can produce more complex fluid effects under energy excitation and mechanical forces, forming more extreme macroscopic structures. The large-scale structures of the universe observed by astronomy prove this point, as shown in Figure 8 - 45(b).

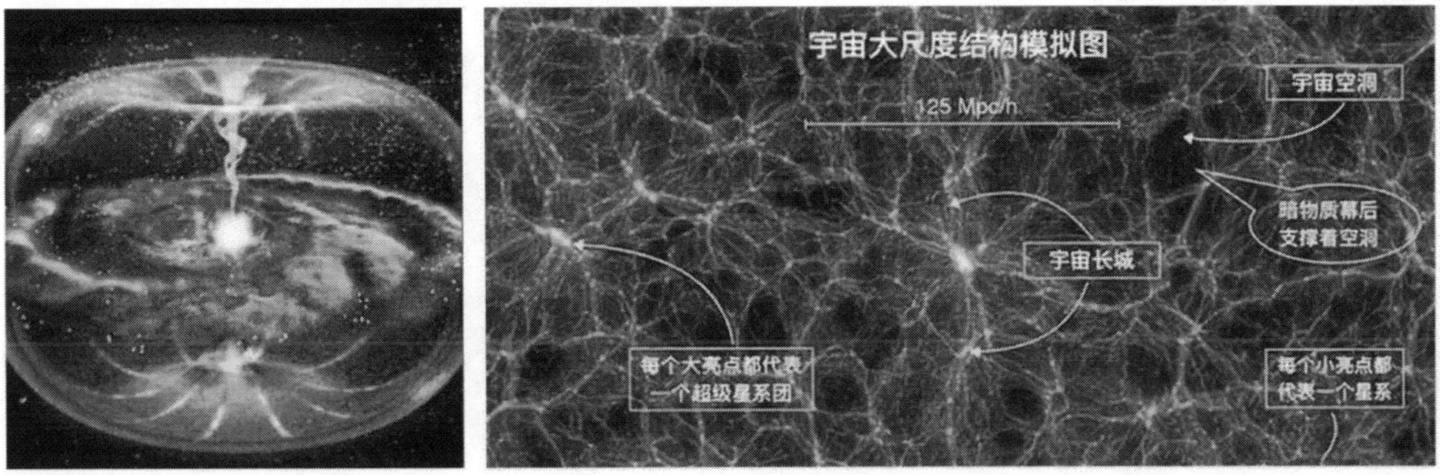

a）星系粒子

b）无数星系粒子形成的宇宙尺度“流体”

Figure 8 - 45 Giant flow field structures formed by galaxy fluids in the cosmic scale

The number of planetary particles included in a galaxy varies depending on the mass of the particles, and we can estimate that the number is on the order of $10^{10} - 10^{20}$.
(8) Cosmic Particles
(1) Cosmic Super Vortex Solitons
As we discussed at the beginning, when the vacuum space superfluid is excited, it forms vortices solitons, which are the most basic type of particles. As the excitation energy increases, the energy of the vortex solitons increases, and their forms are further excited to form vortex rings, vortex balls, etc., which are types of particles or fundamental particles.

On our cosmic level, the super macroscopic fluid composed of countless stars, when excited, forms vortex solitons, which are galaxies. Galaxies can be considered as the most fundamental vortex solitons, or particle-like structures, formed by the fluid of stars.

When the excitation energy is further increased, one can imagine that galactic vortex solitons may form structures similar to vortex rings and vortex balls, leading to general vortex ring galaxies, or even vortex ball galaxy systems.

Vortex ring or vortex ball galaxy systems have energy far exceeding that of the most basic vortex galaxies. By observing the macroscopic structures on the vast scale of the universe composed of countless galaxies, we can consider the universe itself as a super giant vortex ball particle formed by countless galaxy particles and flow fields: the cosmic particle, as shown in Figure 8-46.

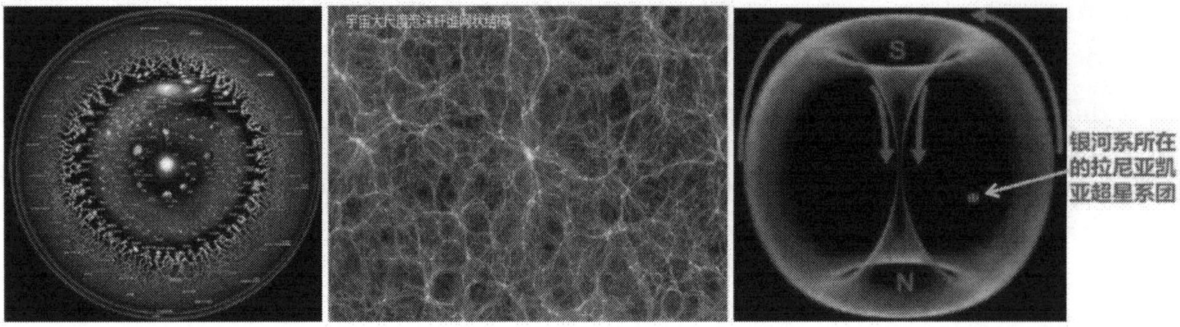

a) 可观测宇宙　　b) 宇宙大尺度丝状结构　　c) 宇宙粒子涡球结构

Figure 62: Figure 8-46: The super giant cosmic particle formed based on the galaxy flow field

(2) Spin of the cosmic particle

As a particle formed by the excitation of a flow field, cosmic particles must also have spin. As previously discussed, the superfluid vacuum space in the universe, when excited by the initial energy, forms vortices, generates vortex ring particles and larger vortex rings, vortex ball particles, and so on, ultimately leading to the formation of countless macroscopic particles and the real material world. This indicates that the superfluid vacuum space is filled with immense energy excitation, and this energy may originate from the spin of the universe itself.

The asymmetry between matter and antimatter particles in the universe, where positive particles greatly outnumber negative ones, and the material world is almost entirely composed of positive particles, may originate from the spin of the universe itself. This spin causes the vortices formed by the superfluid vacuum space to inherently possess asymmetry, with positive vortices being much larger than negative ones, thereby generating more and larger energy positive vortex particles. Through continuous accumulation and chain reactions, this ultimately leads to the result of asymmetry between positive and negative particles.

The countless filamentous structures formed by galaxy clusters on the large scale of the universe we observe may be a microscopic manifestation of the turbulent flow field generated by the internal fluid dynamics of cosmic particles.

Considerations about the overall structure of the universe

As we analyzed earlier, there exists throughout the universe a superfluid vacuum space and a galaxy fluid composed of countless galaxy particles, which are constantly in motion and possess enormous energy.

All particles in the universe are formed under the pressure of the vacuum superfluid, along with the vacuum space

Supercurrents interact to form the current stable state. If there were no walls surrounding the universe, these fluids would spill out into the vacuum space beyond the universe under the influence of huge energy and pressure. The pressure of superfluids in the empty space of the universe would continuously and rapidly decrease, which would directly affect the stability of all matter particles.

If the pressure of superfluids in the empty space drops to a certain limit or completely disappears, all particles that lose the effect of superfluid pressure would quickly collapse and revert back into wave particles in the empty space. The entire material world of the universe would then vanish.

But none of this has happened. According to astronomical observations, in human time terms, the universe's age may have already exceeded several billion years, and there are no signs of any weakening in the pressure of the void space or the stability of matter particles. Therefore, there may be a powerful edge barrier outside the universe, similar to a solid coastline beyond the void, or a barrier like a closed wall, preventing the superfluid of the void space from leaking out.

In other words, the fact that the universe can remain basically stable in its current state suggests that there should be a wall-like structure existing around the universe, ensuring the basic stability of the internal fluids.

However, due to the超强 permeability of the void space superfluid, if this wall is a material existence, its mass density and structural strength must be far beyond what we observe in macroscopic matter, in order to block the outflow of the superfluid.

But if that were the case, this wall surrounding the universe would produce significant mechanical effects, causing noticeable changes in the mechanical observations of all superfluids and galaxies within the universe, which contradicts the physical observation facts.

Therefore, this kind of wall outside the universe should be similar to a vortex ball particle, a field wall, which through high-speed spinning and vortex field allows the universe to become a super giant vortex ball particle, possessing the characteristics of closure, stability, and rigidity similar to particles. All the vacuum space superfluid and galaxy fluid are included inside, which is also one of the reasons we consider the universe as a super giant vortex ball particle.

(4) Regarding dark matter in the universe

As we discussed earlier in the introduction to spatial wave dynamics, the academic community proposed the existence of dark matter in the universe based on two reasons. One is that from observations of the motion of galaxies and large galaxy clusters, their mass and gravitational relationships do not match the actual motion, and from observations of astronomical phenomena such as gravitational lensing and large-scale structure of the universe, it is inferred that there should be a large-scale mass system that we have not yet observed involved.

Now we can speculate that this dark matter in the universe is precisely the superfluid vacuum space we have been discussing. Of course, it may also include many other vortex structures, such as vortices, vortex matrices, vortex rings, and vortex balls, as well as solitons, micro-particles, and fundamental particles that we cannot observe or have not yet observed.

The second reason is that, as discussed in spatial dynamics, the nature of gravity is not yet clearly understood theoretically. Gravity is not a direct mechanical interaction between substances, but rather an effect field formed by the interaction between matter and empty space: therefore, in addition to the gravitational interaction between substances (such as between stars), there is also a gravitational field existing in the region around matter. The academic community's perception of the discrepancy in the relationship between galaxy mass, gravity, and motion is precisely because this regional gravitational field effect has not been considered.

It is precisely for these two reasons that we have the judgment regarding the existence of dark matter in the universe and its mass and energy characteristics.

The number of galaxy particles included in the overall system varies depending on the mass differences of the cosmic particles. We can infer that the number of galaxy-like particles included is on the order of 10^{25} , which is consistent with the number of quasiparticles contained in the initial vortex balls formed during the creation of the vacuum superfluid.

8.9 Cosmic Time Hierarchy

(1) Locality of Time

We have discussed in the basic principles of spatial relativity 3: the principle of time objectivity (Chapter 3.3) that time

Space is local, according to the time locality equation (Equation 3.7):

$$\Delta t_v = \frac{\Delta t_0}{\sqrt{1 - \frac{V^2}{C^2}}}$$

When both observers A and B are in the same vacuum space region Z, and assuming that the vacuum space region Z itself is not moving, the time flow rate of observer A moving at velocity V relative to the vacuum space, denoted as Δt_v , and the time flow rate of observer B at rest relative to the vacuum space, denoted as Δt_0 , follow the Lorentz transformation, that is, $\Delta t_v > \Delta t_0$. A's time flow rate is faster; for B, it might be just one minute, while for A, time could have already passed two minutes, or even more.

The ratio of the time flow rate between the two is related to the speed of A's motion relative to the void space Z. The faster the speed V, the faster the time flow rate of A compared to B, as shown in Figure 8-47.

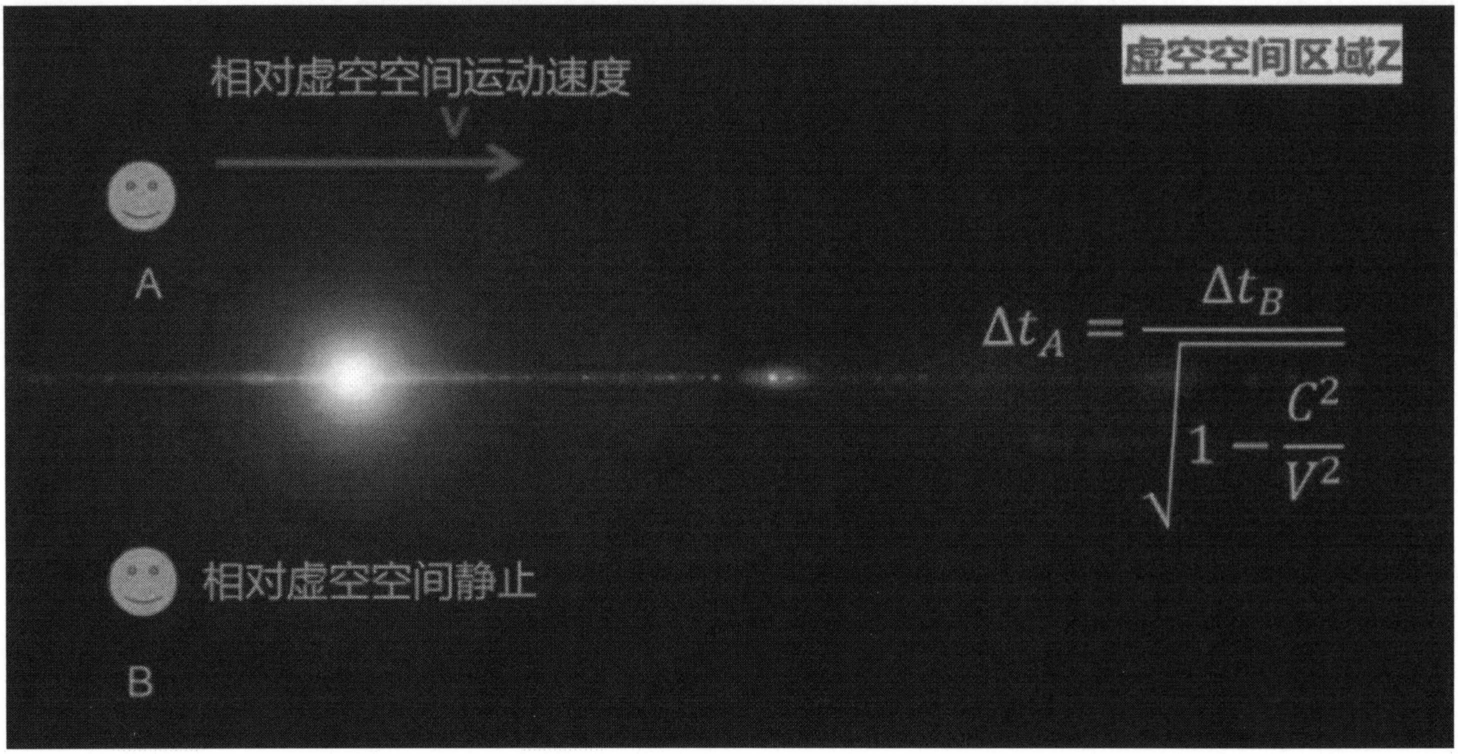

Figure 8-47 Different thermal motion velocities lead to different time flow rates

The root of this difference in the speed of time flow lies in the different thermal motions of A and B, that is, the motion of A and B relative to the void space can be considered as their thermal motion in the void space. The faster the thermal motion, the faster the time flow; the slower the thermal motion, the slower the time flow.

Here, it is assumed that the vacuum space region Z itself is stationary, that is, the vacuum space itself has no thermal motion. When the vacuum space itself also has motion, the hierarchy of time must be considered.

(2) Hierarchies of Time

Let observers A and B be located in two void spaces Z_A and Z_B, respectively, and the relative motion velocity between Z_A and Z_B be V. At this time, even if A and B are stationary relative to their own void spaces, there is still a difference in the speed of time flow between them, because the void spaces Z_A and Z_B themselves are in motion, that is, they have a fundamental thermal motion. Therefore, even if A and B are relatively stationary within their own void spaces, they still possess a fundamental thermal motion, resulting in different time flow rates.

To determine the real motion speed of Z_A and Z_B, the time flow rate of A and B must be considered within the overall background vacuum space Z_0 where both the vacuum space Z_A and Z_B are situated.

According to the basic principles of spatial relativity 5: the new relativity principle (Chapter 3.5), the motion between objects in macroscopic physical phenomena can be approximately simplified as relative motion. However, all objects have absolute motion relative to the space they are in. That is, motion is absolute, especially when it involves light, electric fields, and magnetic fields, where this absolute motion must be taken into account.

Similarly, when two regions of empty space exist in relative motion, there is also absolute motion between them. Since light travels at a constant speed C relative to the uniform and flat empty space it propagates through, if the empty space itself has a motion speed, the total speed of light propagation will also change. A and B can measure the speed of light by referring to the background empty space Z_0, thereby determining the true motion of their respective empty spaces.

Assume the background vacuum space is stationary, and let the relative velocity of ① with respect to ② be ③, and the relative velocity of ④ with respect to ⑤ be ⑥. Then:

$$V_A + V_B = V$$

Let the time flow rate in Z_0 be Δt_0
Then, even if A、 B is stationary relative to its own void space Z_A and Z_B, they still have different time flow rates:

$$\Delta t_A = \frac{\Delta t_0}{\sqrt{1 - \frac{V_A^2}{C^2}}}$$

$$\Delta t_B = \frac{\Delta t_0}{\sqrt{1 - \frac{V_B^2}{C^2}}}$$

That is, in the large void space region Z_0 , the fundamental time flow rate differs among void space regions Z_A and Z_B , which is referred to as the fundamental time flow rate of the void space itself, as shown in Figure 8-48.

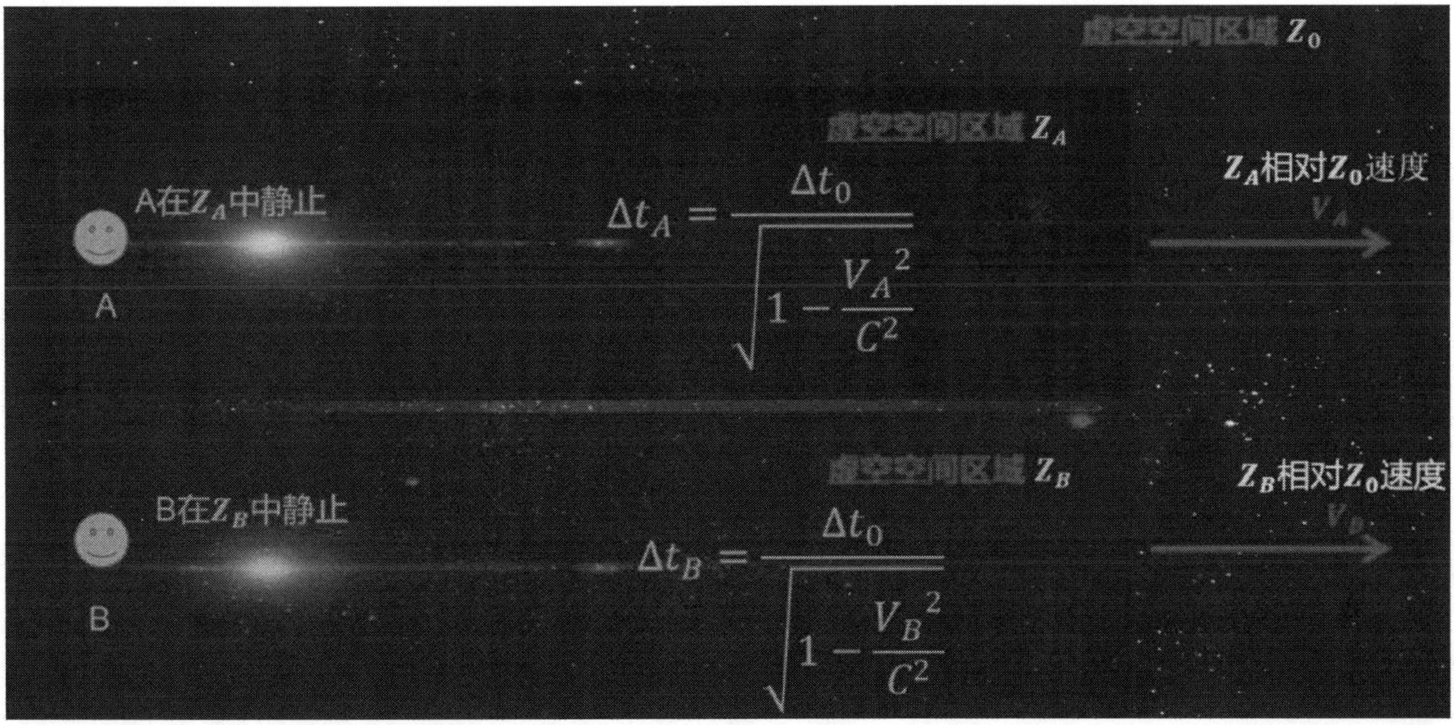

Figure 8-48 Hierarchical Effect of Time Flow Rate

This is the hierarchy of time.
When observers in Z_A and Z_B are moving relative to their own empty space regions, their respective time flow rates are superimposed on the base time flow rate of the space.

(3) Types of Thermal Motion

As analyzed above, different thermal motions of objects lead to different time flow rates, and all forms of material motion can be considered as thermal motion, which are all types of thermal motion.

The motion of the void space itself is a fundamental thermal motion that affects the basic time flow rate of all objects within this void space.

At the same time, the orbital motion of planets, their rotation, and all other movements of matter, such as vibrations, waves, and flows of fluids, are forms of thermal motion, and they all affect the flow of time for matter.

All objects on Earth that move along with the Earth, although the empty space on the Earth's surface is in a stationary state relative to the Earth and surface matter due to the Earth's dragging effect, have their own time flow rate faster than that of the surrounding stellar empty space, because the thermal motion on the Earth's surface is much higher than that of the surrounding cosmic empty space. Therefore, without considering other relative motions of empty space, the time flow rate on the Earth's surface itself will be faster than that in the surrounding cosmic empty space. This is the relationship between the time flow rate and the fluctuation frequency of empty space, as discussed in the basic principle 3 of spatial relativity: the objectivity principle of time (section 3.3), represented by equation (3.5):

$$\Delta t \propto f$$

Among them: ρ is the density of the void space, f is the fluctuation frequency of the void space f
The higher the fluctuation frequency and motion energy of the vacuum space, the higher the overall thermal motion, so the speed of time flow will be faster. Therefore, equation 3.5 can also be written as:

$$\Delta t \propto T$$

Where T is the overall temperature formed by the thermal motion of the vacuum space. On the cosmic scale, this overall temperature of the vacuum space is manifested as the 3 K cosmic background radiation, meaning the overall thermal motion temperature of the vacuum space in the universe is around 3 K.

The thermal motion of the vacuum space can be considered absolute and not arbitrarily relative. For example, two separate gas clouds, one with lower thermal motion and the other with higher thermal motion, this thermal motion itself is absolute and cannot be arbitrarily assumed as relative motion between them. Because different thermal motions will result in different temperatures, leading to different time flow effects.

In summary, the time flow rate of an object depends on the basic thermal motion of the vacuum space it is in and the object's relative motion.

The thermal motion of the void space consists of two parts, the locality and hierarchical characteristics of time flow rate, which essentially reflect the locality and hierarchical characteristics of the void space itself.

(4) Time Hierarchy of the Universe

(1) Time flow at the microscopic level

We observe microscopic particles, such as electrons and photons, which, compared to our macroscopic level, include other particles we have yet to observe, such as bosons, and all exhibit extremely intense thermal motion, including electron spin that may even greatly exceed the speed of light. This makes the flow of time for these microscopic particles extremely fast relative to our macroscopic world. Perhaps one second in our macroscopic world is already an eternity in the world of microscopic particles.

In this situation, if there were life or consciousness in such a microscopic particle world, for example, on an electron, although the electron moves very vigorously relative to us, if one were to move in sync with it, one would not feel this speed of movement. It is similar to how we on Earth, although moving at high speed along with the Earth's rotation, do not feel it at all. If there were life on an electron, they would also experience the same phenomenon.

Because the dragging of the electron on the surrounding vacuum space, the life forms on it are almost stationary relative to the electron and also almost stationary relative to the surrounding vacuum space, and their metabolic rates will also synchronize with the thermal motion in the microscopic world, the time flow rate they experience is greatly reduced, even similar to the time flow rate we perceive on the macroscopic world.

Therefore, although we feel their time passes very quickly, we experience only a day while they have lived for countless years, time flows normally for them. This is the effect of different time scales.

(2) Time flow at the macro level

We humans observe (perceive) that the Earth's speed is very slow, with almost no relative motion to us. Including our observation of the entire Milky Way's movement, it also seems very slow. In reality, both the Earth and the Milky Way are moving

Moving at a very high speed, we do not feel this speed because we are completely synchronized with the Earth and the surrounding empty space.

Just as we do not perceive this speed in the microscopic world.
But, as we analyzed earlier, if the universe is considered as a single particle, it itself would possess a powerful speed. All galaxies and superclusters within the universe are like "basic particles" in the high-energy flow field of this particle. From a macroscopic perspective of the universe as a whole, all galaxies are moving and flowing at high speeds, even seemingly "faster than light." For life within galaxies, one second experienced by an observer of the universe as a whole might have already passed many years, or even an eternity.

Therefore, the thermal motion speed of matter in different levels of the universe, such as the microscopic and macroscopic, differs, leading to different rates of time flow and different perceptions of time's speed. The differing metabolic rates of life forms at various levels also result in different sensations of time.

The ratio of time flow speed between different levels of macroscopic and microscopic worlds may exceed hundreds, thousands, or even millions and billions of times. The story of "a day in heaven, a thousand years on earth" might be happening at any moment.

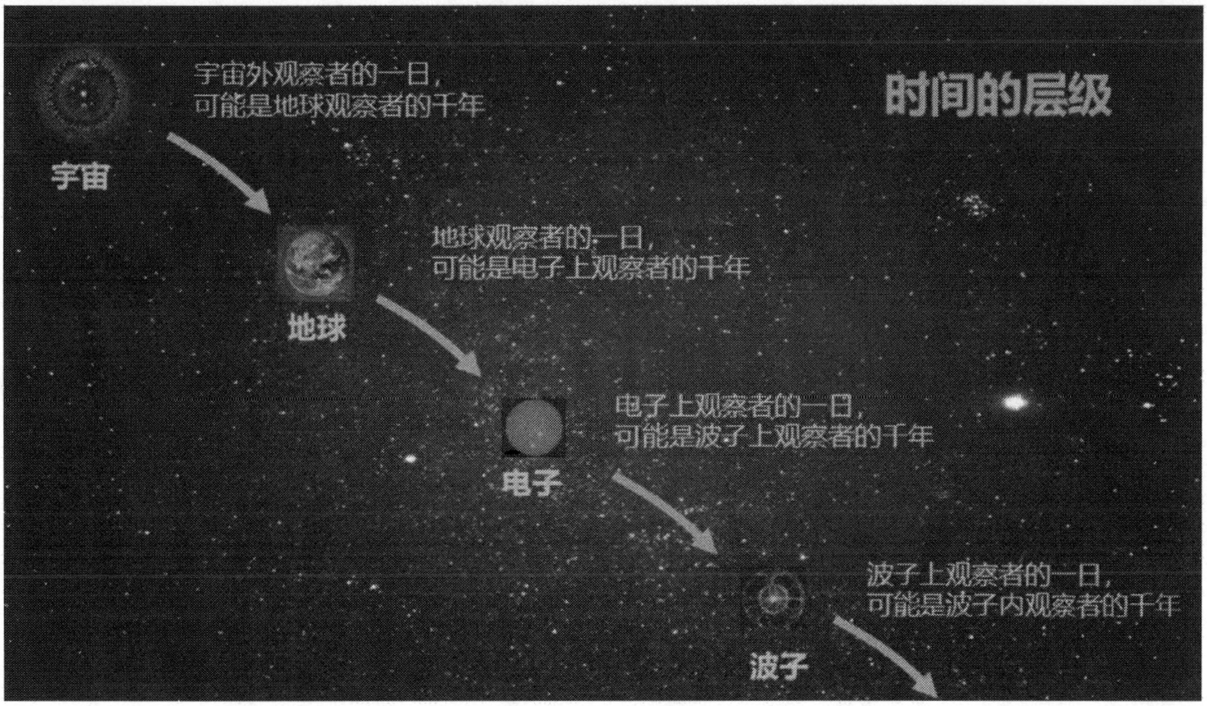

Figure 63: Figure 8-49 Hierarchical Structure of Time

At the same time, within the same level, such as inside the universe, different star systems and planetary regions may also exhibit significant differences in thermal motion, leading to substantial variations in the speed of time flow.

Due to the locality and hierarchical nature of time, differences in the speed of time flow are普遍存在.
For more information on the essence of time and other related topics, please refer to "Extended Reading 12.3: General Introduction to the Principles of Time".

8.10 Principles of Electromagnetic Wave Generation and Propagation [46][47]

(1) Principle of Electromagnetic Wave Formation

In the previous chapter on the general introduction to spatial quantum dynamics, we discussed that the wave-particle duality of all microscopic particles is a phenomenon formed by the interaction between the particle's linear motion and the wave-like nature of the vacuum space superfluid.

The superfluid of empty space typically exhibits longitudinal wave forms due to its almost negligible viscosity. Photons and longitudinal waves of the superfluid of empty space move together at the speed of light. However, due to interactions with the surrounding superfluid of empty space, the photon's path is not strictly straight but deviates around its direction of motion, which is referred to as vibration. This vibration includes both transverse and longitudinal components, resulting in an overall elliptical trajectory.

Therefore, the actual path of a photon is an elliptical spiral, as shown in Figure 8-50.
When the lengths of the major and minor axes of an elliptical trajectory are nearly equal, the vibration trajectory becomes circular, and when the minor axis is extremely small compared to the major axis, the vibration trajectory becomes linear. These different vibration trajectories are the polarization of photons or light waves: elliptical polarization, circular polarization, and linear polarization.

It is precisely this polarization characteristic that allows light waves to exhibit the properties of transverse waves.
This is also the principle behind the formation of electromagnetic waves.
In simple terms, photons are produced in large quantities in the form of quantum vortices and travel at the speed of light. Each vortex corresponds to one or several light rays, which move together with longitudinal waves in space, as the vibration of photons gives rise to the appearance of transverse waves.

(2) Electric and Magnetic Fields in Electromagnetic Waves

Photons in electromagnetic waves spiral in the direction of propagation, meaning there is a vortex along the forward direction. The vortex motion of photons drives the surrounding vacuum space, which is a superfluid, to form a vortex flow field. This vortex flow field is the magnetic field in electromagnetic waves. As the distance from the axis of the propagation direction increases, the strength of the vortex flow field decreases, forming a vortex flow field intensity gradient. This vortex flow field intensity gradient is the electric field.

As shown in Figure 8-50(a), C is the direction of light speed, i.e., the direction of electromagnetic wave propagation, B is the direction of the magnetic field, and E is the direction of the electric field. The three are mutually perpendicular.

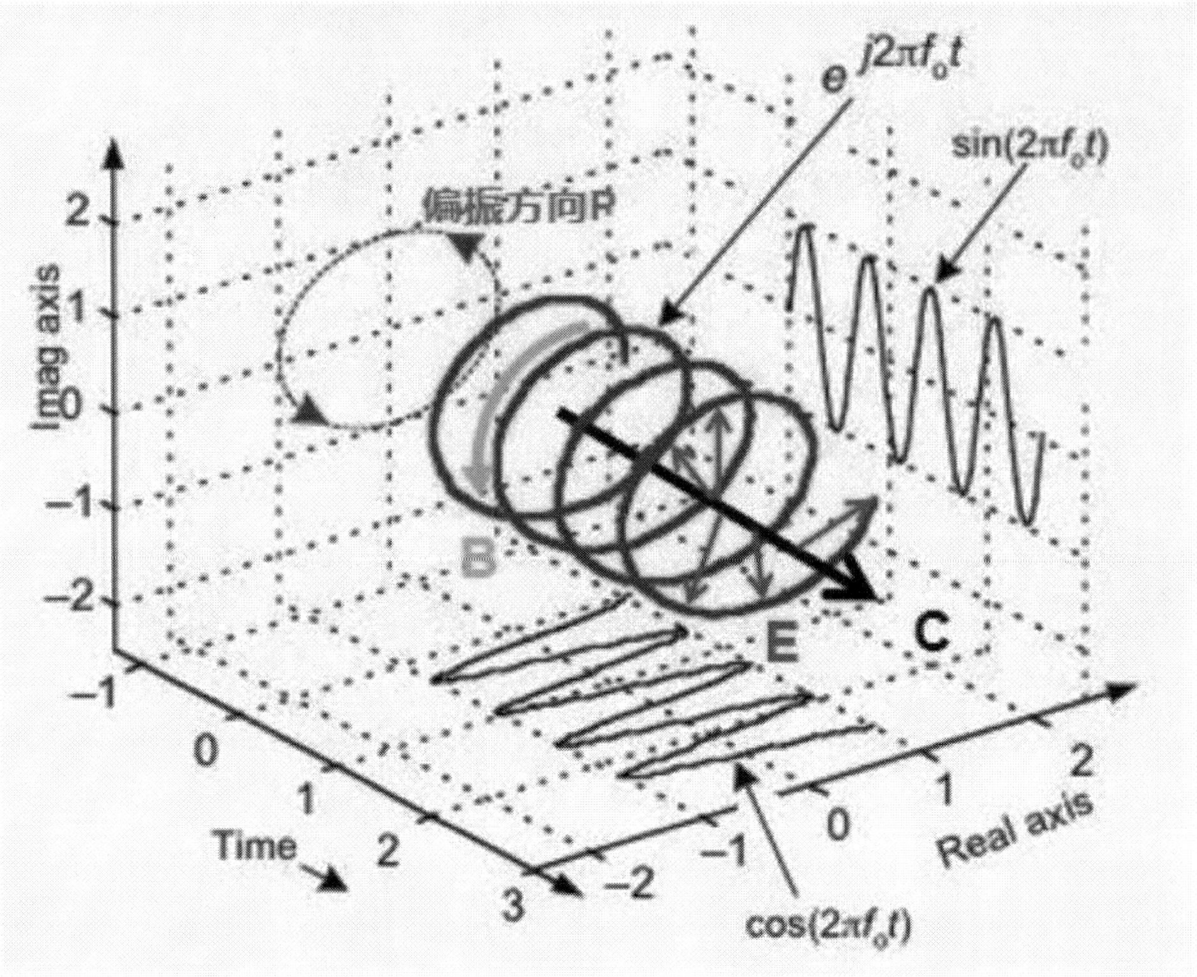

Figure 64: a) Schematic diagram of electromagnetic wave propagation (where C is the propagation direction, E is the electric field direction, and B is the magnetic field direction)

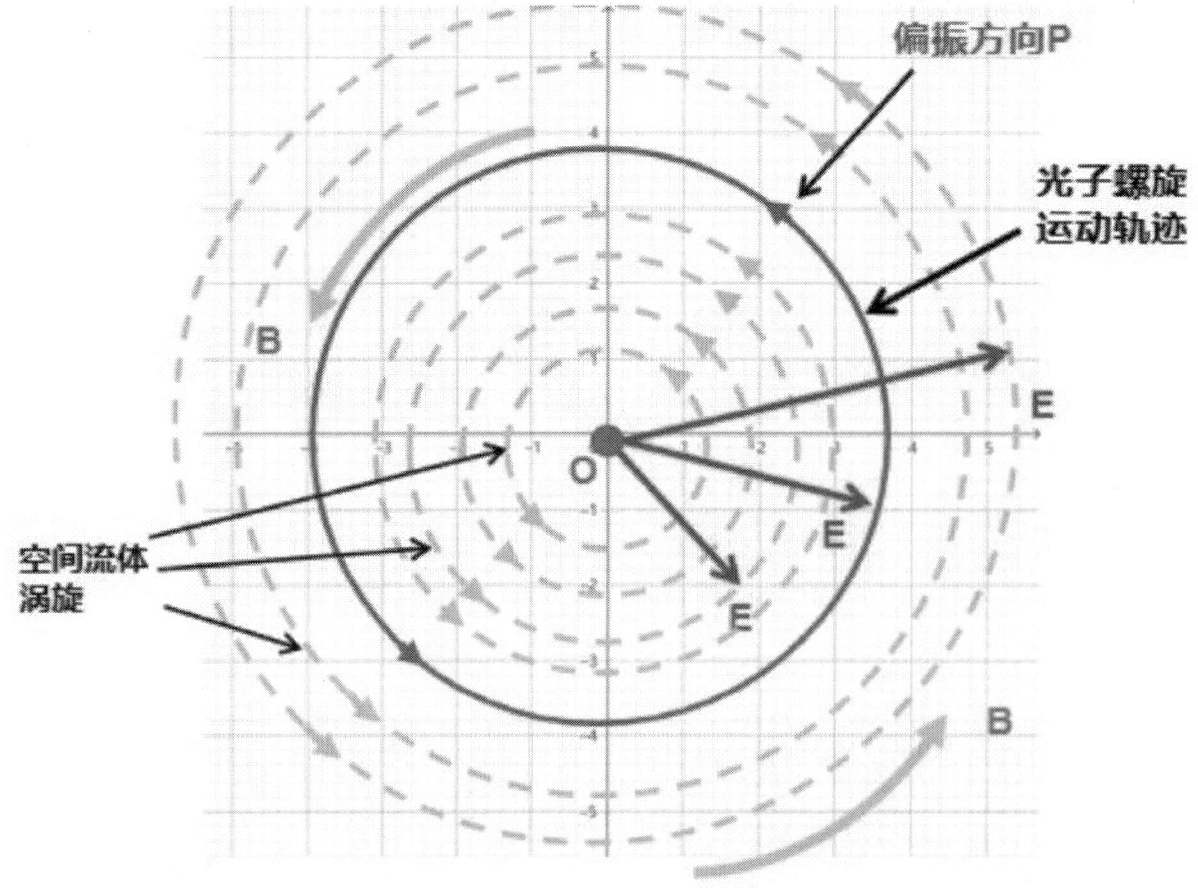

Figure 65: B) Frontal projection of electromagnetic wave propagation (where O is the center of the light wave helix, E is the electric field direction, and B is the magnetic field direction)

Figure 8-50 Electric and Magnetic Fields in an Electromagnetic Wave

As shown in Figure 8-50(b), it is a front projection of the propagation direction of an electromagnetic wave. O is the center point of the electromagnetic wave, and the propagation direction goes outward through the screen. During the helical motion of a photon, it causes the surrounding vacuum space flow field to form a vortex flow field. The strength field of this vortex flow field is the magnetic field, and the gradient field of the strength is the electric field. The direction of the magnetic field B is the same as the polarization direction, which is also the helical direction of the photon. The direction of the electric field E points from the center of the electromagnetic wave vortex to the direction where the photon is located, so as the photon spirals, the direction of the electric field also spirals, forming the polarization direction of the electromagnetic wave.

Because the vortex flow field formed by the superfluid vacuum space spreads out toward the surroundings, forming a diffused electromagnetic field.

By analyzing the principle of electromagnetic wave propagation and the generation of its electric and magnetic fields, it can be concluded that the electric and magnetic fields during the propagation of an electromagnetic wave are not mutually induced, but rather represent the manifestations of the vortex flow field in the vacuum superfluid caused by the helical motion of photons in two aspects: magnetic field and electric field. The flow field strength determines the magnetic field and its direction, while the gradient of the flow field strength determines the electric field and its direction.
(3) Principle of Refraction of Electromagnetic Waves

(1) Wave explanation of refraction: Huygens' principle

The phenomenon of light refraction can generally be explained by Huygens' principle.
Huygens (Christiaan Huygens), a Dutch scientist, proposed this theory in the 17th century to explain the behavior of wave propagation. The core idea of Huygens' principle is: each point on a wavefront can be considered as a new "secondary wave source," and the waves spreading forward from these secondary sources form new wavefronts. In simple terms, wavefronts do not move directly forward, but are formed by the superposition of countless tiny secondary waves, as shown in Figure 8-51.

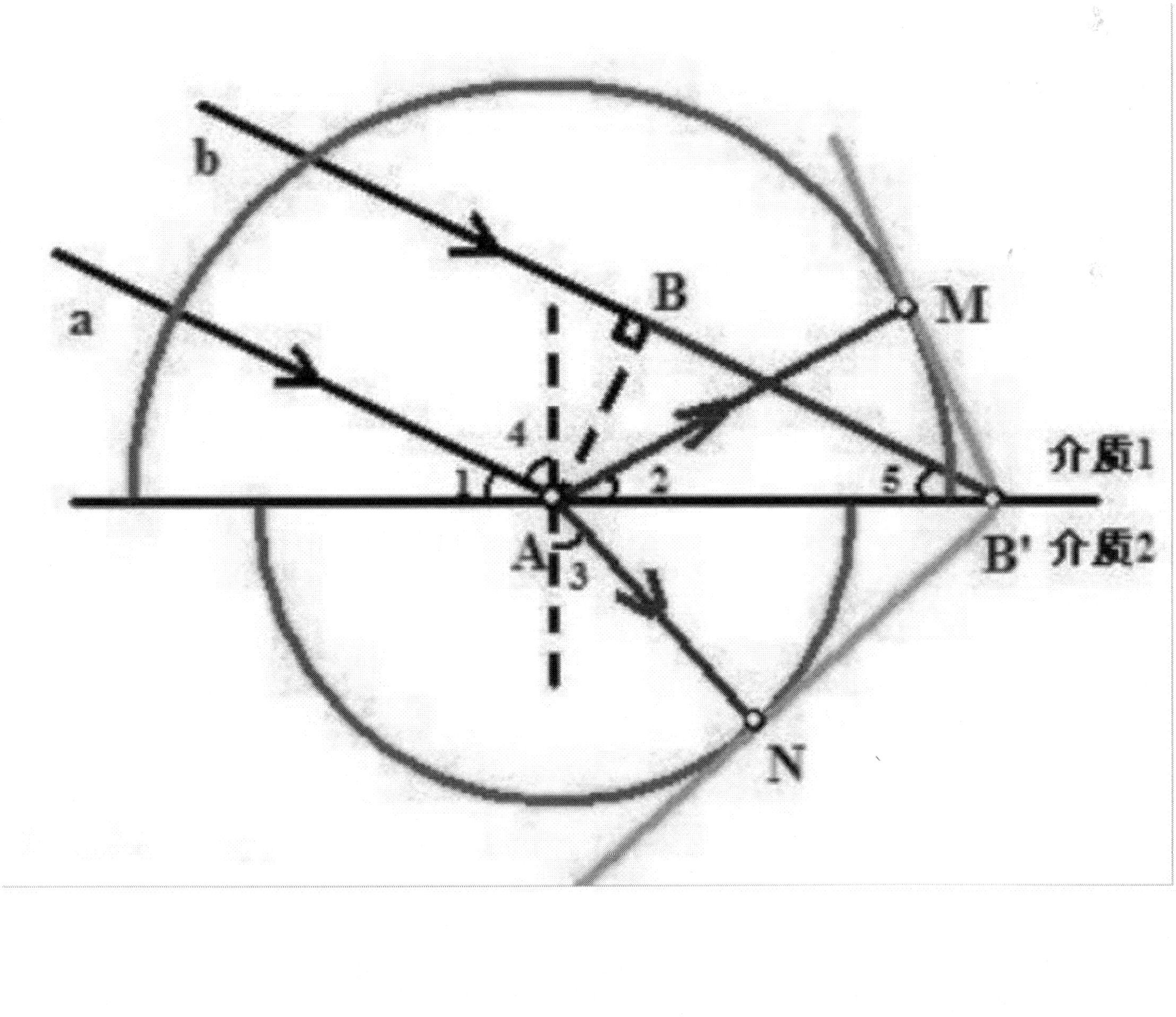

Figure 8-51 Huygens' Principle for Light Wave Refraction

Huygens' principle can well explain the phenomena of light reflection and refraction. Based on this, the French physicist Fresnel (Augustin-Jean Fresnel) proposed a more complete Huygens-Fresnel principle

The interference effect of the secondary source is taken into account, thus more accurately describing the diffraction and interference phenomena of waves.
Huygens-Fresnel principle is one of the foundations of modern optics and provides a good explanation of the refraction of light. However, it does not offer a satisfactory explanation for the phenomenon of different refractive indices for different frequencies of light within the same medium. This is generally explained through an additional polarization phenomenon, that is, when light propagates through a medium, its alternating electric field acts on the charged particles in the medium (such as electrons, ions, or molecular dipoles), causing them to shift or change orientation, forming a polarization effect. This polarization affects the medium's permittivity. Different frequencies of electromagnetic waves produce different polarization effects, thus having different impacts on the permittivity, and the refractive index is directly related to the permittivity:

$$n = \sqrt{\varepsilon_r \mu_r}$$

n is the refractive index of the medium, ε_r is the permittivity of the medium, and μ_r is the permeability of the medium. Thus, the phenomenon that electromagnetic waves of different frequencies have different refractive indices arises. However, this still does not physically explain why the dielectric constant causes changes in light refraction.
(2) Wave and Particle Synthesis Explanation of Refraction
We have explained in the basic principles of space relativity 4: the new light speed invariance principle that refraction of light occurs due to the difference in the density of the superfluid vacuum space in different media, just as water waves will refract when moving through waters of different densities. The superfluid vacuum space is the medium for light propagation, and its density changes in different media will also lead to changes in the propagation of light waves.

As analyzed in the principle of electromagnetic wave formation earlier, the essence of an electromagnetic wave is a combined motion pattern formed by the longitudinal wave of the vacuum space superfluid and the particle motion of photons. The polarization vibration of photons combines with the linear propagation of the longitudinal wave to create the appearance of a transverse wave. The propagation medium of electromagnetic waves is the vacuum space superfluid, and when the density of the vacuum space superfluid changes, the propagation of electromagnetic waves is also affected, including changes in propagation speed and direction.

As shown in Figure 8-52(a), the wavefront ab of the longitudinal wave in the vacuum space propagates from medium L to the new medium L1

When entering L1, the density of the vacuum space superfluid is smaller, so the point a that enters L1 first slows down in L1. Let the space wave speed in L1 be v, and that in L be c. Then according to formula 3.10:

$$n = \frac{c}{v} = \frac{\rho_c}{\rho_v} = \frac{\varepsilon_v}{\varepsilon_c}$$

n is the refractive index of the spatial wave, ρ_c is the density of the vacuum superfluid in medium L, and ρ_v is the density of the vacuum superfluid in medium L1. ε_c、ε_v are the permittivities of media L and L1, respectively, and according to equation 3.9, they are inversely proportional to the densities ρ_c、ρ_v of the vacuum superfluid in media L and L1.

At the same time, when photons enter the L1 medium, the direction of the helical surface of space also changes synchronously, with the change angle matching the above refraction index formula.

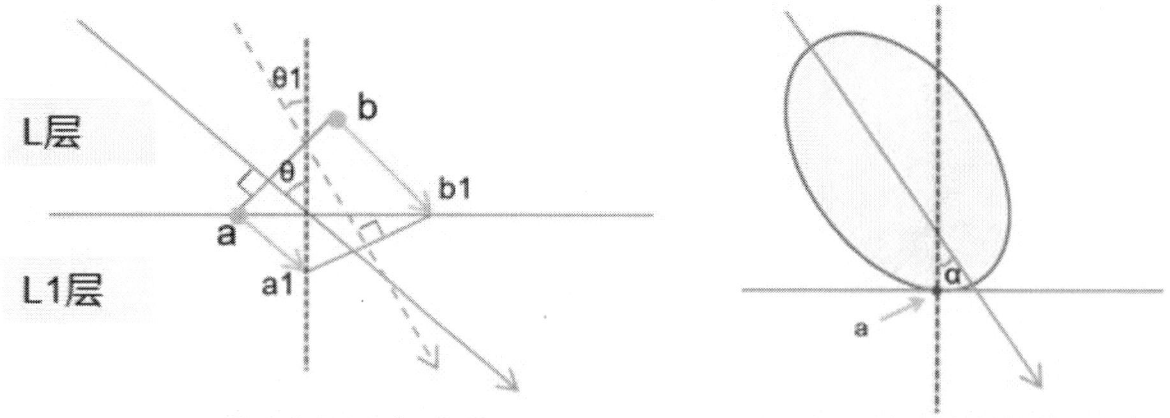

a)纵波波前传播方向折射 b)光子体积向接触法线方向收缩

Figure 66: Schematic illustration of longitudinal wave front refraction and photon volume contraction in Figure 8-52

Additionally, because of $\rho_c > \rho_v$, when entering medium L1, the photon's own volume will also expand, as shown in Figure 8-52 (b). According to our analysis of particle evolution, the photon is an ellipsoidal vortex formed by the superfluid vortex of empty space. Its volume will expand toward the point a where it first contacts the low-density region of the empty space superfluid. This volume expansion will exert a pulling force on different parts of the photon. Since point a is not on the photon's volume axis and forms a certain angle with it, α , this will result in an unbalanced pulling force on the photon, causing its propagation direction to change slightly. In the illustrated direction, the photon will experience more expansion force along the angle α , leading to a change in its propagation direction along the angle α , thus forming a further refraction effect.

That is, when light waves refract, they not only have a refracted angle due to wave behavior, but also an additional refracted angle caused by the contraction or expansion of the photons themselves.

As the frequency of light waves varies, the ellipsoidal shape of photons also changes. Higher frequency indicates higher excitation energy, resulting in greater photon energy. The ratio of the major to minor axes of the volume decreases, making the shape closer to a sphere. Under the same incident angle, the angle α in Figure 8-52(b) will be larger, which in turn increases the additional refraction angle θ_0 . Therefore, the higher the frequency of light, the larger the additional refraction angle θ_0 , and the greater the overall refraction angle of the light wave.

(3) Relationship between refractive index and changes in propagation speed

It can be known from the above analysis that the refraction and change in speed of electromagnetic waves when propagating through different media are two distinct phenomena that occur simultaneously. It is not due to the change in speed causing the change in refractive index, nor is it due to the change in refractive index causing the change in speed. Mathematically, they are equivalent, but there is no causal relationship between them.

The fundamental cause of both changes is due to the variation in the density of the vacuum superfluid in the medium. Of course, the refraction of light is not only affected by the density of the vacuum superfluid in the medium, but also by the arrangement of the crystal matrix in the medium, and many other factors, making the actual application more complex.

(4) The Principle of Electromagnetic Wave Speed

As we discussed earlier, electromagnetic waves are transverse wave states formed by the synchronization of longitudinal waves in the vacuum superfluid and the linear particle motion of photons, due to the polarization of photons. If we take a look, we will find that the formula for the speed of longitudinal waves is similar in various media.

(1) Speed of longitudinal waves in a gas

Longitudinal waves in a gas, also known as sound waves, have a speed formula of:

$$\mathrm{C} = \sqrt{\frac{\gamma \mathrm{P}}{\rho}}$$

Where:
γ is the adiabatic index of the gas (approximately 1.4 for air)
P is the static pressure;
ρ is the density.
(2) Longitudinal wave velocity in a liquid
The formula for longitudinal wave velocity in a liquid (such as water) is:

$$\mathrm{C} = \sqrt{\frac{\mathrm{K}}{\rho}}$$

Where:
K is the volume bulk modulus of the liquid.
ρ is the density.
(3) Longitudinal wave velocity in solids
Longitudinal waves in solids, such as the P-waves in seismic waves, have a velocity formula of:

$$C = \sqrt{\frac{E}{\rho}}$$

Where:
E is the Young's modulus,
The density.
(4) Electromagnetic wave speed
According to the mass-energy equation there is the speed of light formula:

$$C = \sqrt{\frac{E}{M}} = \sqrt{\frac{H}{\rho}}$$

Where:
As the energy density;
As the mass density.
From the above formulas for the longitudinal wave velocity in gases, liquids, and solids, as well as the electromagnetic wave velocity, it can be seen that the longitudinal wave velocity in different media is basically consistent, including the speed of electromagnetic waves, which also aligns with the longitudinal wave velocity. This is consistent with our theoretical analysis of electromagnetic waves, indicating that it is the synthesis of longitudinal wave fluctuations in the superfluid vacuum space and the motion of photon particles. The speed matches that of the longitudinal wave in the superfluid vacuum space.

From this perspective, the longitudinal waves propagating in media such as gases and liquids, where each medium molecule or a large molecular sphere composed of many molecules, also move like photons, exhibiting polarization as a combination of the medium's longitudinal wave motion and the linear motion of particles. Overall, they display transverse wave characteristics, and such large molecular spheres also exhibit quantum states similar to photons.
(5) Propagation Medium of Electromagnetic Waves
From the consistency of the longitudinal wave formula above, we can infer that the parameter γP in the gas longitudinal wave velocity formula, the bulk modulus K in the liquid longitudinal wave velocity formula, and the Young's modulus E in the solid longitudinal wave formula are all similar to the energy density parameter H in the electromagnetic wave velocity formula. They represent different aspects of energy density measurement in various media.

Gases, liquids, and solids are all mediums for the propagation of longitudinal waves, so the vacuum space superfluid should also be a medium for the propagation of electromagnetic waves. The energy density H and mass density ρ in the electromagnetic wave propagation formula correspond respectively to the energy density and mass density of the vacuum space superfluid. This is consistent with our analysis in the basic principles of space relativity, principle 4: the new constancy of light speed, that the vacuum space is the medium for light propagation.

The formula for the speed of light obtained from Maxwell's equations:

$$C = \frac{1}{\sqrt{\varepsilon\mu}}$$

Among them:
ε is the permittivity of the medium;
The magnetic permeability of the medium.
According to equation 8.16, we have:

$$\frac{1}{\sqrt{\varepsilon\mu}} = \sqrt{\frac{H}{\rho}}$$

It can be deduced that the electric permittivity of the vacuum space superfluid ε is proportional to the mass density of the vacuum space superfluid ρ and inversely proportional to the energy density H. At the same time, since H is related to ρ and changes with ρ at a rate much greater than ρ, that is:

$$H = K\rho^2$$

Therefore, the electric permittivity can be considered inversely proportional to the mass density of the vacuum space superfluid, that is:

$$\varepsilon \propto \frac{1}{\rho}$$

The speed of light is proportional to the mass density of the superfluid vacuum space, that is:

$$C \propto \sqrt{\rho}$$

In accordance with our analysis in the basic principle 4 of space relativity: the new light speed invariance principle (equation 3.9).

(5) Principle analysis of the mass-energy equation

In the basic principle 7 of space relativity: the principle of mass-energy conservation, we analyzed that the true meaning of the mass-energy equation $E = MC^2$ is that the energy of a photon with mass M is MC^2 , that is, the energy expression of a photon is the product of its mass and the square of its velocity.

In particle reactions such as nuclear fission, part of the particle is released in the form of micro-particles like photons and wave particles. Photons and these large numbers of wave particles propagate at the speed of light in the form of spatial transverse waves. If the mass lost by the particle is m and is entirely converted into micro-particles such as photons and wave particles, then the total mass of these photons and wave particles can also be considered as m. Since they propagate at the speed of light, they possess kinetic energy $1/2mC^2$. Furthermore, since light waves are transverse waves, they also have the same potential energy $1/2mC^2$, thus their total

Energy is mC^2 , which is also the true meaning of the mass-energy equation.
Now, we know that the vacuum space, as the propagation medium of light, is an ideal non-viscous fluid. The equation for the speed of sound (longitudinal wave speed) in this ideal fluid is:

$$c^2 = \frac{\partial p}{\partial \rho}$$

where $ c $ is the speed of light; $ p $ is the pressure; $ \rho $ is the density.
Also because:

$$\mathrm{p} = \frac{\mathrm{F}}{\mathrm{A}} = \frac{\mathrm{F^*\ d}}{\mathrm{A^*\ d}} = \frac{\mathrm{W}}{\mathrm{V}} = \frac{\mathrm{E}}{\mathrm{V}}$$

where F is force, A is area, d is length, W is work, i.e., energy E, V is volume.
And:

$$\rho = \frac{\mathrm{m}}{\mathrm{V}}$$

That is:

$$\mathrm{V} = \frac{\mathrm{m}}{\rho}$$

Generational 8.18 yields:

$$p = \frac{E}{m}\rho(8.19)$$

Combining equations 8.17 and 8.19, we have:

$$c^2 = \frac{\partial p}{\partial \rho} = \frac{E}{m}$$

That is:

$$\mathrm{E} = \mathrm{mc}^2$$

Equation 8.21 is the well-known mass-energy equation.
It can be seen that the mass-energy equation is a variant form of the equation for the speed of sound in an ideal fluid (longitudinal wave speed), which is consistent with our analysis that light is a longitudinal wave propagating through the vacuum space as a superfluid medium. This is consistent with

our analysis that the vacuum space is an ideal, nearly frictionless superfluid medium.
In equation 8.17, p and ρ can be considered as the pressure and density of the vacuum superfluid when the longitudinal wave speed is C, which is consistent with our previous analysis that the speed of light c and the vacuum permittivity ε are related to the density ρ of the vacuum superfluid.

8.11 The Origin and Evolution Principle of Forces

Based on our previous analysis of the composition and evolution of particles, as well as the nature of electric fields, magnetic fields, etc., we can form a basic understanding of the fundamental principles of various forces.

As we mentioned earlier, there is mass before there is force. All the macroscopic forces we observe are the result of inertial mechanical effects between mass bodies and their mutual interaction with the flow field of empty space.

(1) Electric Field Force

As analyzed in Section 8.3.2, the flow intensity gradient field of the rotating flow field in empty space centered on a particle, formed by its high-speed spin, is the electric field we are familiar with.

When another charged particle Q1 enters the electric field of charge Q2, if the two charges are opposite, meaning their spin directions are opposite, then the direction of the spatial flow field they generate is the same between them and opposite on their respective sides. According to Bernoulli's flow field equation, this creates a spatial negative pressure in the empty space between the two charges, resulting in an attractive force.

Conversely, if Q1 and Q2 are like charges, a repulsive force will be generated between them.
Like like attracts, opposite repels, this is the electric field force we are familiar with.
(2) Magnetic Field Force
As analyzed in section 8.3.3, when charges start to move, countless vortex rings are formed around them, which include both high-speed rotating field flows and high-speed vortices along the flow direction. The field composed of a large number

of such vortex rings is called the magnetic field, and the direction of the field's rotation flow is the direction of the magnetic field. Therefore, the magnetic field is a closed field.

Since the magnetic field is a source-free field, its strength is much smaller than that of the electric field, that is:

$$B = \frac{E}{C}$$

Where: B is the magnetic field strength, E is the electric field strength, and C is the speed of light.
When charge Q enters a magnetic field B, since the charge Q also generates a magnetic field $B1$, $B1$, the magnetic field lines interact with the B magnetic field lines. Due to the difference in relative velocity between the two magnetic field lines on either side of the charge, a field pressure difference is created. This pressure difference acts on the particle, causing it to deflect from its original direction under the influence of the magnetic field B. This force is known as the magnetic force, which is also referred to as the Lorentz force.

If charge Q is stationary in the magnetic field B, it will not produce a magnetic field, and thus will not experience the Lorentz force.
Of course, at this time, the charge Q is still subject to the magnetic field force of B, but this force is very small and insufficient to cause Q to form a noticeable motion, so it appears as if there is no magnetic force.
(3) Weak Force
(1) The Field Principle of the Weak Force
Inside the atomic nucleus or within all particles, the internal field exerts a constraining force on all particles inside the nucleus, which is the weak force, as shown in Figure 8-53.

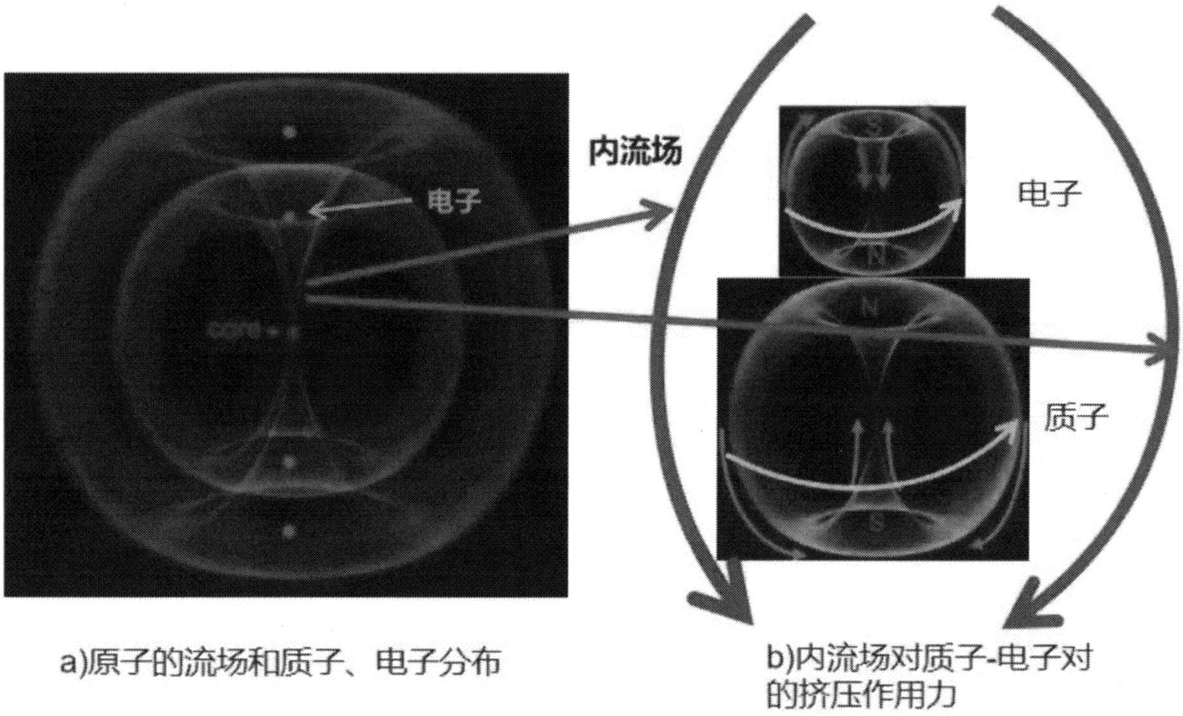

Figure 67: Figure 8-53 The flow field of an atom constrains the electron and proton

As we discussed in section 8.4.4, if an atom is externally excited, the pressure of the internal flow field on the protons and electrons inside the nucleus increases, bringing them extremely close to each other, and thus excites a new flow field between them. If the pressure increases to the limit, the new flow field will excite the formation of a new pair of positive and negative electron neutrinos. The electron neutrino is ejected and released, while the negative electron neutrino combines with the proton and electron to form a new balance, resulting in the formation of a neutron.

This is a typical process of the weak interaction, as follows:

$$\mathrm{p} + \mathrm{e}^- \rightarrow \mathrm{n} + \mathrm{v_e} \ (8.22)$$

When a neutron leaves the atom and becomes a free neutron, the absence of the atomic nucleus's internal field energy suppresses the neutron's stability. The proton, electron, and negative electron neutrino within the neutron will become unstable, and the pressure will cause them to decay quickly, releasing the negative electron neutrino and re-forming a stable proton

$$\mathrm{n} \rightarrow \mathrm{p} + \overline{\mathrm{v_e}}(8.23)$$

Other related processes in nuclear fusion, such as the proton-proton chain reaction in the Sun, involve two protons converting into a deuterium nucleus (D) through the weak interaction, releasing a positron and an electron neutrino:

$p + p \rightarrow D + e^{+} + v_e$ (8.24)

This can be viewed as an enhanced version of the proton-electron neutron formation reaction under stronger weak interactions: two protons, under the constraint of a strong flow field, approach extremely closely, and a strong flow field energy is excited between them, leading to the formation of a positron-electron pair. Then, the positron is released, and one proton and the negative electron approach extremely closely under the weak interaction, forming a neutron through the combination of a proton and an electron, and releasing an electron neutrino (Equation 8.22).

(2) Chirality of the weak interaction

It should be noted that a proton and an electron will form $S - S$ polar pair and $N - N$ polar pair, among which $S - S$ polar pair has a relatively weak repulsive force and is more stable under the constraint of the nuclear flow field, so when arranging the electron distribution in atomic energy levels, the electrons will preferentially arrange according to the S-S polar pair. Only after all S polar ports are occupied will the arrangement proceed.

Perform N orbital pairing to form $N - N$ spin pairs, so each layer of orbitals always completes the arrangement of electrons with the same spin before arranging electrons with opposite spins, which is the Hund's rule of electron arrangement.

While the $N - N$ pole pairs are less stable due to strong repulsive forces, they are more likely to respond under stronger nuclear flow field constraints, forming a powerful flow field between them and subsequently undergoing weak interactions that lead to neutron formation.

Other weak interactions also follow similar principles, so the weak interaction has a rather obvious characteristic: chirality, meaning it only acts on left-handed fermions or right-handed antifermions.

(4) Strong Force

Protons and neutrons together form the nuclear matrix of an atom. Within the atom's internal flow field, the protons and neutrons in the nuclear matrix are strongly constrained by the flow field forces, which keep them stably bound together. This force causes the flow fields of protons, neutrons, and other particles to merge, meaning that parts of their outer flow fields and parts of their inner flow fields interact and close upon each other: part of proton A's inner flow field flows into proton B's inner flow field, and then flows back from proton B's outer flow field to proton A's outer flow field, forming a unified flow field loop.

This unified flow field creates a more powerful flow field constraint between multiple protons, allowing numerous high-energy protons and neutrons to be stably bound together, forming a stable and powerful atomic nucleus.

The constraining force formed by this unified flow field is the strong force, as shown in Figure 8-54.

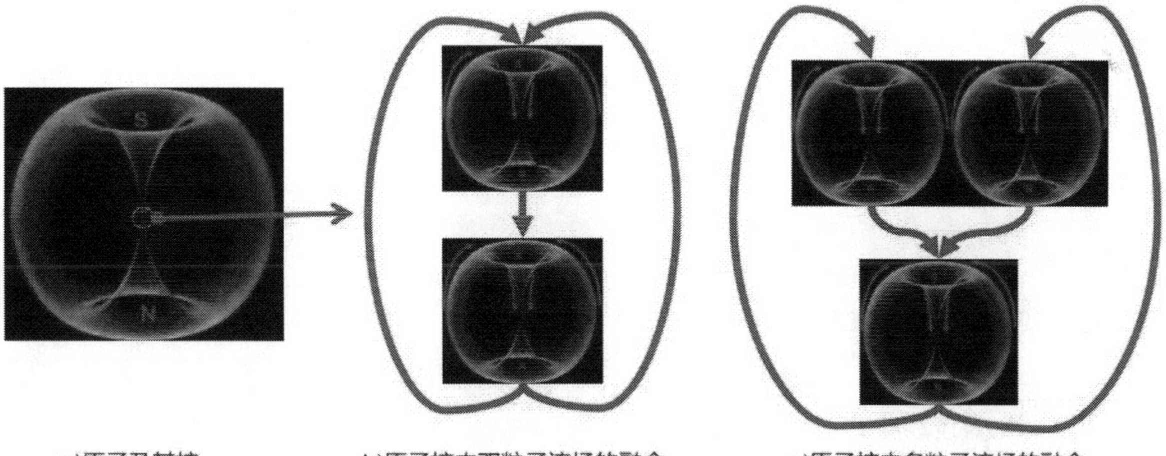

Figure 68: Figure 8-54 Multi-particle fields inside the atomic nucleus merge to form the strong force effect

The strong force between quarks in a proton, and others like it, also follows the same principle.

It can be concluded that the strong force is much more powerful than the weak force. Modern experiments show that the strong force is about 10^{15} times stronger than the weak force, which aligns with our expectations.

(5) Gravity

We have previously discussed the formation principle of gravity in the introduction to spatial dynamics (Section 5.3.4). The thermal motion of a mass body causes fluctuations in the surrounding vacuum space superfluid, leading to changes in flow velocity. When two or more mass bodies meet, the changes in flow velocity of the vacuum space superfluid between and around them result in fluid dynamic effects. Specifically, the flow velocity of the vacuum space superfluid between the mass bodies increases, while the flow velocity in other directions around them decreases, creating a negative pressure in the flow field between them and thus generating an attractive force.

This is essentially the same principle as that of electric and magnetic fields, which are also manifestations of negative pressure phenomena caused by the influence of the superfluid motion velocity in different directions of surrounding empty space, and are all expressions of Bernoulli's principle in fluid dynamics across various flow field phenomena.

The difference lies in that electric and magnetic fields are vortex flows, thus having direction, whereas gravity is caused by the velocity change of the superfluid in space due to the energy vibration of matter.

Therefore, there is no direct gravitational force between celestial bodies and equal mass objects, but rather a force resulting from the pressure difference of the superfluid in the surrounding space, which manifests as mutual attraction. Hence, there is no such thing as gravitons, just as there are no magnetic monopoles.

In fact, as we discussed earlier, the rotation and revolution of Earth and other planets also produce corresponding electric and magnetic field effects similar to the spin of particles, although the effects are extremely weak due to their slow rotational and orbital speeds, making them almost undetectable.

It should be noted that we also discussed in the introduction to spatial dynamics that gravity, being a mechanical effect of the vacuum space flow field, is not a linear force between mass bodies, but rather a range force that permeates the surrounding space, that is, an existing range gravitational field. When a third mass body appears, it not only affects the two mass bodies but also influences the gravitational field between them

There is a gravitational effect between objects, and they are also subject to an additional range gravitational field, resulting in an increase in gravity.
This kind of range gravitational field is one of the reasons for the observed galactic gravitational anomalies and the consistent deviations in experiments detecting the gravitational constant G.

Because the longitudinal wave propagation speed of the vacuum space superfluid is the speed of light, the speed of gravitational changes is also the speed of light.

In a broad sense, gravity can take into account all the electric and magnetic effects within a mass body, as well as its thermal motion, and even the unified fluid dynamics effects around the mass body caused by its rotation and revolution.

In simple terms, the gravitational force of an object is proportional to the object's mass and thermal motion energy, and inversely proportional to the square of the distance, that is:

$$F \propto \frac{M\Omega}{r^2}$$

Where: F is the gravitational force of an object, M is the mass of the object, Ω is the thermal motion intensity of the object, and r is the distance from the object.

The thermal motion intensity of an object can be equated to the thermal motion temperature of the superfluid in the surrounding vacuum space, that is, the vacuum temperature, so the above formula can be written as:

$$F \propto \frac{MT}{r^2}$$

Where, T is the temperature of the superfluid in the vacuum space around the object.
Considering the empirical gravitational formula:

$$F = G\frac{M}{r^2}$$

Combining equations 8.26 and 8.27 gives:

$$G \propto T(8.29)$$

That is, the gravitational constant G is proportional to the intensity of thermal motion of an object or the temperature of the superfluid vacuum surrounding it.

That is, when the effect of an object's thermal motion is not considered, the gravitational constant G would not be a constant but a variable value.

The experimental determination of the gravitational constant G has always been a challenge in the physics community [48], as results from different experiments often show significant discrepancies, which are directly influenced by variations in temperature conditions.

(6) Summary of the principles of various forces

Electric field force: The mechanical effect between active vortex flow fields, i.e., the mechanical effect between electric fields and electric fields.

Magnetic field force: The mechanical effect between sourceless closed vortex flow fields, i.e., the mechanical effect between magnetic fields and magnetic fields.

Electric field is a source-based magnetic field (with particle spin continuously exciting), and is a vortex flow field intensity gradient field; magnetic field is a source-free electric field, and is a vortex flow field intensity field. Electric field and magnetic field are measurements of the two directions and variables of the vortex flow field, and they are perpendicular to each other.

From the perspective of dimensions, the vortex field strength described by electric and magnetic fields should not be the same type of strength, just as momentum and kinetic energy, which describe the intensity of an object's motion, are not the same type.

Weak Force: The constraining force of the particle field on internal particles.
Strong Force: The strong constraint force formed by the fusion of particle fields between particles.
Gravity: The field mechanics effect caused by the thermal motion of matter triggering the formation of a superfluid field in the surrounding vacuum space. The field mentioned here includes all fields formed by the wave, flow, vortex, and other movements of the superfluid in vacuum space.

It can be anticipated that if there are other forces in the universe that we have not yet observed, these forces would not exceed the two principles of particle inertia and the overall force of the flow field, and their essence would be the same.

8.12 Origin, Composition, and Evolution of the Universe

(1) Origin and Evolution of the Material World of the Universe
Based on our previous analysis, the origin and evolution of the universe can be divided into the following stages:
(1) Vacuum Space
At first, there was vacuum space with no matter at all.
(2) Emergence of Vacuum Space Superfluid
The most fundamental vacuum space superfluid, composed of countless wave particles, emerged in the vacuum space, spreading throughout it.
(3) Energy disturbance, emergence of initial particles
The superfluid vacuum space is excited by energy disturbances, exhibiting thermal motion and forming vortex flow fields, leading to the creation of fundamental solitons such as individual quantum vortices, vortex pairs, vortex matrices, vortexons, and even vortex rings.

At the same time, the overall vortex motion of the superfluid vacuum space in the universe also forms the universal particle spin and flow field, resulting in the formation of cosmic particles.

Exciting the superfluid energy of the vacuum space may exist in two scenarios:
The first possibility is that superfluid appears in the vacuum space along with this energy, meaning that when superfluid enters the vacuum space, it exhibits flow, vortices, and other thermal motions, thereby beginning to excite fundamental solitons. In this case, the universe can be viewed as a whole, starting as a complete vacuum and later being infused with superfluid to form a thermal field, which then begins to generate internal matter particles. That is, the theory of the primordial universe: first there is the universe and vacuum, and then the superfluid in the vacuum space forms particles and matter.

The second possibility is that the vacuum space already contains the superfluid, and later this superfluid is disturbed by energy, leading to thermal motion and the emergence of fundamental particles such as solitons. In this scenario, it is precisely this disturbance that creates the overall superfluid flow field in the universe space, thereby giving rise to the formation of cosmic particles. That is, the theory of the post-

Cosmology: First there was vacuum, vacuum superfluid, then the flow field of the vacuum superfluid as a whole formed composite particles and cosmic particles.

(4) Formation of composite particles and matter

The most fundamental solitons and basic particles merge to form more complex composite particles, or the superfluid vacuum space creates higher quality and energy composite particles under greater energy disturbances, further forming macroscopic matter particles, macroscopic matter, planets, etc.

(2) Ancient Chinese Cosmological Philosophy

After analyzing the process of the universe's creation and evolution, recalling our ancestors' thoughts on cosmic space and time, we find many similarities.

The hypothetical superfluid vacuum space we imagine filling the entire universe is actually similar to the chaotic gas envisioned by our ancient philosophers. At the beginning, the universe was filled with this chaotic gas, where heaven and earth were not yet differentiated, meaning the universe was initially filled with such chaotic gas, or in other words, the superfluid vacuum space, but since it had not yet been excited, no material particles had formed, resulting in a state of complete chaos. It was only later, as it was excited, that the chaotic gas gave rise to matter and created a world with clear distinctions between yin and yang, which is basically similar to the process of the universe's generation and evolution we have analyzed.

On the other hand, ancient Chinese Daoist philosophy also deeply explored the process of the formation of the universe, with the concept of the "Four Great Beginnings" being a particularly representative idea.

The "Four Great Beginnings" first appeared in the "Liezi" and later in the Han dynasty's contemporary text "Zhouyi Gan Zuo Du" of the谶纬神学 (omens and divine interpretations), representing the process of transition from the boundless void to the Taiji (Supreme Ultimate) and the birth of heaven and earth. It is divided into four stages: Taiyi, Taichu, Taishi, and Taisu.

There is an explanation in the "Liezi":
"Taiyi, before the appearance of qi," meaning that in the universe there was only vacuum space, and there was no superfluid space, that is, the chaotic qi.
"Out of the void, the first appearance was of chaotic energy," meaning that in the vacuum of the universe, the chaotic energy began to emerge, which is what we refer to as the superfluid of the void space.
"Tai Shi" refers to the instant formation of form, meaning the beginning of the formation of the initial form from chaotic energy, which is the first emergence of solitary wave particles in the vacuum space superfluid, also the initial prototype of particles.
"Tai Su" is the beginning of matter's substance but not yet formed into a body, meaning the initial form generated from chaotic energy continues to gather and form substantial matter, which is the further fusion and excitation of the most primitive solitary particles in the vacuum space superfluid to create composite particles with higher energy. Composite particles have a particle nucleus, and from the initial wave-like form of the vacuum space superfluid, the solid particle form finally emerges as substantial matter.

At the same time, these high-energy composite particles formed yin and yang properties due to differences in their spin motion, thus forming the Taiji stage, which is what we refer to as particles acquiring electric charge and becoming positive and negative.

It is precisely because of the yin and yang properties that when they combine, they form neutral and more stable particles, leading to the complex and vast family of material particles, and thus forming the magnificent and mysterious material world we see today. This is the philosophical concept of "All things carry yin and embrace yang; the interaction of opposites produces harmony," as expressed by our ancient sages, where the fusion of yin and yang gives rise to all things.

It can be said that the ancient concept of wuji (nothingness) evolving through four stages—taiyi, taichu, taishi, and taisu—leading to taiji, completely describes the entire process of the universe's birth and evolution, which coincides with our analysis of the superfluid vacuum space and the formation and evolution of the universe.

Our ancient sages, perhaps limited by their lower technological level, their wisdom and philosophical thinking about the universe are no less than ours.
(3) The Five Major Problems in Current Cosmology
The Big Bang theory of the origin of the universe currently faces several significant issues.
(1) Singularity Puzzle

In the Big Bang theory of the universe, the "singularity" is a point with an infinitely small volume, infinitely high density and temperature, considered as the origin of the universe. According to mainstream modern scientific theory, the universe was born about 13.8 billion years ago from a Big Bang, which began from a singularity. This singularity had infinite density and temperature, containing all the matter and energy that formed the entire universe. However, the specific origin and nature of the singularity remain many unsolved mysteries in the scientific community.

A singularity is a point in mathematics that cannot be processed, representing an unmanageable limit state of a mathematical object under certain conditions. In physics, a singularity is described as a point with infinite small volume, infinite density and temperature, a state that renders all current physical laws, including general relativity, ineffective. The existence of singularities challenges our understanding of physical limits, and the origin of singularities remains without conclusive answers in the scientific community, becoming an unresolved mystery in the Big Bang theory.
(2) The Flatness Problem
The flatness of the universe is a property of cosmic space, referring to its large-scale geometric structure in physics. Flatness means that the curvature of a space (the force causing geodesic deviation) is zero.

Measurements of the angular power spectrum of the cosmic microwave background radiation show that our observable universe is incredibly flat, with no observable curvature, Ω very close to 1, only 1.005 ± 0.007, differing by only a few thousandths.

Why is the universe so flat? According to the Big Bang theory, the non-flatness of the universe does not violate any physical laws. However, the universe we observe is almost completely flat. This seems like a coincidence, as the Big Bang could have produced a universe with any curvature.

But using coincidence to explain this is clearly unsatisfactory, and it would even lead people into metaphysical thinking, making them believe in a creator, which clearly goes against the spirit of science.
(3) The Monopole Problem
The problem of magnetic monopoles seems to be an issue in the field of high-energy particle physics, but the big bang theory as a description of

Theories describing the origin of the universe should explain the problem of magnetic monopoles. However, current theories clearly fail to address this issue.
(4) Symmetry breaking problem
The universe big bang theory has not provided a reasonable explanation for phenomena such as the breaking of matter-

antimatter symmetry.
(5) The problem of discrepancies in the measurement of the Hubble constant
The discovery of the Hubble constant is a crucial foundation of the big bang theory, but there are significant discrepancies between measurements obtained through the "late universe" method and the "early universe" method. The academic community has yet to offer a reasonable explanation for this.

These issues are actually major flaws in the big bang theory, undermining its scientific validity and rationality.

In our model of the vacuum space superfluid and cosmic particles, these issues can be given reasonable explanations.

According to the cosmic particle model, there is no singularity, so the singularity problem does not exist. The universe, as a giant particle formed by the disturbance of a superfluid, has a continuous flow field. However, we only observe a small part of it, which appears flat. Based on our analysis of the magnetic field, there is no problem of magnetic monopoles. The asymmetry between matter and antimatter arises from the overall vorticity direction of the vacuum space superfluid in the universe, as discussed in our analysis of the matter-antimatter principle. This asymmetry in the flow field leads to the observed asymmetry between matter and antimatter.

As for the Hubble constant anomaly, the universe was not created by the Big Bang, so observations naturally lead to anomalies.

Recently, István Szapudi from the University of Hawaii at Manoa and his collaborators proposed in a paper published in the Monthly Notices of the Royal Astronomical Society that adding a small amount of rotation to the cosmic model can perfectly resolve the Hubble constant measurement anomaly without violating any physical laws or current principles.

There are observations. This is consistent with our theoretical analysis of the particle model of the universe.
In addition, for several important pieces of evidence supporting the Big Bang theory, such as the redshift phenomenon, cosmic background radiation, and the abundance of helium elements, our model of vacuum space superfluid and particle universe can also provide good explanations. These topics have been discussed in the sections "Redshift and Blueshift of Light" in Chapter 5.9 and "On the Big Bang Theory" in Chapter 5.10 of the Introduction to Spatial Dynamics.

(4) Entropy and Emergence: The Essence of Order and Chaos

(1) Entropy

Entropy is one of the parameters that characterize the state of matter in thermodynamics, initially a thermodynamic function, later evolving into a measure of the disorder of a system.

In 1865, the German physicist Clausius proposed the entropy increase law, stating that heat transfer in an isolated system always leads to an increase in entropy, which remains unchanged only in reversible processes. This is also known as the second law of thermodynamics.

Later, Clausius regarded the universe as an isolated thermodynamic system and derived two conclusions from the second law of thermodynamics: the energy (internal energy) of the universe is constant; the entropy of the universe is always approaching its maximum value, which later developed into the "heat death" theory. That is, as an "isolated" system, the entropy of the universe increases with time and eventually leads to a state of heat death.

The theory of the universe's heat death has sparked continuous debate within the scientific and philosophical communities since its inception. Due to its implications regarding the future of the universe and the fate of humanity, its influence has extended far beyond these fields, becoming one of the most vexing cultural puzzles in modern history.

However, many scholars have also pointed out that the entropy increase law is a law applicable to the field of thermodynamics, with strict limitations on its applicability, and it is not suitable for promotion to more complex areas such as the universe, which goes beyond its scope.
(2) Order and Chaos

On the other hand, we also need to clarify that the so-called order and chaos of a system are merely human perceptions. They are defined by us as order and chaos, and they do not represent an objective reality in the natural world.

All physical phenomena in the objective physical world are natural processes and results of material motion, and are equal material phenomena, or to say, equal in rights.

It can be considered that there is a probabilistic issue between different phenomena, but not a priority issue. In other words, the order or disorder we define does not have a priority difference, and for the motion and evolution of material systems, there is no distinction between the two.

(3) Emergence is not special

The cosmic vacuum space superfluid composed of countless waves, when excited, generates a large number of particles, leading to the formation of a vast and complex objective physical world. This can be considered as an emergence phenomenon within the vacuum superfluid. However, these particles themselves do not represent order, nor do the most fundamental waves represent chaos.

Similarly, the emergence of life and intelligence in the evolution of the objective physical world, up to the formation of our current human world, is also a form of emergence in the material world. It does not imply that intelligent beings such as humans have a higher priority than non-living entities like soil, stones, water, and air. Both are normal outcomes of the material world evolving under different conditions, and neither is special.

Including the large-scale AI models we create today, they are also an emergence phenomenon in another domain.
The emergent result is a common outcome and phenomenon in the countless evolutionary processes of the objective physical world. However, the frequent occurrence of emergent phenomena and their ability to persist for a long time actually indicates their stability, which suggests the possibility of the long-term existence of matter, life, and civilization.

Therefore, the so-called order and chaos are actually equal states of matter. As long as the overall evolutionary environment of the universe does not undergo significant changes, there is no reason for the existence of the material world and human civilization, which are considered "order," to necessarily evolve into "chaos." According to the evolutionary history of the material world, it might even be the opposite.

(5) Understanding of Vacuum Space

Vacuum space is a space that contains no matter and can accommodate all things. There are many speculations about the properties of vacuum space, but currently, many theories do not actually refer to a vacuum as we understand it—completely devoid of any matter—but rather to the superfluid of the empty space we have been discussing all along.

As quantum field theory considers the vacuum as the ground state of quantum fields, the vacuum is not "empty," but rather the state where all quantum fields are at their lowest energy level. According to quantum field theory, space is filled with various fields (such as electromagnetic fields, electron fields), and the excited states of these fields manifest as real particles, while the ground state corresponds to the "no real particles" state. Here, the so-called quantum fields refer to the superfluid vacuum space we are discussing, which is a form of matter, not the true "vacuum" that contains no matter at all.

Others, such as Dirac, who proposed the theory that vacuum is an "ocean" of filled negative energy states to resolve the negative energy solution problem, are also the same.

Since the Special Theory of Relativity denied the existence of the ether, the theoretical community has experienced a lot of confusion regarding the nature of vacuum. The above statements are all compromises made to maintain consistency with relativity, attributing material and mechanical properties to vacuum.

In fact, the superfluid vacuum space can be considered as the ether that was denied by special relativity, but it is a further developed ether superfluid with dynamic and mechanical properties, composed of countless wave particles.

These aspects have not fully explored the nature of complete vacuum.
It seems that when we talk about absolute vacuum, people cannot understand its origin and nature, such as why there is a vacuum space that can contain everything, what the essence of vacuum space is, where it comes from, and so on.

In fact, if we consider all matter to exist, then the vacuum space that contains everything is non-existent. That is, vacuum space is not matter, nor is it an objective reality; the concept of containing everything does not apply to it.

Everything exists naturally, and it has nothing to do with vacuum space. Vacuum space can be considered as a concept we assign, just like time, mass, length, and quantity, which are all concepts we use to understand the objective world.

One of my concepts is that space is not an objective reality; there is no such object as a "vacuum space" in the universe.
In this way, it may seem difficult to understand. However, just like the concept of "0," people initially understood numbers from 1、2、3 and similar quantities without the concept of "0" itself. People could not imagine what "0" represented, just as they could not initially comprehend concepts like infinity and infinitesimals. These concepts did not exist at first, but emerged gradually as people's understanding of mathematics deepened.

We can consider vacuum space as a concept, similar to "0", representing nothingness, rather than an objective reality. That is: there is no vacuum in the universe.

(6) Outside the universe, inside the particle

(1) Beyond the Universe

In the previous chapter discussing the origin and evolution of cosmic matter time, we discussed two mechanisms for the emergence of the universe. One is the innate cosmology: the universe and vacuum existed from the beginning, and later, a superfluid of empty space emerged in the vacuum. The superfluid then produced matter particles through motion and energy excitation, eventually evolving into the real world.

The second is the post-natal cosmology: the universe vortices are formed as particles in the fluid vacuum space filled with particles due to energy excitation, while the excitation of the vacuum space superfluid creates a large amount of matter, galaxies, etc. Galaxies and supergalactic clusters act as "micro-particles" to form the flow field of cosmic particles.

For the a priori cosmology, we cannot explain how the superfluid vacuum space suddenly appeared and how it was disturbed by energy to form the material world. However, the a posteriori cosmology can well explain it, as the universe itself is located in a vast ocean of superfluid vacuum space, and it is a giant particle formed by the excitation of the superfluid vacuum space.

Of course, this cosmic giant particle, in the vast superfluid ocean of outer space, is itself just a microscopic particle, similar to the wave particles that form matter.

The cosmology of the day after tomorrow is more in line with the philosophical view of materialist science, and we can consider this as the formation of the universe and

The truth of the evolutionary process. In this case, the outside of the universe should be filled with countless similar "universes" or "particles," even particles with greater energy. Countless "universe" particles form a vast and immense "cosmic matter" world, similar to how countless fundamental particles in our universe form the complex and vast reality world.

Therefore, outside the universe, there should be a world composed of countless "Pang substances," and the universe is merely a negligible "particle" in that world. We can call this world the upper realm or the great world.
(2) Inside the wave particle
Bosons are the fundamental particles that make up the superfluid vacuum space we imagine to be filled with the universe. They are the most microscopic particles that compose our world.

Bosons have mass, volume, and dynamic properties, so the superfluid vacuum space also possesses mass, density, and pressure. All matter in our world, including ourselves, is constantly subjected to the pressure of the superfluid vacuum space. This pressure may be stronger than atmospheric pressure or water pressure, but we are almost unaware of it, just as we are always under atmospheric pressure yet cannot feel its immense force.

On the other hand, according to the sound speed formula in the section on the principle of the speed of electromagnetic waves (Chapter 8.10), under the condition of neglecting temperature and interparticle collision effects, it can be considered that the faster the wave propagation speed of the fluid, the stronger the rigidity of the particles composing the fluid and the smaller their mass. That is, the square of the longitudinal wave speed is inversely proportional to the mass of the medium particles and directly proportional to the rigidity of the medium particles:

$$v \propto \sqrt{\frac{K}{m}} = \sqrt{\frac{H}{\rho}} (8.30)$$

Equation 8.30 is the relationship equation for longitudinal wave velocity, where v is the longitudinal wave velocity in the medium, m is the mass of the particles composing the medium, K is the rigidity of the medium particles, ρ is the overall density of the longitudinal wave medium, H is the overall rigidity density of the longitudinal wave medium, corresponding to ideal gas, liquid, H is the elastic modulus, corresponding to solid H is the Young's modulus, corresponding to superfluid in empty space, H is the energy density. It is evident that the rigidity of the medium particles is from the macroscopic

The manifestation corresponds to the overall energy density of the medium itself.
The smaller the mass of the medium particles, the less resistance the wave encounters, and the stronger the rigidity, the shorter the time for the force of collision to be transmitted, so the faster the speed of sound wave propagation.

The speed of sound waves (longitudinal wave speed) in the superfluid vacuum space is the speed of light, far exceeding that of all gases, liquids, and solids, so it can be considered that the rigidity of the wave particles is much greater than that of all macroscopic matter particles.

At the same time, all macroscopic material particles themselves consist of countless wave particles, so the wave particles also have much less mass than the macroscopic material particles.

Exactly for these two reasons, the longitudinal wave speed of the superfluid vacuum space composed of countless wave particles far exceeds the longitudinal wave speed in the fluid medium formed by all macroscopic particles.

On the other hand, the speed of longitudinal waves in the vacuum space superfluid is the speed of light, which is finite, indicating that the wave particles are not entirely rigid. That is, the wave particles themselves should possess some elasticity. Based on this, we can conclude that the wave particles are not the most fundamental microscopic particles, but rather have an internal structure, which gives rise to their elasticity. This elasticity means that the transmission of collision forces between wave particles requires time, thereby establishing a speed limit of the speed of light for longitudinal waves in the vacuum space superfluid.

So, we can infer that there is also a more fundamental and complex microscopic world inside the boson, where countless more fundamental submicroscopic substances interact in complex ways to form the boson particle.

This microscopic world inside the boson can be called the lower world, or the small world.
(3) The End of the World
Beyond the universe, there is the upper world, and within the boson, there is the lower world. Adding our own world, the three worlds are actually connected as one.

But what lies beyond our three worlds? Above the upper realm and below the lower realm, are there more worlds?

Does this material world have an end? What should be the limit of speed?
Such questions should be answered by science, but we are currently unable to access the upper realm beyond the universe or to explore the lower realm within the particle, and thus we cannot understand the fundamental rules of these worlds or their basic composition. As a result, we are also unable to provide fundamental explanations.

However, when our vision can reach such a world and understand the physical rules and laws governing it, we will certainly be able to find the answer and draw a conclusion about the ultimate form of this material world.

Some people say that the most mysterious and incomprehensible aspect of this world is that it is actually understandable. We should recognize that a world must operate and evolve according to certain rules, and since these rules can be observed and understood, the world itself must be understandable as long as we can observe it.

8.13 Five Possible Models of the Universe Space

Is the universe bounded? If it is, what lies beyond the boundary? If not, how can it be unbounded?

What is the smallest particle of matter in the universe? Can it be divided infinitely? If we keep dividing, is there an end? If not, what would it be? Why can't it be divided further?

What is space in the universe? Why is there such a strange existence called "space" that can contain everything?

These puzzles about the universe and matter have confused countless people, leaving generations of scientists and philosophers puzzled and feeling lost.

We are trapped in such a dilemma, on one hand due to our incomplete understanding of the universe, having not yet grasped its deeper laws and principles, and on the other hand, it may be because of our own limitations in experience and perspective

Limitations have led to many misconceptions in thinking.
Next, we attempt to consider the ultimate model of the universe from several different perspectives, resolve the above confusions, and provide clear possible results and answers.

(1) Linear Universe Space Model

As we analyzed earlier, the universe itself is a super-massive particle composed of countless galaxy clusters. Beyond our own world, that is, outside the universe, there exists a larger world, which can be called the upper world or the big world. In that big world, the universe is merely like a fundamental particle, meaning there are countless, even "infinite," similar universe particles.

At the same time, the smallest particle in our world, the wave particle, is not entirely rigid, so it should also have an internal structure. That is, even the smallest particles contain a world within them, which we can call the lower world or the small world.

According to the linear universe space model, these universe spaces are linearly arranged. That is, beyond our own world, there may be a larger big world and a smaller small world. Beyond the big world, there might also be an even larger world, and beyond the small world, there might also be an even smaller world, as shown in Figure 8-55.

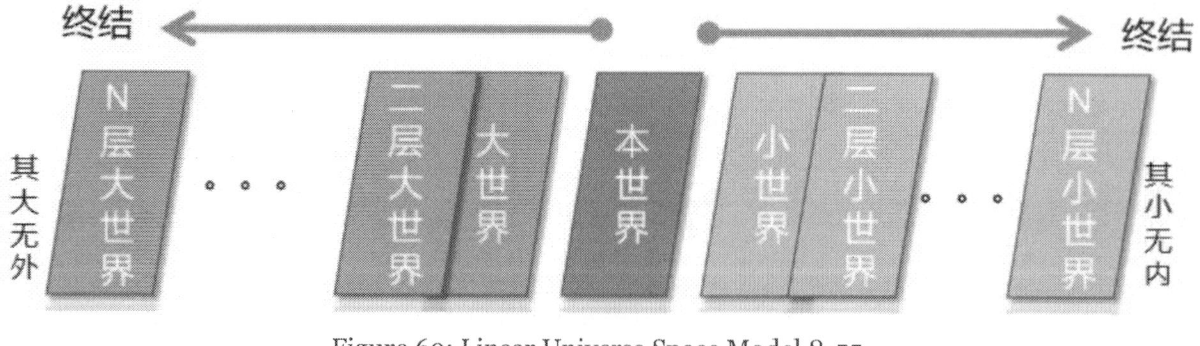

Figure 69: Linear Universe Space Model 8-55

A world is a layer, perhaps there are only three layers in total, or maybe it can reach up to thirty-three layers. The ancient concept of "thirty-three heavens" or "thousand worlds" refers to this. However, such a linear model of cosmic space has its limits. That is, when a world reaches a certain layer, its rules of operation and material laws become restricted, preventing it from expanding further or being divided into smaller layers. The entire universe space has an end

The spatial model reaches its endpoint, which is the situation of "boundless in size and infinitely small in interior."
These operational rules or laws within these large or small worlds are unknown to us inside the universe. Once we observe the rules of matter movement in these large or small worlds, we will understand the reasons and principles behind them and no longer feel confused.
(2) Emergent-type universe space model
If we were to identify the most fundamental characteristics of the world we live in within the universe, they should generally include matter, motion, and so on. It is precisely these characteristics that constitute our vast and complex material world.

However, there is another important characteristic, which is determinism. The motion of all matter is deterministic, meaning it follows the law of causality. Even in the quantum world that we have not yet fully observed, its motion principles should fundamentally be deterministic. Including the most complex chaotic systems, none have escaped the deterministic law of causality; they are simply difficult to predict, not contradictory to determinism.

Therefore, the determinacy of motion is one of the most fundamental and important characteristics of our universe's space. If there are larger and smaller worlds outside the universe, these worlds should also possess the characteristics of matter and motion, otherwise they would not be considered worlds. However, the determinacy of motion laws may not be the same in all such worlds.

The larger and smaller worlds outside the universe might have phenomena of indeterminate motion of matter, or even be entirely random worlds, where there is no causality, and almost no laws at all, completely chaotic.

From such a completely chaotic small world to the emergence of our deterministic present world, this is an emergent phenomenon, where determinacy arises from randomness.

Or in other words, the transition from our completely deterministic world to a completely chaotic big world outside the universe is also an emergence, that is, randomness emerging from determinism, as shown in Figure 8-56.

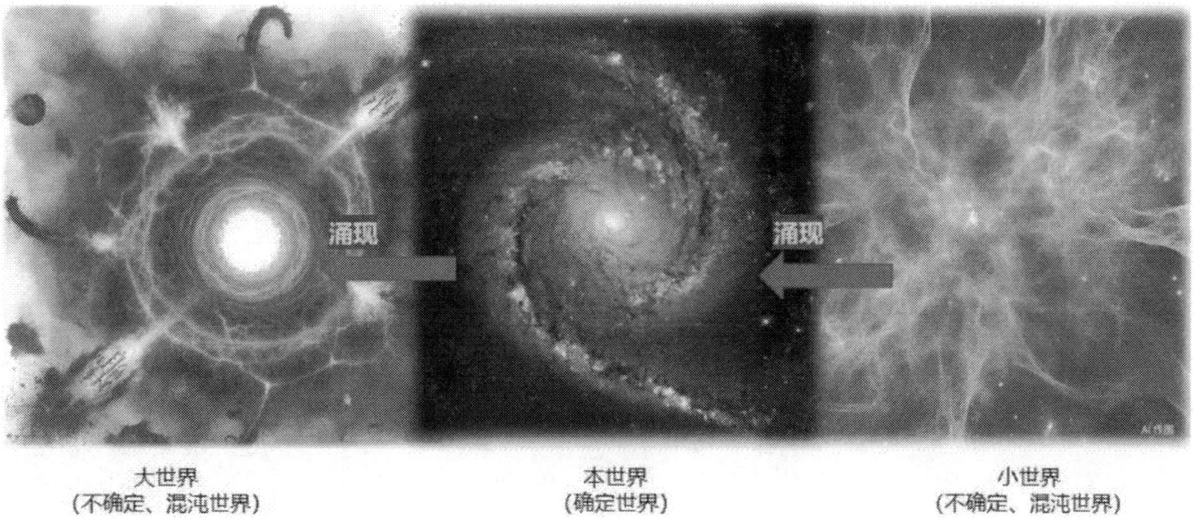

Figure 70: Figure 8-56 Emergence Model of Cosmic Space

In a world as chaotic as the big world and the small world, where there is no causality and even no regularity, the physical rules governing its operation are entirely different from those in our world. If life exists there too, it would be entirely the result of emergence from chaos. From the moment they come into being, they would be fighting against chaos, creating and maintaining determinism, and eventually being reabsorbed into chaos again.

Such a big world may contain many universes like ours.
In such a world, there are no familiar concepts of particle subdivision or physics beyond the world, because without these deterministic physical laws and material worlds, such a world might be the end and the origin of the universe space.

(3) The Cyclic Universe Space Model

Our real-life experiences have deeply ingrained concepts such as size, inside and outside, presence and absence, which also influence our understanding of the entire universe space, leading us to endlessly question what is outside the universe, what is inside the smallest particle, and what an empty space is, among other things.

But these concepts may become misconceptions in our understanding of the essence of cosmic space.
As for cosmic space itself, there may be no distinction of inside and outside, or big and small. What is "infinite" in size becomes small, and what is "infinite" in smallness becomes large. The outside is essentially the inside, and they are fundamentally one. Space is itself a substance.

Itself.
The inner world of our universe, when reaching outside the universe, becomes the upper world, or the great world. From the great world to an even greater external space, it might be our lower world, or the small world. From the small world to the outside, we return to the inner world of our universe. There may be more or fewer worlds nested between the great and small worlds, but overall, they form a circular structure, or the cycle of world space.

In such a world, what is inside or outside the universe loses its meaning, and what lies within the smallest particles of matter also has an answer. Space is the very essence of matter itself. "There is matter, so there is no space; there is no space, so there is matter." Infinity is the same as infinitesimal.

This is the toroidal universe space model, as shown in Figure 8-57.

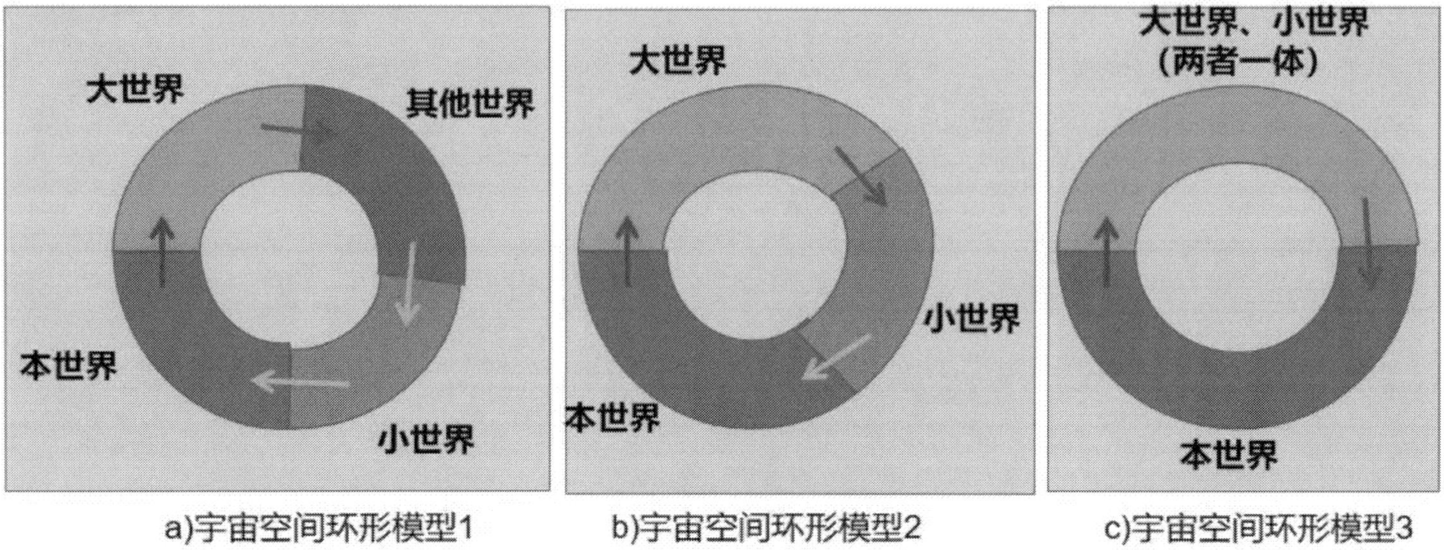

Figure 8-57 Cosmic Space Ring Model

Many scholars in academia have speculated that the universe may be bounded yet infinite, or that it could be modeled as a Möbius strip, Klein bottle, or similar structure, implying a unified cyclic system with no end or boundary. However, the exact nature of such cycles remains unclear, and we can speculate that a toroidal universe is one possible model.

Like the other two models mentioned above, the universe in the cosmic space toroidal model can also have many, even "infinite," universes, similar to the small worlds inside our smallest particles "wave particles," meaning there exist numerous ones

The parallel universe to ours. Here, parallel refers to physical parallelism and coexistence, similar to the parallel existence between different planets in the universe, not the concept of parallel timelines or alternate worlds found in science fiction.

(4) Open Universe Space Model

The above-mentioned universe models we discussed consider the universe as a particle in the form of a super-large vortex or other similar shapes, meaning such a universe is closed, like a particle with a particle wall that isolates it from the outside world.

But the universe may not be a completely closed particle; it could be in the most basic soliton form, such as vortices. In this case, the universe's superfluid would no longer be an isolated, closed space disconnected from the external world, but rather an open part of the larger external world.

Figure 71: Open Universe Model 8-58

In such an open universe model, if we could leave the universe, that is, the edge of the vortex, we might discover different forms of matter, similar to how in our macroscopic world, leaving water would result in different forms of matter such as atmosphere and land, as shown in Figure 8-58.

The universe space of the several cosmological models discussed above, aside from the basic closed particle model, cannot be ruled out as possibly being such an open model.

(5) Isolated Universe Space Model

The isolated universe space model is the most extreme model, meaning the universe is the only existence, with no larger big world or smaller small world. When people step out of the edge of the universe, they emerge from the smallest wave particles and return to the universe, or leave from different edges of the universe and return through different wave particles, or directly return to the universe from different spatial positions.

In other words, the universe has no distinction between inside and outside; outside the universe is inside the universe. There is no difference between large and small; the largest is the smallest. There is no separation between matter and space; space is the essence of matter, and matter is the attribute of space. They are two aspects of the same entity.

In summary, the several cosmic space models mentioned above, including the linear, emergent, circular, and open models, all belong to the later cosmological theory we discussed earlier (see Chapter 8.12), that is, the universe is a closed system formed later, and thus they are all later universe models. The fifth model, the isolated universe space model, belongs to the earlier cosmological theory we discussed, that is, the earlier universe model, meaning the universe exists from the beginning.

Since our current exploration and understanding of the universe space are still in a very初级 stage, our models of the entire universe space can only be more based on imagination and speculation grounded in the real world. However, regardless of what the real universe space is like, it should belong to one of the above models or their variations.

Of course, there are many more mysteries in the universe that await our continued exploration.

8.14 The Essence of Quantum Entanglement and Quantum Interference

(1) Quantum Entanglement
(1) The essence of quantum entanglement
Traditional quantum theory holds that two entangled electrons remain correlated even when separated by vast distances; that is, when one is manipulated (for example, through quantum measurement) and its state changes,

When one particle undergoes a change, the other will instantly experience a corresponding change. This ghostly action at a distance violates the principle of locality as stated in special relativity. This is also the reason why Einstein, Boris Podolsky, and Nathan Rosen proposed the EPR paradox in 1935 to question the completeness of quantum mechanics.

In quantum entanglement theory, if we do not measure, we cannot know the individual states of the two electrons. If we measure electron 1 and determine its state, then the state of electron 2 is also determined simultaneously.

This is considered as the entanglement of electron 1 and electron 2 in traditional quantum theory, where the measurement of electron 1 instantaneously affects electron 2, determining its state instantly.

Although many experiments related to the EPR paradox seem to confirm this phenomenon of quantum entanglement, as discussed in the introduction to space quantum dynamics (Chapter 7.4), these experiments can only serve as necessary conditions to prove the existence of quantum entanglement, not sufficient conditions. The so-called entanglement in quantum mechanics is actually determined at the moment the two quantum particles are created, and has nothing to do with non-local interaction. There is no non-local correlation, and this so-called non-local entanglement is merely a mathematical description within the statistical framework of quantum mechanics, not a physical objective reality.

The essence of quantum entanglement is a symmetry formed between two or more quantum particles at the moment of their creation, due to the mechanism of their generation or interaction. This symmetry causes the different particles to be in specific quantum states. At this moment, all the quantum states have already been determined. When we measure the state of one particle, we can immediately know the state of the others through this symmetry relationship.

As analyzed in the principle of antimatter (Chapter 8.6), when the superfluid vacuum encounters a particle obstacle, it forms a vortex particle pair behind the particle. One of these vortex particles has a positive spin, and the other has a negative spin. We can consider these two particles to be in an entangled state. If we test one particle and find it has a positive spin, we can immediately know the other has a negative spin, as shown in Figure 8-59.

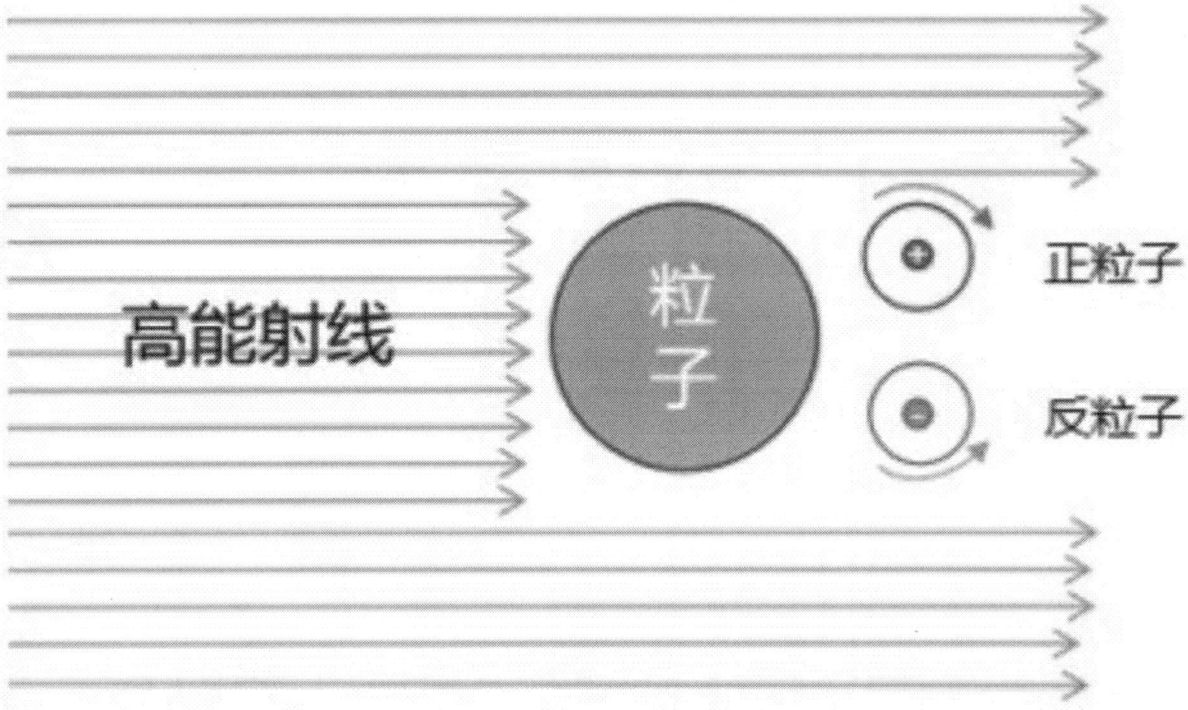

Figure 72: The "entangled state" of particles with opposite spins formed after high-energy rays impact a particle

Figure 8-59 Formation of "entangled states" of matter and antimatter particles

In other words, the so-called entanglement of these two particles is due to the symmetry of their production mechanism, which gives them corresponding specific "quantum states." This so-called "quantum state" is determined at the moment of their production, and no measurement made on one particle afterward will have any so-called "nonlocal effect" on the other.

The same principle applies to other types of entangled states, which are formed through production or interaction mechanisms, resulting in a distribution of particles in specific "quantum states." These states are determined at the moment of production and remain unchanged afterward, regardless of measurement, and do not influence each other.
(2) Quantum entanglement phenomena in the macroscopic world
Understanding the essence and principles of quantum entanglement, we can easily replicate and demonstrate this "mysterious" quantum entanglement phenomenon in the macroscopic world.

For example, if we push a plate or another object forward in a pond, this will create two vortices in the pond, with their vortex directions opposite to each other. In this way, we prepare two vortices that are in an "entangled state," as shown in Figure 8-60.

We can determine that one vortex is positively spinning and the other is negatively spinning. If we do not measure (i.e., observe) them, they will remain in an "entangled state," and we will not know which one is positively spinning and which one is negatively spinning.

However, once we measure one of the vortices (i.e., observe it), we immediately obtain the spin "quantum state" of that vortex and simultaneously know the spin "quantum state" of the other vortex, no matter how far apart they are. "As if there is a non-local instantaneous interaction, causing the wave function of the other vortex to collapse instantly, and its quantum state to be determined at once."

This is the entire process of quantum entanglement and measurement.

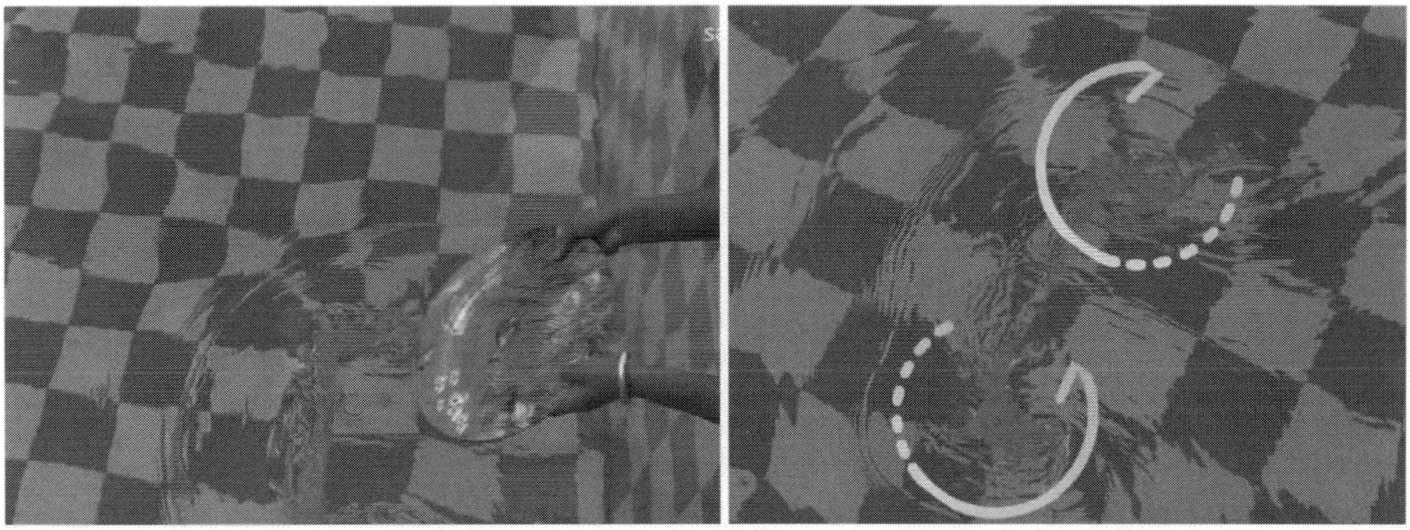

Figure 8-60 Macroscopic Quantum Entanglement Phenomenon

For example, if we light two firecrackers and throw them out, and only one goes off, we know that one has exploded and the other hasn't, but we don't know which one is which. They are in an "entangled" state. When we measure one firecracker and determine whether it is in the "exploded" or "not exploded" quantum state, the quantum state of the other firecracker is instantly determined, no matter how far apart they are, as if there is a "instantaneous action at a distance."

Of course, this so-called wave function collapse and instantaneous action at a distance do not actually exist; they are merely interpretations.

A computational model of the mathematical description system of quantum mechanics, not an objective physical reality. The so-called quantum entanglement is actually a distribution on a specific "quantum state" formed between quantum particles due to symmetry and other characteristics in the process of quantum generation or the mechanism of acting on quantum particles. This distribution of quantum states is determined at the moment of generation and remains unchanged regardless of subsequent measurements.
(2) The essence of quantum double-slit interference
(1) The essence of quantum double-slit interference
The double-slit interference experiment is considered one of the most mysterious and confusing phenomena in quantum mechanics, casting a shadow of mysticism over the entire field. The famous physicist Richard Feynman once stated in his book "Feynman Lectures on Physics" that the quantum phenomena demonstrated by the double-slit experiment are impossible, absolutely impossible, to explain in any classical way, and it embodies the core ideas of quantum mechanics.

But as we analyzed in the introduction to quantum dynamics (Chapter 7.2), in fact, what is manifested in the double-slit interference in the quantum realm is the apparent wave-particle duality formed by the synthesis of microscopic particles and spatial wave fluctuations.

Microscopic particles move linearly in their own way, but due to the influence of the superfluid vacuum space, they do not move strictly in a straight line. Instead, they vibrate in left-right and up-down directions as they move forward. For photons or electrons, this vibration is known as polarization. This polarization causes their propagation to exhibit the characteristics of transverse waves, thereby demonstrating the wave-particle duality.

At the same time, in the Young's double-slit interference experiment, when photons, electrons, and other particles pass through the double slits, the longitudinal waves of the superfluid vacuum space that travel with them form interference patterns behind the slits, which affect the propagation paths of the photons, electrons, and other particles. These particles choose their propagation paths according to the principle of minimum energy consumption, meaning they take the path that requires the least amount of energy. Since the interference of the longitudinal waves of the superfluid vacuum space remains constant, this also allows the propagation paths of photons, electrons, and other particles to have a general directional preference, plus the initial one when passing through the double slits

Differences in conditions lead photons, electrons, etc., to have different probabilities of choosing specific paths, ultimately accumulating on the display to form an interference pattern, as shown in Figure 8-61.

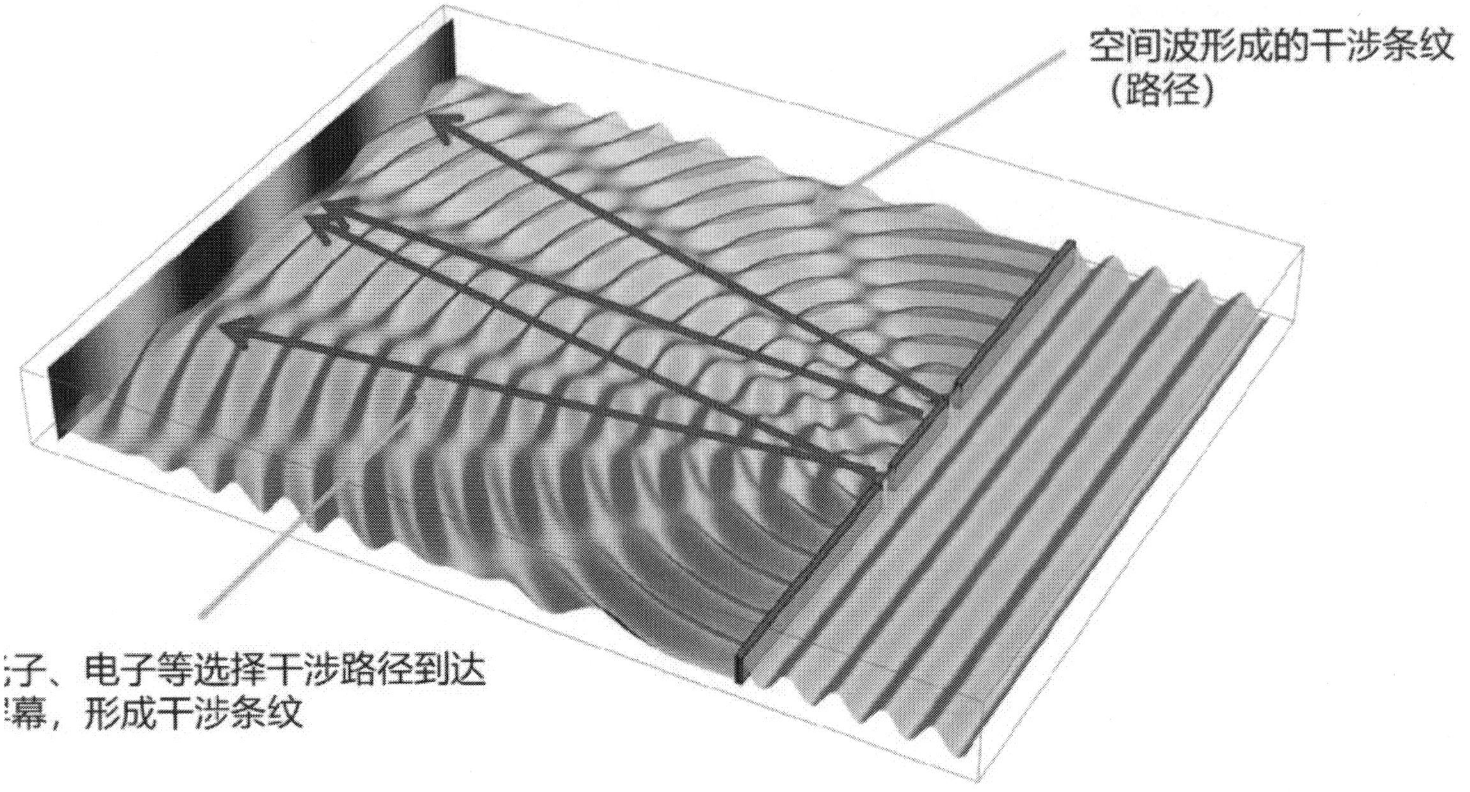

Figure 8-61 Particles propagate along the interference paths of spatial waves, forming an interference pattern

This seems, from the experimental results, to be a mysterious "single photon or single electron passing through both slits at the same time and interfering with itself," but the reality is clearly different.

Double-slit interference of single photons or electrons is essentially the result of photons or electrons moving along the interference paths formed by spatial wave interference, and the final pattern observed is that of spatial wave interference, not the result of "self-interference" of the photons or electrons.

When we observe the double-slit interference through instruments, the act of observation disturbs the spatial wave after the double slits, destroying the interference state and restoring the disordered state. In this case, the motion of the photons is no longer influenced by the spatial wave interference state, reverting to a basic linear state, which causes the interference pattern to disappear.

As we discussed in Section 7.2, there are no mysteries in the universe; any observed so-called mystery can be reasonably understood and explained through analysis of the underlying principles. The mystical and metaphysical views and interpretations imposed on quantum mechanics due to phenomena such as wave-particle duality, double-slit interference, and quantum entanglement are clearly

It is inappropriate.
We should have firm confidence in the materiality and determinism of the physical world.

(2) Quantum double-slit interference in the macroscopic world

Understanding the fundamental principle of quantum double-slit interference, which is that particles choose paths according to the principle of minimum energy consumption, and propagate through space following wave interference paths, ultimately forming an interference pattern. According to this principle, all particles, even macroscopic matter, can completely reproduce this "quantum double-slit interference" experiment.

As the British statistician Galton once designed a Galton board, where each black dot represents a nail fixed on the board, and they are equally spaced. The horizontal position of each nail in the upper layer is exactly in the middle of two nails in the lower layer.

Drop a small glass ball with a diameter slightly smaller than the distance between two adjacent pins into the Galton board entrance. As the ball falls, it hits the pins and has a probability of $1/2$ to roll left or right, then encounters the next layer of pins. This process continues until the ball reaches one of the slots at the bottom. If a large number of such balls are dropped continuously from the entrance, they will pile up on the bottom plate forming a density function graph that approximates a normal distribution, which is high in the middle, low at both ends, and symmetric like an ancient clock shape, as shown in Figure 8-62.

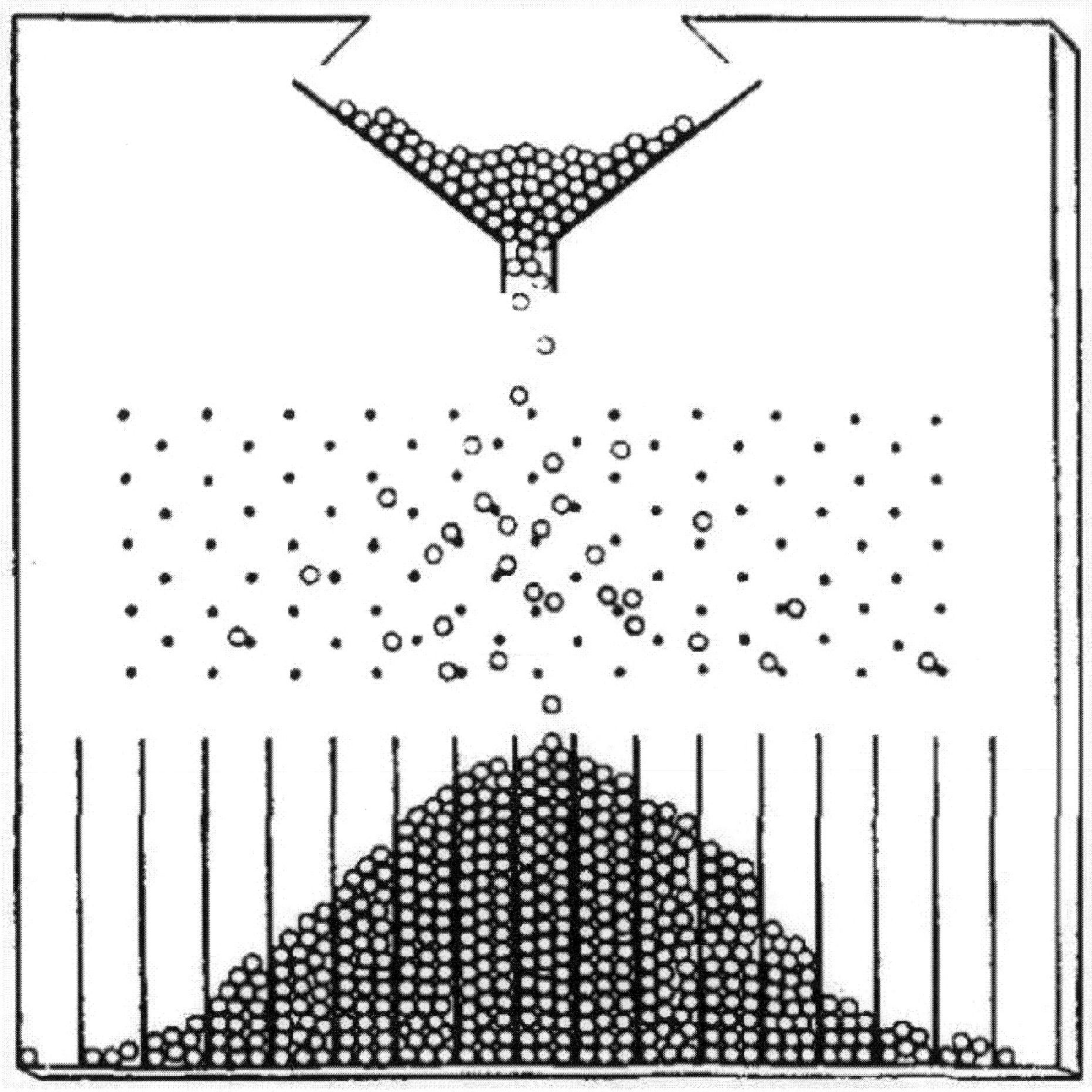

Figure 73: Figure 8-62 Normal distribution formed by the falling of balls in the Galton board

The principle of the Galton board is actually the same as the double-slit interference principle we mentioned above, as both are special distribution patterns formed by particles during their movement and propagation due to path selection.

According to the same principle, we can also modify the Galton board to create an interference path design, by placing the pins that block the path along the interference lines of the wave ripples, that is, along the paths where the wave amplitude is zero. In this case, as the balls fall, they will choose the path of least energy consumption, which is the path without pins, or the path without interference. As a result, the distribution formed at the bottom will not be a normal distribution, but an interference fringe distribution, as shown in Figure 8-63.

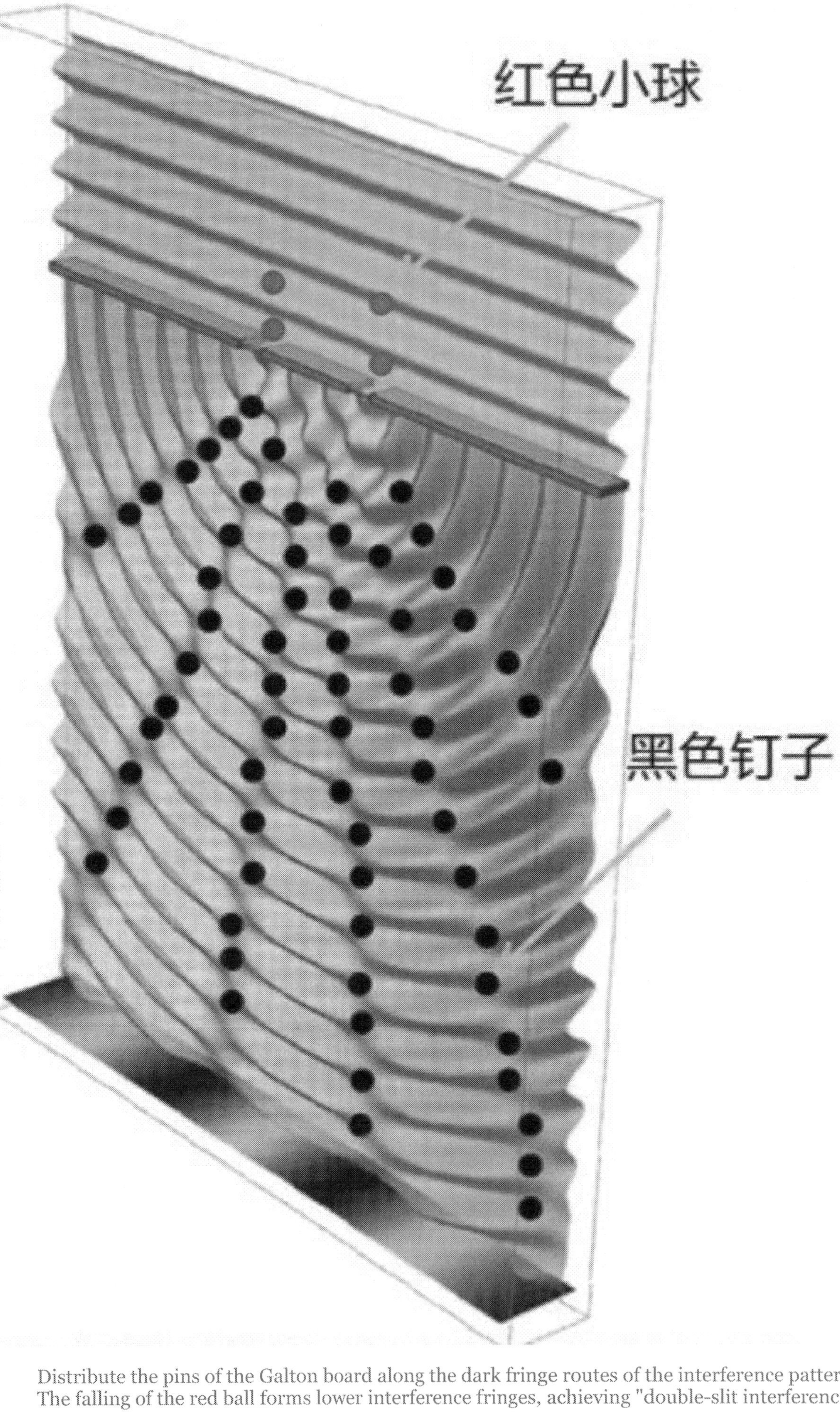

Distribute the pins of the Galton board along the dark fringe routes of the interference pattern,
The falling of the red ball forms lower interference fringes, achieving "double-slit interference"

Figure 8-63 "Double-slit interference" in the macroscopic world

This is the quantum double-slit interference in the macroscopic world, following the same principle as that of photons, electrons, and other microscopic particles.

8.15 The Emergence of Determinism and Randomness

We have discussed earlier in the cosmological model that our universe's material world operates according to determinism

and causality. In our basic principles of space relativity 1: the principle of objective physical laws (Chapter 3.1), it was also mentioned that according to the objectivity of physical laws, the objectivity and uniqueness of causality necessarily follow. This is also the fundamental principle from which we derive the determinism of the universe's material world.

But can randomness appear in such a deterministic material world? The academic community has actually been making efforts in this area all along.

(1) Current research on randomness

The study of random numbers is an important field in current randomness research. Currently, there are mainly two types of random number generation: one is pseudo-random numbers, which are generated through mathematical formulas and algorithms. These are considered not truly completely random, hence called pseudo-random numbers.

In contrast, the academic community has been continuously seeking for truly random numbers. The core of true randomness lies in utilizing the randomness of physical processes to generate random numbers that are unpredictable by humans. The main methods include:

Circuit noise: such as random voltage fluctuations caused by the thermal motion of electrons in conductors.
Quantum effects: such as photon detection based on the uncertainty principle of quantum mechanics, or the random time intervals of radioactive decay.

Environmental noise: such as receiving electromagnetic noise from the atmosphere through an antenna, i.e., atmospheric noise; or the subjective time intervals from user operations like keyboard input and mouse movement, which can also serve as a source of randomness.

Chaotic systems: such as using nonlinear components in chaotic circuits (e.g., chaotic oscillators), relying on their extreme sensitivity to initial conditions to generate randomness.

Etc.
But as we discussed earlier, this so-called randomness that relies on physical phenomena is essentially the result of material processes. In the real world, all material operations and interactions follow determinism and causality. The reason they appear to exhibit所谓的 randomness is merely due to our limited observational and computational capabilities

The lack of completeness is not truly random, as in the case of the so-called quantum uncertainty principle. Similarly, chaotic systems can only be said to be unpredictable, yet they still conform to determinism and causality. Therefore, this randomness is not genuine randomness.

(2) The Three-Body Problem

The three-body problem is a classic challenge in celestial mechanics and mathematics, studying the motion of three point masses under the influence of universal gravitation. Its complexity arises from nonlinear dynamics and chaotic properties, and there is still no complete analytical solution to date.

Like other chaotic systems, the three-body system also conforms to the determinism and causality of the physical world, and is inherently predictable. However, the ability to predict is limited by observational and computational capabilities, that is, the precision of initial condition data acquisition and the capacity of the mathematical calculation system, which are essentially engineering issues.

In fact, if single-body and two-body systems can still be analyzed and considered from a particle perspective, starting from three-body systems and beyond, such as four-body and multi-body systems, they gradually enter the realm of fluid mechanics. The complexity of the three-body and multi-body problems lies in the fact that they represent the transitional stage from particle mechanics to fluid mechanics.

Fluid dynamics includes not only the linear motion of individual particles but also wave motion. As we have mentioned several times before, photons are particles themselves, but when moving through the superfluid vacuum space, they are influenced by the superfluid and produce wave-like behavior, eventually manifesting as transverse wave motion, resulting in polarization and forming complex characteristics such as electromagnetic vortices in the vacuum space.

The motion of objects in the Three-Body problem has also begun to show the influence of this direction, and it is precisely for this reason that it has become one of the research directions in the study of three-body or multi-body problems.

Precisely because of the complexity of the Three-Body problem, which seems to contain infinite possibilities, everything in the world may arise from the different changes in the three-body system. This might be one of the sources of the ancient

Chinese saying "three gives rise to all things."
(3) Emergence of randomness that breaks the law of causality
(1) Determinism of physical laws: Causality

In the previous section on the five possible models of cosmic space (Chapter 8.13), we discussed that a deterministic material world may give rise to uncertainty.

Just as our current world operates based on interactions between substances, it must necessarily conform to determinism and the law of causality. How can randomness emerge in such a world?

Can our human initiative break through this determinism and causality, truly affecting the operational laws of the objective material world?
(2) Methods to Break Causality
To truly break the determinism of material operations, we must rely on non-material forces outside the material world.

The mathematical system, which we are all familiar with, is another purely conceptual system outside the material operational system. For example, we can use any irrational number, and when making a decision, we can use a random number generator to select a digit at the corresponding position after the decimal point of an irrational number, as the basis for the next step. In this way, the digits in the irrational number are completely unrelated to the operation of the physical world, because it is a purely conceptual existence that does not depend on the operation of the material world.

Decisions made based on such concepts of numbers continue to drive the operation of the material world, thereby breaking the control of causality. The previous action and the next decision are no longer entirely deterministic interactions between matter following mechanical effects, but instead introduce new factors beyond mechanical interactions between matter, so the causality governing the operation of matter no longer applies between these two actions.

This process still conforms to the principle of determinism, but it does indeed break the physical causality: there is no physical causal relationship between the two steps.

When human subjectivity begins to rely on such conceptual systems to influence the operation of the material world, it frees humans from the control of causal laws in the material world, allowing the world's operational laws to be understood as

Human subjective consciousness transcends and exerts reverse influence and transformation.
Therefore, the operation of the objective material world strictly follows the law of causality, but human subjective initiative can break the law of causality, meaning that people can take control of their own destiny.

The philosophical view of determinism that arose due to the emergence of Newton's classical mechanics and caused despair for everyone is clearly wrong.

The emergence of quantum mechanics made people feel as though they had found randomness, but this randomness is not truly random. Therefore, the uncertainty principle based on quantum mechanics is also inaccurate. However, it aligns with our philosophical view that "human subjective initiative can break the law of causality."
(3) Concept System Types
To break the material causality formed by the physical world operating according to mechanical laws, we need to use a concept system outside of the material world. The irrational numbers we mentioned above are exactly such a concept system.

Concept systems outside the physical world have many types. Overall, they can be divided into those invented by humans and those discovered by humans.

The conceptual systems invented by humans, such as color, mass, length, etc., which are formed based on how the physical world operates, including the act of human invention itself, all follow the mechanical laws of the physical world. Therefore, although these human-invented conceptual systems are non-material in nature, their creation and application are inseparable from the operation of the material world, and can even be said to be the result of the material world's functioning.

This type of conceptual system, which we may call a conceptual system based on the physical world, is unable to disrupt the material causal laws of the material world.

When you try to make important decisions by means such as drawing straws or flipping a coin, gambling a chance and changing your fate, these actions are actually operations within the physical world and are still under the control of causality. They do not change your fate; they are your fate itself.

Another concept system is one that humans have discovered, meaning that this concept system exists inherently and is not created by humans.

Conceptual systems that are material or have no relation to any material mechanical interaction, which operate independently of the workings of the material world, are completely pure conceptual systems. For example, as we mentioned above, irrational numbers have no relation to human subjective will or any mechanical laws of the physical world; they are not affected by causality and are purely conceptual systems unrelated to the physical world.

When such pure conceptual systems enter the physical world, they can break the chain of physical causality, leading to uncertainty in the physical world.

Because such a process is not influenced by any physical laws, it is also not subject to the intervention of causality.

This conceptual system, or a collection of such conceptual systems, including other similar logics, rules, etc., is the one discussed in the basic principles of spatial relativity 1: the principle of objective physical laws (Chapter 3.1), which refers to a virtual world independent of our material world.

The effects from the virtual world onto the material world (real world) are, for the material world, causeless.

The result.

The operation of the material world has transferred into the virtual world, which is the cause of the material world with no result.

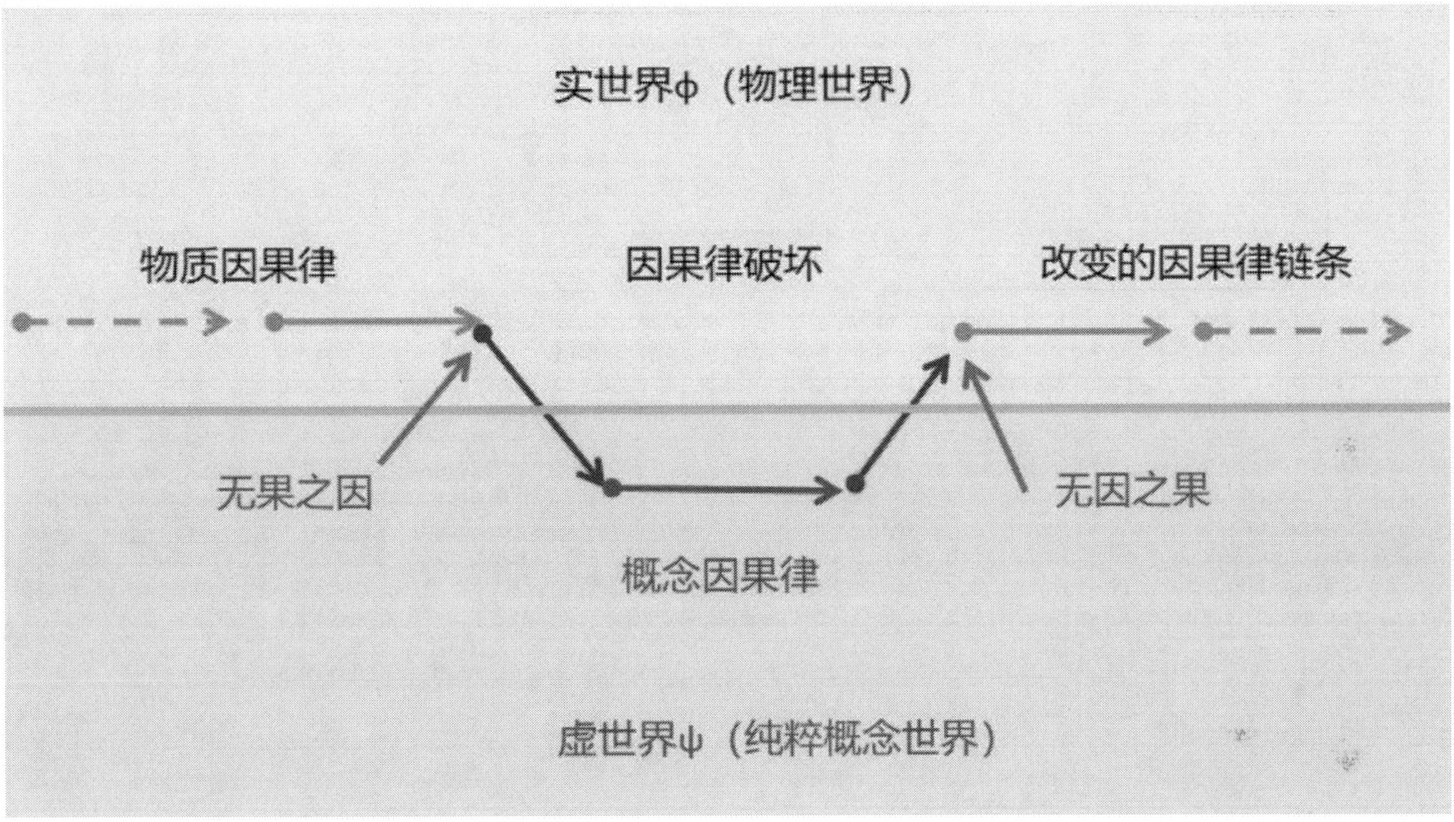

Figure 8-64 The Destruction of Physical Causality

When the operation of the material world exhibits effects without causes and causes without effects, the causality of the material world is disrupted, leading to physical randomness.

This is the emergence of uncertainty and randomness in a deterministic physical world, as shown in Figure 8-64.
Of course, when considering the entire world including the real world and the virtual world as a whole, the laws of causality and determinism remain intact. However, for the real world, the physical causality and determinism that rely entirely on mechanical laws have already been disrupted.

This allows humans or sentient beings to break through the causal laws of the physical world, gaining the ability to transform the physical world. They also possess the power to change their fate by altering the causal development chain through the modification of causality.

Concept systems similar to irrational numbers exist in many forms in real life, and there are even more types that humans have yet to discover. These concept systems, used unconsciously by humans, have already broken the control of physical world causality, enabling humans to take control of their own destiny.

(4) Origin of the Conservation of Mass and Energy Principle

Because the material world is deterministic and follows the laws of causality, it adheres to the principles of mass and momentum conservation, and energy conservation. But what is the fundamental reason behind the conservation of momentum and energy?

Does the existence of mass in the vacuum superfluid space itself play a role? Such a vacuum superfluid would exert uniform external forces, resistance, etc., on all objects within the space. Is this the foundation upon which physical laws such as momentum and energy conservation are based?

When a force is applied to an object, the object experiences a constant acceleration, with its speed increasing uniformly. What is the mechanism behind the acceleration and the change in speed? Why does the object's speed not immediately become infinite?

If there were no vacuum superfluid space, and the object were in a complete vacuum, would the motion state and laws of the object under force change? Would the speed increase significantly?

After experiencing a force, an object's speed does not increase significantly, and there is a limitation. Is this due to the incomplete rigidity of particles or matter? If particles or matter were completely rigid, the effect of the force would be instantaneous. Would the instantaneous force during a collision be infinite, leading to a more significant speed effect?

In classical physics and engineering, the law of conservation of energy is an empirical law derived from experimental observations, similar to many descriptive theories, and it is known to be true without understanding the underlying reasons.

The Noether's theorem, proposed by mathematician Emmy Noether in the early 20th century, states that continuous symmetries correspond one-to-one with conservation laws. This theorem applies to all physical laws based on the principle of action. According to Noether's theorem, the essence of energy conservation can be considered as a manifestation of time translation symmetry. However, due to the complexity of energy issues in the universe, the form of energy changes with the development of theories, and whether it necessarily conforms to time translation symmetry remains uncertain.

The essence and universality of the law of conservation of energy still require further exploration and verification as scientific understanding continues to develop.

8.16 Relationship with the Standard Model

The Standard Model is the fundamental theory that describes the behavior of subatomic particles and their interactions, mainly covering three fundamental forces: electromagnetism, the strong force, and the weak force. It successfully integrates all known fundamental particles and unifies most phenomena in particle physics through a mathematical framework, aligning with most experimental observations, making it a successful theory in particle physics.

The Standard Model is derived from the analysis of higher-level particles, while the general theory of spatial particle dynamics serves as a supplement and correction to the Standard Model. It explains the formation of fundamental particles from the origin of the most basic particles, analyzing them from multiple dimensions such as spin, vibration frequency, orbital motion, matrices, flow fields, and their interactions with the superfluid vacuum space. This leads to the formation of various mechanical effects, providing a complete picture of the nature, origin, composition, and evolution of particles, and forming an explanation based on physical principles. These aspects are missing in the Standard Model.

At the same time, spatial particle dynamics also corrects some issues of the standard model in certain areas.

An introduction to spatial particle dynamics provides the main supplements and corrections to the standard model:

(1) Supplement to the Particle Standard Model

(1) The Essence and Origin of Particles
Based on the fundamental principles of space relativity 2: the Principle of Objective Space and fundamental principle 6: the Principle of Relative Space, this provides a principle-based explanation and description of the mechanism of particle generation in the vacuum space superfluid and the intrinsic nature of particles.
(2) Explanation of Particle Composition and Evolution Path
An analysis of the structural composition of particles and the evolutionary path from initial particles to composite particles up to macroscopic particles is provided, supplementing the missing mechanisms of the formation and evolution of more microscopic particles beneath the basic ones in the standard model.
(3) The essence of charge and the essence of matter-antimatter
The origin and mechanism of matter-antimatter asymmetry are explained according to a unified principle for issues that the standard model cannot account for, such as the essence of charge and the essence of matter-antimatter.
(4) The origin and mechanism explanation of matter-antimatter asymmetry
Provides a theoretical explanation for the matter-antimatter asymmetry and the problem of symmetry breaking in the universe that the standard model cannot account for, in line with experimental observations.
(5) Origin and Mechanism of Gravity
Supplements the standard model's inability to explain the origin of gravity.
(2) Modification of the Particle Standard Model
(1) Origin and Mechanism of Electric Fields, Magnetic Fields, and Electricity
The essence and distinction of electric fields and magnetic fields are presented, offering a unified theoretical description of the nature and formation mechanisms of electric force, magnetic force, weak force, and strong force, which aligns with experimental observations and astronomical observations.
(2) Understanding of Particle Mass

The standard model's assumption that particles such as photons and neutrinos have no mass has been corrected. According to space particle dynamics, which considers the essence, composition, and evolution of particles, all particles have mass. Current experimental observations have detected that neutrinos possess extremely small masses, consistent with our theoretical predictions.

(3) Understanding of Particle Internal Structure

The Standard Model assumes that particles such as electrons have no internal structure and are point-like entities. Space Particle Dynamics provides clarification and correction, stating that all particles we observe have internal structures and are composed of more microscopic particles.

(4) Understanding of Particle Spin

Due to concerns that the spin speed of particles like electrons might exceed the speed of light, the Standard Model and other theories have avoided addressing the fundamental nature of spin, considering it merely an intrinsic property of particles. Space Particle Dynamics has made corrections to this understanding.

The speed of light is not the ultimate limit of particle motion; all interactions conform to the new principle of locality, and the spin of electrons can exceed the speed of light. This spin is an important mechanism for the formation of particle charge.

8.17 Relationship with Fluid Cosmology (SVT)

Half a century ago, the idea that "spacetime is a fluid" had already appeared in academic circles, and this theory is known as the "superfluid vacuum theory" (SVT). In recent years, a theory that may be a "liquid superfluid" (liquid-superfluid) has also been re-proposed by many scholars, such as Dr. Roberto Bellati from the International School for Advanced Studies (SISSA) in Italy, and scientists like Mauro Maccione from the Ludwig-Maximilians-University in Munich. They have provided new interpretations of this theory, which share the same understanding as spatial relativity and spatial particle dynamics.

Both theories believe that there exists a fundamental superfluid throughout the vacuum space, akin to an ocean of emptiness in the universe. All stars and matter originate from this superfluid and drift within it.

However, fluid cosmology differs from space relativity in aspects such as the asymmetry of matter and antimatter in the origin of matter, the formation principle of gravity, the formation and evolution of the universe, the formation and evolution of electromagnetic fields, and the mechanism of formation and propagation of electromagnetic waves.

Space Relativity, space particle dynamics, and fluid cosmology can be considered two theoretical systems based on the same theory of the vacuum space superfluid, with different interpretations in some fields.

Introduction to Spatial Soliton Dynamics

9.1 Definition of Quantum Solitons

As we discussed earlier in the introduction to particle dynamics, the origin of particles is the vortices, vortex pairs, or more energetic vortexons, vortex rings, vortex balls, etc., formed by the excitation of the vacuum space superfluid. These fundamental particles share the common motion characteristic of having spin, and high-energy vortex balls and other particles will possess complete vortex flow fields.

Composite particles and macroscopic matter particles formed by these fundamental particles will naturally contain these vortex motions within them. In addition, they will also exhibit overall linear motion and vibrational motion resulting from interactions with the surrounding vacuum space superfluid or other particles. These combined motions form the overall composite motion of the particles, also known as thermal motion.

All particles exhibit thermal motion. For composite particles and macroscopic matter, the various thermal motions of a large number of internal particles interact with each other, eventually forming a balance and reaching a state similar to resonance, resulting in overall thermal motion.

All particles originate from the superfluid of the void space and move and drift within it, interacting continuously and complexly at the quantum level with the superfluid of the void space or other particles and matter. They are individual entities with coherent motion structures, which we can refer to as quantum solitons.

All particles and matter in the universe are formed from quantum entities and continuously undergo quantum interactions, so they are all quantum solitons. These quantum solitons are like countless islands in a quantum ocean, forming the material world structure of the universe.

The medium particles, called waveons, that form the superfluid of the void space are the smallest quantum solitons. Quantum solitons can be considered as the whole of matter and its induced quantum field, wave field, etc.

9.2 Characteristics of Quantum Solitons

All quantum solitons have the following characteristics:
(1) Mass, volume, motion
All quantum solitons are particles formed by the excitation of superfluid energy in empty space, and macroscopic matter composed of these particles is also formed from the most fundamental bosons, so they all possess the characteristics of mass, volume, and motion.

Bosons can be considered as the most fundamental quantum solitons with the smallest mass, volume, and energy.
(2) Internal Structure
All quantum solitons have internal structures, including the most basic bosons. Even macroscopic particles, matter, and composite particles above are composed of a large number of quantum solitons, forming a composite quantum soliton system.
(3) Quantum Fingerprint
All quantum solitons exhibit thermal motion, as the differences in their internal complexity and interactions with surrounding quantum solitons and superfluid vacuum space result in significant variations in each quantum soliton's quantum thermal motion. If in the macroscopic world no two leaves are exactly the same, then in the microscopic world, no two quantum solitons have exactly the same quantum thermal motion.

In other words, the thermal motion characteristics of each quantum soliton are unique, similar to each quantum soliton's quantum fingerprint, differing from one another, with each being distinct.

It could even be said that all the particles within each quantum soliton, including the most fundamental isolated particles, are themselves quantum solitons and possess unique quantum fingerprints.

The motion of each quantum soliton changes over time, but its overall quantum fingerprint can remain unchanged for a long period.

(4) Holistic Solitary Flow Field

Macroscopically, a quantum solitary as a composite system is generally formed by a large number of quantum solitons. The thermal motions of all these quantum solitons interact with each other, eventually forming a stable overall resonant state. This resonant state is manifested as the thermal motion of the collective flow field formed by the quantum solitary and the surrounding vacuum superfluid, including waves, flows, vortices, etc. This flow field is referred to as the solitary flow field of the quantum solitary.

Since these solitary flow fields are the result of internal resonance, they can be considered as the result and manifestation of the coherent motion of all the sub-quantum solitons within. They contain the information of all the sub-quantum solitons, which means they include the holistic information of the quantum solitary.

This holistic information is not merely the surface information like that in a hologram, but also includes the internal structure and the information of the internal sub-solitons. This is what is known as the holographic solitary flow field of the quantum solitary.

The holographic solitary flow field is the foundation of the quantum fingerprint that forms quantum solitary waves, and can be considered the core of the quantum fingerprint.

(5) Quantum Interaction

Each quantum entity is immersed in the quantum ocean of the vacuum space superfluid, so it continuously interacts with the vacuum space superfluid at the quantum level through its holographic flow field. The movement of the vacuum space superfluid affects the quantum entities, and the thermal motion of the quantum entities also affects the vacuum space superfluid, forming different thermal motions such as vortices, flows, and waves in the vacuum space superfluid, which represent the exchange of thermal energy between the two.

In addition, each quantum soliton interacts with others through their respective effects on the superfluid thermal motion of the vacuum space, even through direct collisions. This interaction not only occurs between quantum solitons, but also continuously takes place between sub-quantum solitons within the quantum soliton.

It can be said that the thermal motion interaction or collision interaction between quantum solitons and between quantum solitons and the superfluid vacuum space are all forms of interaction among quantum solitons.

All quantum solitons are undergoing quantum interactions with each other when they are within each other's interaction range.

(6) The speed of quantum interaction

This quantum interaction of quantum solitons is mainly formed through the waves and flow of the superfluid vacuum space, or through the movement of quantum solitons. The waves of the superfluid vacuum space are longitudinal waves with a speed of light, while the speed of quantum solitons' movement may be less than the speed of light, or may exceed, even greatly surpass, the speed of light.

Just as in any fluid, the movement speed of an object may be less than the speed of longitudinal waves in the fluid, or exceed, even greatly exceed, the speed of longitudinal waves.

(7) System Evolution of Quantum Solitons

In addition to the most basic wave particles, all quantum solitons emerge in the vacuum space superfluid, and the interactions between these quantum solitons drive the formation of more complex and massive quantum soliton systems, thereby giving rise to a complex macroscopic material world.

Each material in the macroscopic material world is a complex quantum soliton giant system, containing countless sub-quantum soliton subsystems, making the thermal motion of these macroscopic quantum soliton systems more complex and powerful.

And countless macroscopic substances further interact and evolve, continuing to emerge various quantum isolated giant systems such as planets, atmospheres, oceans, and land. Such quantum isolated giant systems continue to evolve and give rise to various life-like entities. It can be said that all life forms, including humans, are also quantum isolated giant systems.

As quantum isolated systems become more complex and evolve, the interactions they possess also become more complex and intensified. For example, life forms including humans develop subjective consciousness, thus enabling more complex interactions with their surrounding environment.

9.3 Interaction Between Quantum Solitons

As we discussed earlier, the interaction between all quantum solitons mainly takes two forms, one being direct

The collision interaction, one is an indirect interaction through thermal motion in the superfluid of empty space.
The electromagnetic force, gravity, strong and weak forces between substances discussed earlier are all forms of indirect interaction.

Direct collision interactions between particles, such as annihilation of matter and antimatter, belong to collision interactions.
Direct collision interactions, in addition to the direct collision of particles themselves, can also involve partial emission and collision interactions within the particles.
(1) Quantum Orientation and Communication
A particle is formed and composed of countless complex movements of wave particles, such as spin and vortices. In addition to forming the motion of the particle, these wave particles also have high-speed spin movements themselves. This powerful spin gives the wave particles an axial and directional energy, similar to a gyroscope. That is, these wave particles may maintain their direction constant during the process of forming the particle's motion. For example, if they initially point along the X-axis (idealized discussion), they will continue to point along the X-axis in the future, unless changed by collisions between particles.

Other particles with high-speed spin will also have similar orientational effects.
Therefore, if a particle, at the moment of its creation, has its internal high-speed spinning wave particles or sub-particles pointing to a certain point, they will continue to point to that point in the future. This is like a natural gyroscope compass, which can serve as a means of communication in the universe.
(2) Quantum Communication Distance and Speed
The speed of longitudinal waves in the superfluid vacuum is the speed of light. However, as the most fundamental wave particles or energy-rich solitons, they can move at speeds far exceeding the speed of light when acting as particles under high-energy excitation. Just as the movement speed of substances in water can far exceed the speed of water waves, and airplanes in air can surpass the speed of sound, communication and positioning across vast distances in the universe at superluminal speeds with high precision can be achieved through these wave particles or micro-particles.

Here, the micro-particles refer to solitons with much less energy than photons, which are fundamental vortex solitons or vortex

equivalent. Since the energy and mass of photons make their movement in the superfluid vacuum subject to significant interactions from the superfluid itself, they can only move at the speed of longitudinal waves, i.e., the speed of light.

Only smaller microparticles can better break through the limitations of longitudinal wave velocity, enabling motion at speeds far exceeding the speed of light. At the same time, due to their smaller size and mass, they can penetrate more effectively than photons or electromagnetic waves, similar to neutrinos, which can easily pass through most objects. Even ordinary magnetic shielding fields cannot prevent this, and interaction only occurs when encountering corresponding quantum fingerprint fields, similar to entangled particles or matter-antimatter annihilation. Only corresponding particles and quantum fields will interact.

9.4 The Principle of Superconductivity

Since the discovery of superconductivity by Onnes in mercury (Hg) at 4 K in 1911, after many years of research, it was recognized in 1933 that superconductors have another fundamental property: perfect diamagnetism, known as the Meissner effect. Based on this foundation, the BCS theory was completed in the 1950s, after more than forty years of development.

The BCS theory has influenced condensed matter physics and even the entire field of physics for more than half a century. However, in recent years, the anomalous properties of the superconducting and normal states of high-temperature superconductivity have challenged and questioned the authority of traditional theories. While continuing to explore higher T_c (critical temperature) superconducting materials, an increasing number of scholars have recognized that the microscopic mechanism of high-temperature superconductivity is quite different from that of BCS superconductors, but consensus [49] has not yet been reached.

Based on the theory of spatial relativity and quantum soliton theory, we can provide the following framework for understanding superconductivity.
(1) Effective mechanical range of holographic flow field
As we analyzed earlier, all particles such as electrons, atoms, molecules, etc., are quantum solitons, that is, combinations of the material particle and its quantum holographic flow field. In addition to collision effects between quantum solitons,

there are also interaction effects through their quantum holographic flow fields. All forces we observe now, including electric field force, magnetic field force, gravity, and even the strong and weak forces inside particles, are actually fluid dynamic effects formed by the mutual interactions of quantum holographic flow fields between different quantum solitons.

As the distance from a particle increases, the energy of its holographic quantum field also decreases, and so does the force exerted by the field. When the distance r increases to a certain extent, the field's energy is no longer sufficient to produce a noticeable mechanical effect on other particles, and we refer to this distance as the effective mechanical distance of the particle field r_c. See Figure 9-1.

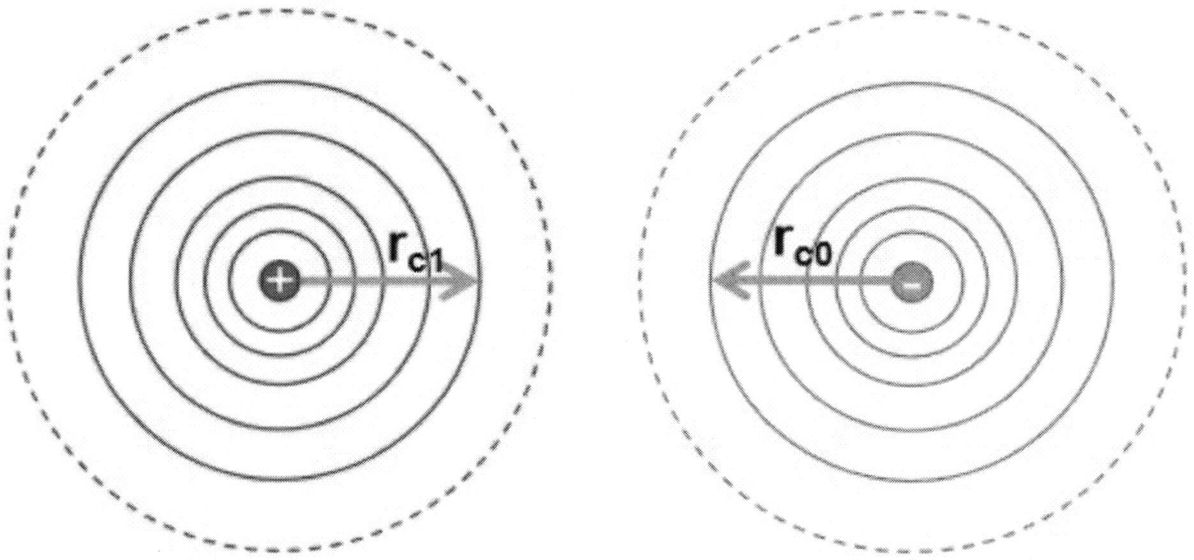

Figure 74: Figure 9-1 Effective Mechanical Distance of the Quantum Soliton Holographic Flow Field

(2) The Superconducting Principle of Conductors
When electrons move within a conductor, they interact with the holographic flow fields of all surrounding particles. These interactions manifest as corresponding mechanical effects. From a macroscopic perspective, this results in electrons experiencing various forces, which act as resistance during their motion. For the conductor, this corresponds to its resistive properties.

The greater the hydrodynamic effects of other quantum entities on the electron, the greater the resistance exhibited by the conductor. Conversely, the smaller the hydrodynamic effects of other quantum entities on the electron, the smaller the resistance exhibited by the conductor.

$$F \propto R(9.1)$$

Where: F is the total resistance experienced by electrons as they move through a conductor, and R is the resistance of the conductor.
The holographic flow field of a quantum soliton is the result of its interaction with the surrounding vacuum space superfluid, manifesting as different movements such as flow, wave, and vortex of the vacuum superfluid centered on the quantum soliton. These movements collectively represent the thermal motion of the quantum soliton and the vacuum space superfluid.

Therefore, when the temperature is lower, the overall thermal motion of the surrounding vacuum space superfluid and all quantum solitons is lower, resulting in a lower effective mechanical influence range of the holographic flow fields of all quantum solitons.

$r_c \propto T(9.2)$

Thus, the resistance experienced by electrons in this environment is also smaller.

$F \propto T(9.3)$

Where: F is the total resistance experienced by electrons moving in a conductor, and T is the temperature of the conductor. According to equations 9.1 and 9.3:

$R \propto T(9.4)$

The resistance of a conductor is proportional to the temperature.

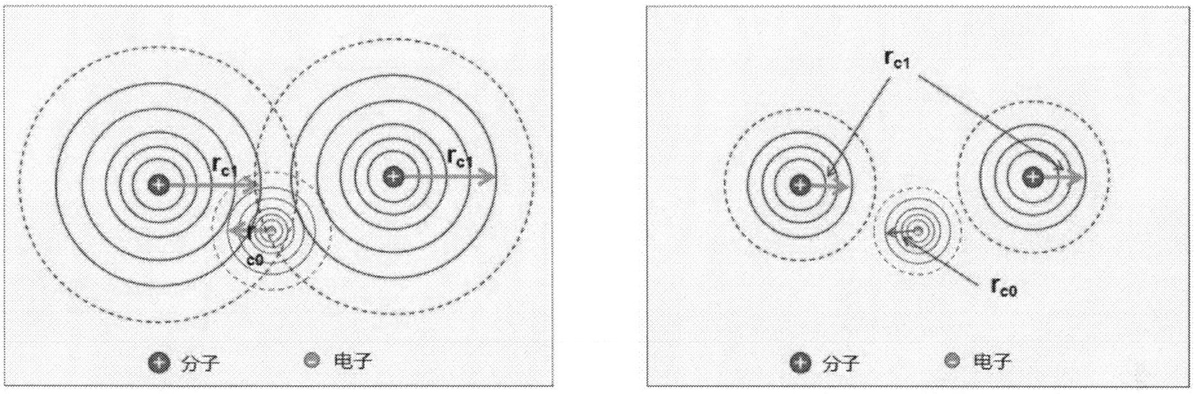

a)正常温度时导体中电子和分子的流场力学范围

b)超导时导体中电子和分子的流场力学范围

Figure 75: Figure 9-2 Effective mechanical range of molecular and electron fields in a conductor at different temperatures

It can be expected that as the temperature decreases, the effective range of the fields of various particles in the conductor will decrease, and the corresponding field mechanical effects will also diminish. When the temperature drops to a certain limit, the effective range of the field will decrease to a point where..

To a certain extent, the average distance between electrons and other atoms, molecules, etc., is greater than their respective flow field effective mechanical radii, at which point the force acting on the electrons drops almost to zero, showing zero resistance

When $T < T_c$

$$r_{c0} < \frac{L}{2} - r_{c1}$$
$$F_c = 0$$
$$R_c = 0$$

Where:
T is the conductor temperature;
Cut-off temperature;
r_{c1} as the average flow field effective mechanical radius of conductor medium molecules;
The effective mechanical radius of the conductor's electron field;
The cutoff force acting on moving electrons in a conductor;
R_c is the conductor's cutoff resistance;
L is the average distance between two medium molecules in a conductor.
Equation 9.5 is the effective mechanical effect cutoff radius equation for the superconductor medium particle flow field. Therefore, the superconducting effect is essentially due to the temperature decrease causing the distance between molecules inside the conductor to be greater than the sum of the effective mechanical radii of the two holographic fields, resulting in the difference being larger than the effective diameter of the electron's holographic field. This leads to the mechanical effects of the surrounding molecular fields on the moving electrons in the conductor being reduced to nearly zero, as shown in Figure 9-2.
(3) Isotope Effect
According to our previous analysis, the holographic flow field of a quantum soliton is a manifestation of its thermal motion. The thermal motion of quantum solitons triggers movements such as superfluid fluctuations, flow, and vortices in the surrounding vacuum space, exhibiting an overall flow field effect. Therefore, the effective mechanical half-radius r_c of the holographic flow field is related to the motion energy of the quantum soliton,

Or related directly to mass,
because the vibrational frequency of an atom (phonon energy) is inversely proportional to the square root of the atom's mass, that is:

$$\omega \propto \frac{1}{\sqrt{M}}$$

Where: ω is the vibrational frequency of the atom, and M is the mass of the atom.
Therefore, it can be known that the greater the mass of a quantum soliton, the greater its thermal motion energy, and the larger the corresponding holographic flow field's effective mechanical radius r_c becomes

$$r_c \propto \sqrt{M} \quad (9.6)$$

Equation 9.6 is the relationship equation between the holographic flow field effective mechanical radius of a quantum soliton and its particle mass, where M is the mass of the quantum soliton.

According to equation 9.5, we can further derive that, when the distance between molecules in the conductor medium $ L $ remains constant, the superconducting critical temperature of the conductor is inversely proportional to the effective mechanical radius of the holographic flow field of the conductor medium, denoted as $ @o\# $

$$T_c \propto \frac{1}{r_{c1}}$$

According to equations 9.6 and 9.7, we can infer that, at the same temperature, the larger the mass of the medium particles in a conductor, the lower the corresponding superconducting critical temperature will be, assuming the spacing between medium particles remains unchanged.

$$T_c \propto \frac{1}{\sqrt{M}}$$

If the medium in a conductor is the same isotope of a particle, then the superconducting critical temperature of the conductor will decrease as the isotope's mass increases.

In history, in May 1950, scientists at the U.S. National Bureau of Standards, such as Serrin (B. Serrin), discovered through precise measurements of the superconducting temperatures of various isotopes of metallic mercury that the superconducting critical temperature is actually inversely proportional to the square root of the isotope's mass (as shown in Figure 9 - 3) [50] . Baden of Bell Labs keenly realized the isotope effect in superconductivity.

The physical essence: the square root of atomic mass is exactly related to the atomic vibration energy, meaning that superconductivity is necessarily connected to atomic lattice vibrations. Combined with the then existing phenomenological theory and experiments of superconductivity, which had already shown that superconductivity is a second-order phase transition in the material's electronic system, it was almost certain that the "backstage driver" of superconductivity is very likely to come from the interaction between electrons and the atomic lattice.

The phenomenon discovered by Seling is the isotope effect in superconductivity.

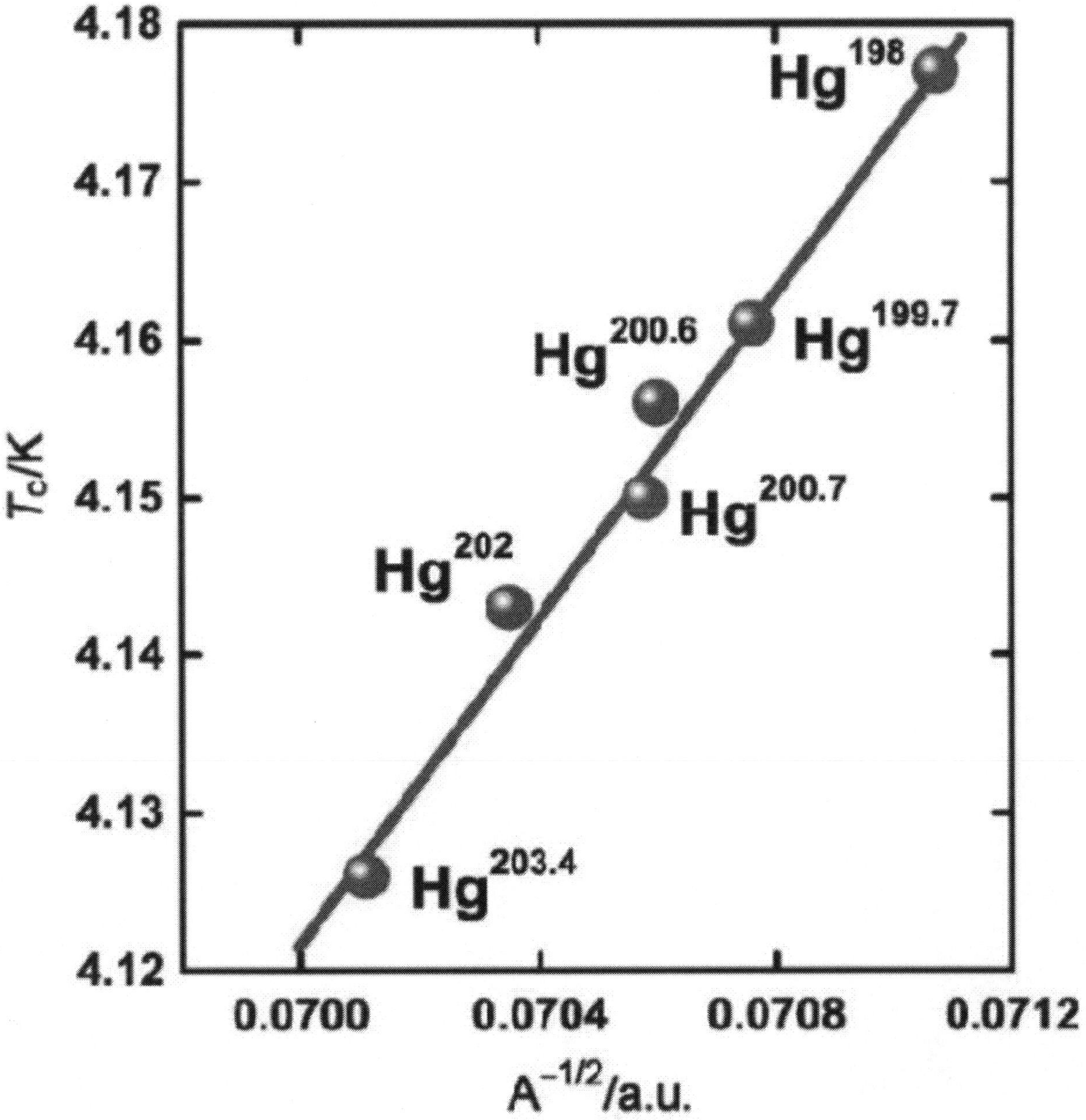

Figure 76: Isotope Effect on Superconducting Critical Temperature in Metallic Mercury (Figure 9-3)

The essence of the isotope effect in superconductivity is that as the mass M of the medium particles increases, the effective mechanical radius r_c of the holographic flow field of the medium particles also increases, leading to a decrease in the superconducting critical temperature T_c.

Of course, the increase in the mass of conductor medium particles will also bring about some changes in other properties of the conductor's crystalline structure, which will in turn affect the manifestation of superconducting effects, and in some cases even lead to anomalous isotope effects, requiring specific analysis.

(4) Meissner Effect

As discussed earlier, the reduction in the holographic flow field effect radius of the medium molecules in a conductor with decreasing temperature is essentially due to the suppression of molecular thermal motion energy, which leads to a decrease in the superfluid flow field strength around them.

Of course, this is only one aspect, because the superfluid in the void space, as a fluid or gas, is also composed of countless wave particles and has its own thermal motion. When the temperature decreases, the thermal motion of the superfluid in the void space will also be suppressed and reduced, while the superfluid will also experience an overall increase in density, that is, the thermal expansion and contraction effect.

$$\rho_c \propto T$$

Where ρ_o is the density of the superfluid in the vacuum space of the conductor, and T is the temperature of the conductor.

The reduction of the holographic flow field energy of the conductor medium molecules, as well as the decreased thermal motion and increased density of the surrounding vacuum space superfluid, is also an important factor. The greater the density of the vacuum space superfluid, the smaller the effective mechanical effect range of the same molecule's holographic flow field, that is:

$$r_c \propto \frac{1}{\rho_c} (9.10)$$

Among them, r_c is the effective mechanical radius of the holographic flow field of conductor medium molecules. When the temperature drops to the superconducting critical temperature, the density of the vacuum space superfluid increases significantly, and the motion of quasiparticles is greatly reduced. The ability of conductor's medium molecules to drive the motion of surrounding vacuum space superfluids through their thermal motion is also significantly weakened, leading to a substantial suppression of the holographic flow field, and thus a reduction in the effective range of mechanical effects. According to the equation for the cutoff radius of the medium particle flow field mechanical effects in superconductors (Equation 9.5), the conductor exhibits superconducting effects.

The same principle applies. As we analyzed earlier, a magnetic field is a vortex flow field of the vacuum space superfluid (see Section 8.3). Therefore, when the density of the vacuum space superfluid increases and its thermal motion weakens, and simultaneously, according to the second fundamental principle of spatial relativity: the principle of spatial objectivity (see Section 3.2), the conductor's medium molecules' ability to influence the vacuum space...

The distortion and dragging effects of the superfluid in space, when acting together, cause the superfluid in the void space to form a similar state to the semi-frozen condition along with the conductor. In this state, the vortex flow field of the magnetic field will encounter greater resistance, leading to a weakening of the magnetic field effect.

When the conductor's temperature drops to the critical temperature, the density of the superfluid in the internal void space increases significantly, and thermal motion decreases greatly. As a result, the magnetic field flow that was originally propagating through the conductor will be expelled, similar to how water in a flowing river would be forced to detour when the water in the center cools down significantly or even starts to freeze. The magnetic field follows the same principle.

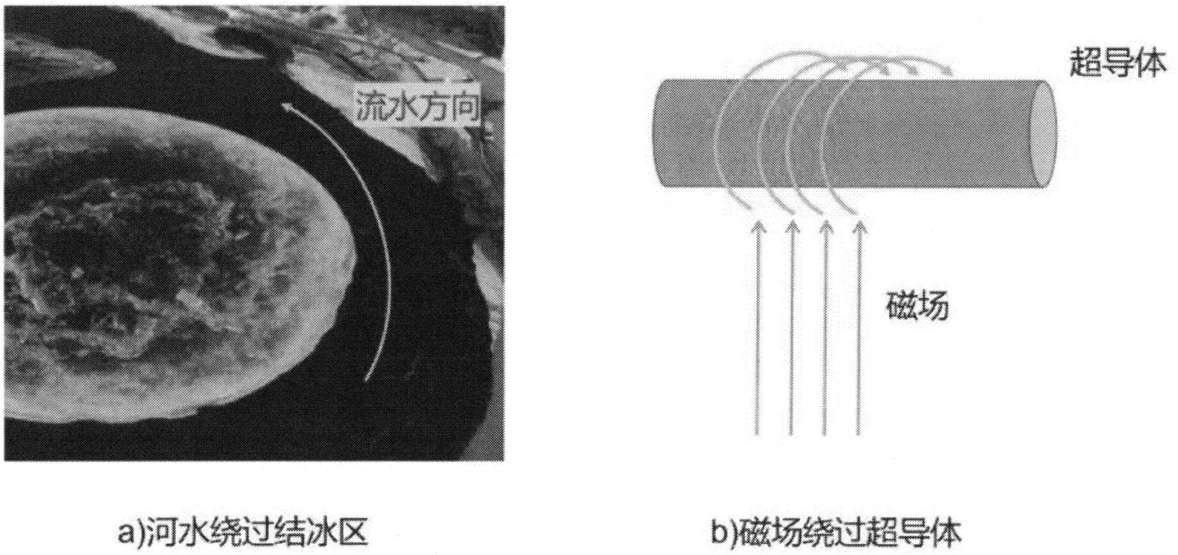

Figure 77: Figure 9-4 - Schematic illustration of the principle of magnetic field bypassing a superconductor

At this moment, if an external magnetic field approaches, it will also be "expelled" from the conductor for the same reason, following the same principle, as shown in Figure 9-4.

And since electrons have much greater mass and energy, the strength of the electric field is much greater than that of the magnetic field ($B = E/C$, C is the speed of light), electrons can still propagate smoothly in conductors. Moreover, due to the reduction in the effective mechanical effect radius of the particle field, they can exhibit superconducting effects, which is not contradictory to magnetic field shielding.

Superconductors are very easy to generate current, so when a magnetic field enters a superconductor, it easily excites induced currents that cancel out the magnetic field, preventing the magnetic field from entering the interior of the superconductor. It seems as if the magnetic field encounters a field barrier. This shielding effect of the magnetic field within the superconductor is also the result of the "ice-like" phenomenon of superconductors.

Historically, in 1933, Walther Meissner and his student Robert O

Robert Ochsenfeld discovered during magnetic field distribution measurements on metal spheres that when tin or lead metal spheres were cooled into the superconducting state in a magnetic field, the magnetic field lines appeared to be suddenly cleared from inside the sphere, as shown in Figure 9-5(a). Since they could not directly measure the change of the internal magnetic field of the superconductor, they inferred indirectly from the opposite changes in the internal and external magnetic fields that the magnetic induction inside the superconductor was zero, and the magnetic field lines would bypass the superconductor, as shown in Figure 9-5(b).

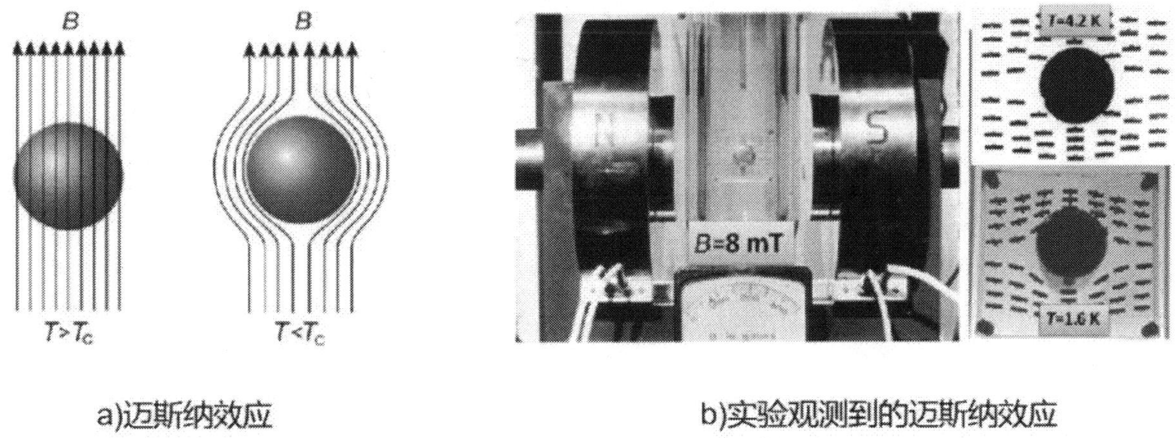

Figure 78: Figure 9-5 Meissner effect: Perfect diamagnetism of superconductors

This is the famous Meissner effect in superconductivity.
Therefore, the essence of the Meissner effect is that when the conductor's temperature drops below the critical temperature, the internal vacuum space superfluid and the conductor's thermal motion significantly decrease, forming an effect similar to freezing. As a result, the magnetic field, which is the vortex flow field of the vacuum space superfluid, is restricted in its flow and is expelled from the conductor.

Of course, just like a strong water column can penetrate through ice, if the magnetic field is strong enough, it can still penetrate the superconductor and enter its interior, i.e., the London penetration depth of the superconductor [51] , as shown in Figure 9 - 6.

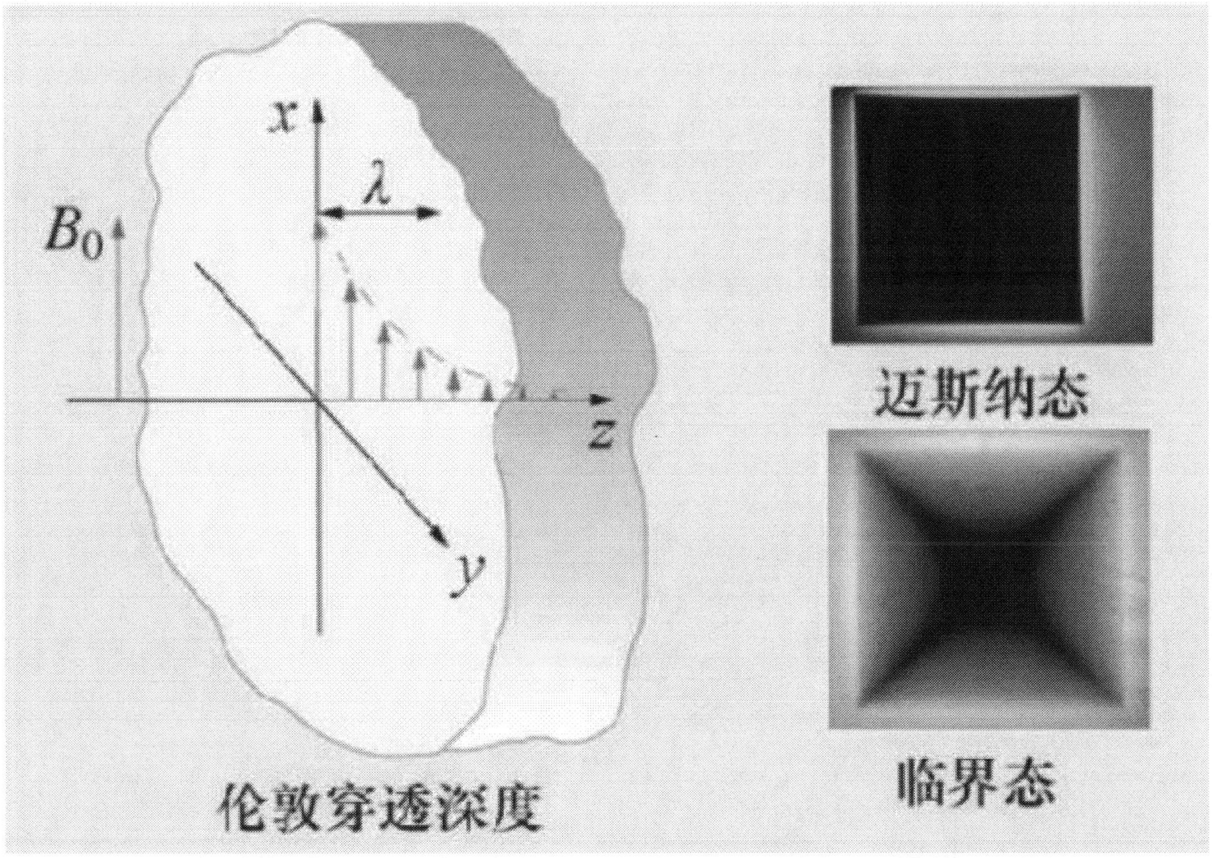

Figure 79: Figure 9-6 London penetration depth of the magnetic field in superconductors

(5) Design direction of superconducting schemes

According to the equation for the effective mechanical effect cutoff radius of the particle flow field in a superconductor (Equation 9.5), when designing superconducting conductors, one can consider arranging the molecules inside the conductor into a fine tubular crystal structure. This would allow the sum of the effective mechanical effect radius of the conductor molecules and electrons to be smaller than the radius of the tubular space channel, enabling electrons to move with less resistance through the tubular channels.

Similar to the internal structure of natrolite crystals, there are countless fine fiber-like channels in the same direction, serving as high-speed pathways for electrons.

At the same time, the radius of this pipe cannot be too large, allowing the electric field effect formed after the molecule loses electrons to effectively act on the electrons and drive their movement, as shown in Figure 9-7.

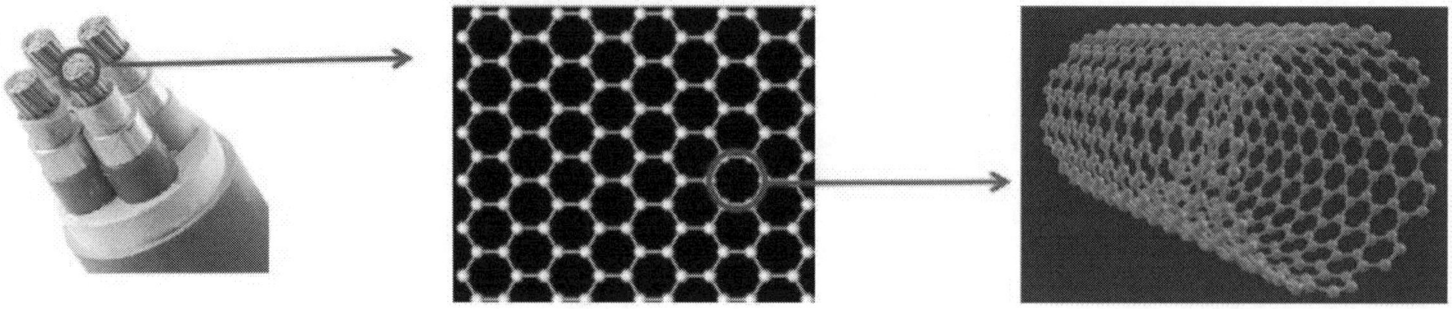

Figure 9-7 Schematic Diagram of the Tubular Crystal Channel Superconductor Structure

The superconductor design must consider three conditions: conductor conductivity performance, internal crystal channel structure, and the radius of the crystal channels must meet the requirements of the molecular flow field mechanical effect radius and the molecular electric field mechanical effect radius.

If the conductor's conductivity performance is not sufficient, it can be achieved by injecting an electron current into the internal crystal channels.

9.5 Interaction of Life Quantum Solitons

Living organisms are a more advanced form of existence that emerges from the evolution of quantum entities, and they have developed more capabilities, allowing for more complex interactions between quantum entities.
(1) Perception and memory of information from other quantum solitons
The life quantum soliton itself has intelligence, capable of observing the surrounding environment and responding to it. Similarly, it can sense the quantum field information from other surrounding quantum solitons and distinguish the quantum fingerprints of different quantum solitons, even forming memories of these quantum fingerprints.
(2) Searching, locating, and actively interacting with other quantum solitons
The life quantum soliton can search, locate, and find other quantum solitons in the surrounding quantum space based on the quantum fingerprints it has perceived and remembered. Once located, it can actively interact with the quantum soliton through the direct and indirect methods discussed earlier.
(3) The spirit and consciousness of life quantum solitons
The spirit and consciousness of living beings themselves are also one of the forms and results of thermal motion of quantum solitons. Therefore, for higher-level life quantum solitons, their spirit and consciousness are also part of their quantum fingerprint. In other words, the spirit and consciousness of living beings can also be transmitted to the surroundings through quantum fluid fields, wave fields, and other means, and directly or indirectly affect other quantum solitons around them.

This is an important characteristic that distinguishes the life quantum soliton from other quantum solitons.
(4) Quantum Interaction of Life Quantum Isolates
As mentioned earlier, life quantum isolates can perceive, identify, and remember the quantum fingerprints of other quantum isolates, and can locate and find corresponding quantum isolates based on the stored quantum fingerprints in memory. These capabilities belong to the unique quantum interaction methods of life quantum isolates.

Life quantum isolates can also interact with each other through this method.

The ability to identify, remember the quantum fingerprints of other life quantum entities, and to locate and interact with them remotely.

The operation of a living organism can be viewed as the mutual interaction among countless subsystems within the body, which are referred to as sub-quantum entities. These interactions between quantum entities include both direct contact interactions and indirect quantum interactions, meaning that indirect quantum interactions can also exert influence and effect on all quantum entities within the organism.

Including the spirit and consciousness of living beings, remote quantum interaction can also influence and intervene in the reaction of all sub-quantum entities within the body.

When life quantum entities can influence and intervene through remote quantum interaction by sensing and identifying the quantum fingerprints of other life quantum entities' subsystems.

This is referred to in the field of modern neuroscience as mind-matter interaction or telepathy. Currently, research in neuroscience mainly focuses on the interaction and sensing between the mind and instruments, that is, brain-computer interfaces. Research on direct brain-to-brain interaction and sensing has not yet begun.

The remote mind-matter interaction force between life quantum entities includes this remote interaction force between non-living entities or all entities. We can refer to this concept from quantum mechanics and call it entity entanglement force. The range of action and penetration ability of entity entanglement force are similar to those of neutrinos, which are significantly stronger than electromagnetic waves and photons.

(5) The collective consciousness field of life quantum entities

When a large number of quantum entities gather, their quantum holographic soliton fields will influence each other, leading to a resonant effect as a whole. This causes all the quantum holographic soliton fields of the quantum entities to undergo corresponding changes and be affected accordingly, forming an overall, collective quantum holographic soliton field. This allows these quantum entities to form a larger quantum entity as a whole.

All macroscopic matter, such as particles, matter, gases, liquids, and so on, is like this, and the same applies to living organisms.

A large number of life quantum entities gather and interact, and their quantum field and wave field produce a collective resonant effect, forming a loosely integrated whole quantum entity.

The thoughts and consciousness of all life quantum entities also follow this principle, so a gathering of living beings will have an overall consciousness field. This collective consciousness arises from the influence of all individual consciousnesses and in turn affects each individual's consciousness and spirit.

However, the formation and influence of this collective consciousness mainly occur in the subconscious realm, and the conscious mind is unable to perceive or feel it.

This is true for all life, including humans.
This shared consciousness field plays a more significant role in the interaction, communication, and evolution among life forms without language, such as animals, and even in the ancient periods of human history when there was no language or writing.

It continues to subtly influence the entire life community and will play an increasingly important role. For more information on quantum interactions between life quantum entities, see Chapter 12.4: An Introduction to Telepathy in the Extended Reading section.

9.6 Thoughts and Quantum Internet

As mentioned above, thoughts and consciousness can form quantum flow fields, wave fields, etc., and have their own unique quantum fingerprints, which can interact with other quantum flow fields, wave fields, etc., forming holographic soliton fields.

Therefore, if the next-generation internet has the capability to recognize and access quantum flow fields, with fixed quantum fingerprint characteristics, then the thoughts and consciousness of living beings can interact through quantum interactions, enabling recognition, positioning, and communication, thus achieving remote access to the internet.

This has significant implications for the development of the next-generation quantum internet and the mind internet.

9.7 Academic-related research

Quantum solitons and their corresponding quantum fluid fields, wave fields (holomorphic soliton fields), quantum fingerprints and distant

Programmatic interaction capability also exists in many similar studies in academia, such as the so-called field Ψ , torsion field [52] , and other theoretical research. These studies are all based on the ability of remote communication and interaction formed by quantum fluid fields or wave fields. However, most of these theories are still in the exploratory stage and have many inaccuracies. They are in the same direction and approach as our quantum soliton system and soliton field, but there are significant differences in specific understanding and details.

There is an old Chinese saying: "All things have spirit." We can understand this as meaning that all things, that is, all quantum entities, have corresponding holographic solitary flow fields, which allow them to interact and influence each other through these fields, forming an overall system of giant quantum solitary entities. This giant quantum solitary system can, in turn, influence all quantum entities, that is, influence all things, and no one can escape it, just as the existence of the legendary "Dao." This "Dao" is part of the overall determinism and causality.

Further research and exploration are needed for quantum solitons and their corresponding holographic soliton fields, as well as for the next generation of quantum internet and even mind internet.

10 Unified Physics Establishment and development of a fundamental theoretical framework

10.1 Requirements for Establishing a Unified Foundation Framework of Physical Sciences

In current research across different fields of theoretical physics, due to the lack of a unified foundational framework and worldview, there are numerous conflicts and contradictions in the theoretical systems between different fields in terms of philosophical perspectives, making it impossible for them to be reconciled and integrated. Therefore, although the academic community has made many efforts and attempts, it has yet to form a complete framework for a unified physical foundation theory.

Through our previous analysis, to establish a unified theoretical foundation framework for physical science, the following points must be clearly defined:
(1) The fundamentality of the motion and mechanical properties of empty space
Vacuum space is an objective real entity, a superfluid composed of countless fundamental particles (which can be called wave particles) that possess relative motion and mechanical effects. Since wave particles have mass, volume, and motion properties, vacuum space itself also possesses mass, density, and motion, exhibiting complex patterns such as flow and wave.

It is precisely these properties of vacuum space and wave particles that form the origin of fundamental interactions and mechanics in the physical world, giving rise to our vast and complex reality. The motion and interaction of all microscopic particles, macroscopic objects, planets, and so on occur within the context of vacuum space and are influenced by it. This is the most fundamental theoretical basis for unifying the micro and macro aspects of the universe.

At the same time, all microscopic particles, macroscopic matter, and matter-antimatter are forms of particles or accumulated particle structures generated by the superfluid of empty space under different energy excitations and motion states

The asymmetry of matter and antimatter at the cosmic level originates from the asymmetric vortical motion of the superfluid vacuum itself.

Now research in all fields of physics lacks a correct and complete understanding of this point, leading to the inability to establish a unified fundamental theoretical framework.

(2) Unification of micro and macro realms

Vacuum space is the same in the microscopic and macroscopic worlds, which forms the basis for the unification of microscopic and macroscopic world theories. For example, the refraction of light in the macroscopic world and that in the microscopic world both follow the same principle, as the change in the path and wavelength of light is caused by the density of vacuum space.

Based on this unified understanding, it is possible to establish a unified framework for both macroscopic and microscopic physical theories.

(3) The Unity of Wave-like and Particle-like Properties

All microscopic particles, macroscopic objects, and celestial bodies in the universe are subject to particle-like linear motion. However, since they all exist in the void space, they inevitably experience the effects of the movement and mechanics of the void space. The movement of the void space includes flow and wave-like motion similar to superfluids. Therefore, the motion of all objects will exhibit wave-like properties, which is the underlying reason for the de Broglie matter wave principle.

It is only because the effects of the void space are more pronounced on objects with smaller mass, that microscopic particles display significant wave-particle duality. In contrast, the wave-like properties of macroscopic objects and celestial bodies are relatively weak. Nevertheless, the motion of macroscopic objects, such as the rotation of stars and galaxies, is significantly influenced by the void space. Many explanations of cosmic phenomena must take into account the effects of the void space.

(4) Unity of mechanical principles

There is mass first, then force. The origin of all forces comes from the collision forces between mass bodies, that is, the conversion of momentum and kinetic energy. This force is manifested macroscopically as the pressure of the empty space, and electromagnetic force and gravitational force are the macroscopic results of this force.

At the microscopic level, various complex particle interactions are essentially also the result of the mass interaction forces between particles and between particles and the empty space.
This is the fundamental framework shared by all mechanics theories, and it must be based on this framework to construct a complete physical foundational theoretical framework system.

(5) The Unity of Determinism and Indeterminism

All mechanical actions conform to determinism and follow the law of causality. When we cannot grasp all the information and factors of mutual interactions, or actively abandon some information or factors, the observed results will deviate from the expected ones within a certain range. This deviation manifests macroscopically as probability and uncertainty.

At the microscopic level, if we do not consider the role of empty space, the motion of quantum particles would exhibit many probabilistic behaviors that are difficult to understand. At the macroscopic level, if we do not take into account many influencing factors, everything would also show a randomness akin to flipping a coin, deviating from determinism.

Therefore, behind the apparent randomness, there must be hidden deterministic factors and laws that we have not discovered or properly handled. In this sense, determinism and indeterminism are unified; it is either a limitation in our understanding of the underlying laws or an active choice in describing events.

Only by being clear can we firmly adhere to the materialist core values of science, otherwise we are easily misled by mystical and esoteric ways of thinking.

(6) Unity of Different Theoretical Systems

For the same event or law, different interpretations can be formed from different perspectives, leading to various theoretical systems. These theoretical systems express the event or law differently based on the phenomena they observe. As long as these theories are consistent with the observed phenomena and form a self-consistent system, they can be considered reasonable. Since they describe the same event or law, the underlying unity of these different theoretical systems is essential, meaning they are expressions and descriptions of the same phenomenon from different aspects.

Depending on the type of theoretical description, there can be principle-based descriptive theories, statistical descriptive theories, and several other types

Mathematical descriptive theories, for example, although they have unity in describing phenomena and laws, all such theories must be supported by a deterministic principle descriptive theory.

Principle descriptive theories are based on fundamental principles of determinism, objectivity, and materialism in the physical world, making them the most basic theories. All other statistical, geometric, and mathematical theories that explain the objective physical world are upper-level mathematical theories and must have corresponding principle descriptive theories to support them. Without such support, they are considered incomplete.

The above six points are essential conditions for establishing the foundational framework of a unified physical theory. Current theories such as special and general relativity and quantum mechanics do not fully satisfy these conditions, leading to incomplete aspects in each theory and making it difficult to achieve a consistent unification between them.

10.2 Main Issues of Special Relativity

(1) Denies the objective reality and motion of empty space. Although Einstein made corrections in general relativity to acknowledge the objective reality of empty space, he still did not accept its motion, as this would completely undermine the foundation of special relativity. This is also the fundamental reason for the subsequent several errors in special relativity;
(2) Misuse of the constancy of the speed of light hypothesis. The speed of light is constant in the vacuum space it resides in, which is fine, but assuming that the speed of light is constant for all observers is an error, confusing the distinction between physical laws and observed phenomena. Physical laws are objective and unchanging, while observed phenomena vary depending on the observer and their state of observation;
(3) Misuse of the principle of relativity. Einstein's special relativity claims that the motion of objects is absolutely relative and that it is impossible for them to move relative to the vacuum space, which is precisely due to the misunderstanding of the principle of relativity caused by the first error. The motion of objects is relative, but since vacuum space is an objective reality, the motion of light, electromagnetic waves, and gravity is relative to the vacuum space. Therefore, when involving these factors, it is essential to consider the relative motion to the vacuum space.

Considering the relative motion of each object with respect to the empty space, otherwise a large number of paradoxes will appear. Many famous paradoxes in special relativity are related to this issue.
(4) Misunderstanding of the logic for determining simultaneity. Einstein's foundational logic for determining simultaneity in special relativity is that different inertial frames may judge the simultaneity of the same event differently due to their own motion. However, he overlooked the fact that any inertial frame can have an infinite number of observers who can accurately determine the simultaneity of two events. In other words, simultaneity is objective and does not change with different inertial frames; it is only the observation of the phenomenon that differs, requiring observers to analyze and correct for these differences.

It is precisely due to the aforementioned important errors that the special theory of relativity is a theory that distorts the objective physical world, and it is an incorrect explanation of the objective world.

Throughout the 20th century, there was no philosophical reconciliation between relativity and quantum mechanics, as they remained in fundamental philosophical conflict in many aspects, and this contradiction could never be resolved. This is partly due to the errors and defects inherent in the theory of relativity mentioned above.

10.3 The Main Problems of General Relativity

(1) It is merely a mathematical theory of a geometric nature at the upper level, without a foundational descriptive theory, and does not conform to physical reality.
(2) The motion and mechanical properties of empty space are not recognized, and the essence of mass distorting space is not understood.
(3) The equivalence principle has flaws, leading to incorrect interpretations of phenomena such as "gravitational redshift."
(4) There is no physical explanation of gravity, causing deviations in many theoretical interpretations in cosmology such as black holes.

10.4 The Main Issues of Quantum Mechanics Theory

(1) The failure to consider the influence of empty space on microscopic particles leads to difficulties in explaining phenomena such as wave-particle duality, double-slit interference, and quantum tunneling.
(2) Equating statistical probability descriptions with the actual physical world denies determinism and causality, causing confusion in the philosophical view of physics.
(3) Influenced by the old concept of locality, it causes difficulties in providing physical explanations for phenomena such as particle spin and quantum entanglement.
(4) It fails to explain the relationship between fields and empty space, and in explaining photon energy, does not take into account the role of spatial wave energy.

10.5 Main Issues of Classical Mechanics Theory

(1) Describes the phenomenon of gravity but fails to explain its source.
(2) The motion and mechanical properties of empty space are not considered in the analysis and description of classical mechanics laws.

10.6 The Role and Development of the Theory of Space Relativity

Due to the above reasons, the special theory of relativity, including the general theory of relativity developed upon it, has existed in serious philosophical contradictions with quantum theory, which have never been resolved. Its wide-ranging influence has significantly affected the thinking and development of theoretical physics.

"Space Relativity" initially forms a unified theoretical framework through the discussion of seven basic principles: the principle of objective physical laws, the principle of spatial objectivity, the principle of temporal objectivity, the new principle of invariant speed, the new principle of relativity, the principle of spatial relativity, and the principle of conservation of mass and energy, as well as sections on the general overview of space electrodynamics, space hydrodynamics, space classical mechanics, and space quantum dynamics

The issues and contradictions of the theory, including:
(1) Establish the correct concept of empty space, by updating the principle of relativity and the principle of invariant light speed, to correct the erroneous space-time interpretation of special relativity.
(2) In response to the geometric interpretation of gravity in general relativity, a corresponding fundamental principle-based theory is proposed, forming a unified interpretation framework for cosmic forces.
(3) Resolve the philosophical incompatibility between general relativity and quantum mechanics, and overcome the difficulty of the geometricized mechanics of general relativity in being quantized.
(4) Resolve the defects of the old ether theory and establish the correct concept of empty space, which can be called the new ether theory.
(5) Clarify the distinction between mathematical descriptions and computational systems of quantum mechanics, relativity, etc., and the physical reality and objective world, resolve the difficulties in explaining quantum phenomena such as probability, quantum entanglement, wave-particle duality, and double-slit interference, and provide a theoretical description based on deterministic fundamental principles.
(6) Explain and supplement the difficulties in the standard model and particle physics theories regarding the nature of antimatter, the asymmetry between matter and antimatter, the origin, structure, and evolution mechanisms of particles, and the essence of charge. It also corrects and improves the understanding of the nature of spin, mass, electromagnetic fields, and the generation and propagation mechanisms of electromagnetic waves.

In simple terms, spatial relativity supplements and refines theories in various fields of physics by taking into account the role of the superfluid vacuum space. For example, it considers the effect of the superfluid vacuum space on microscopic particles, supplementing and refining quantum mechanics; it considers the effect of the superfluid vacuum space on macroscopic objects, supplementing and refining general relativity, special relativity, and classical mechanics; and it considers the role of the superfluid vacuum space in the formation and evolution processes of particles, supplementing and refining particle physics, etc.

The above work has formed a foundational theoretical framework for both macroscopic and microscopic physical theories, enabling the integration of the aforementioned various macroscopic and microscopic theoretical systems into a complete and unified theoretical system.

11 Conclusion

Since Newton published the "Mathematical Principles of Natural Philosophy," foundational physics in various fields such as classical mechanics and later special, general relativity, and quantum mechanics has made significant progress, driving tremendous advancements in human civilization. However, due to insufficient understanding of the objectivity and motion of empty space and its mechanical effects, there are still issues of incompleteness and inconsistency in the foundational physical theories of various human domains, even leading to fundamental and structural contradictions and errors. These problems have prevented the unification of several physics fields for a long time, making it difficult to form a complete cosmological worldview and a unified physical foundation framework.

Space Relativity, building upon the accumulated thoughts of predecessors, clarifies the vague understanding and cognition of empty space, fully considers the material nature and mechanical properties of empty space, and on this basis, completes the construction of a unified cosmological worldview and a foundational physical theoretical framework. It has carried out consistent revisions and reconstructions for the aforementioned four fundamental physics domains, initially forming a theory of cosmic unification.

Considering that theories such as special and general relativity, quantum mechanics, etc., have undergone long-term, even centuries of development, and have deeply and widely influenced society, science and technology, and other fields, revising these theories is not something that can be done by one person or a small group, nor can it be achieved in a short period of time. At the same time, spatial relativity is currently only in the early stage of establishing a basic framework, which is still very rough, and many aspects are only qualitative descriptions, lacking quantitative analysis and precise mathematical formulations. To complete the consistency revisions and integration of various physical theory systems, a lot of specific work still needs to be done. All of this requires the participation of more forces in the future, even unified social efforts, and joint efforts, in order to continuously improve and develop the framework of a unified physical theory, and to achieve the construction of a unified physical theory.

12 Extended Reading

12.1 Eight Intrinsic Problems in the Theories of Special and General Relativity

Summarizing our previous analysis, the main problems with the special and general theories of relativity are:
(1) Misuse of the principle of the constancy of the speed of light
The special theory of relativity's principle of the constancy of the speed of light includes three meanings:
(1) The speed of light in a uniform flat vacuum space is constant at C, which is approximately 300,000 kilometers per second;
(2) The speed of light is independent of the motion of the light source;
(3) The speed of light is independent of the motion of the observer.
The main error lies in point three: the speed of light is indeed constant when observed from a reference frame of relative empty space, and it is independent of the observer's motion. However, this does not mean that the relative speed of light observed directly is also constant at C.

When an observer moves with different velocities in empty space, the light propagating in this empty space will be observed to have different relative speeds, not a constant C.

Because the propagation of light relies on the background reference frame of empty space, its speed is only constant relative to empty space, not relative to the observer. The speed of light in a uniform and flat empty space is constant at C, which is a physical law. Different observers in different states of motion directly observe different light speeds, which is an observational phenomenon. The special theory of relativity's postulate of the constancy of the speed of light confuses the distinction between physical laws and observational phenomena.
(2) Misuse of the principle of relativity
The understanding of the principle of relativity in special relativity is that all motion is relative, and two relative motions

The motion of objects, their actual velocity can be arbitrarily assumed; one can consider A moving and B stationary, or B moving and A stationary, or both moving, as long as the total relative velocity remains consistent.

In reality, all motion occurs within the surrounding vacuum space, so their motion speed relative to the vacuum space is unique and cannot be arbitrarily assumed. The sum of their individual motion speeds relative to the vacuum space equals their relative motion speed.

Especially when it comes to situations involving light, electromagnetic force, and gravity, this point must be considered, because the propagation of electromagnetic waves and gravitational waves relies on the vacuum space, and their motion speeds are all relative to the vacuum space. Leaving this out will inevitably lead to errors.

(3) Misunderstanding of Vacuum Space

Special relativity treats empty space as an illusory, conceptual reference frame rather than an objective reality, which is also the fundamental reason for its misapplication of the constancy of the speed of light and the principle of relativity.

In fact, the vacuum space is an objective reality, a real existing background reference frame, and the vacuum space itself has relative motion. The propagation and movement of electromagnetic waves, gravitational waves, etc., all rely on the vacuum space, and the so-called constant propagation speed C they refer to is also relative to the particular vacuum space they are in.

When the void space is in motion, the speed of light will inevitably be affected, resulting in relative changes.

The hypothesis of the constancy of the speed of light did not take this into account, leading to the mistaken belief that "the speed of light is constant relative to different observers in motion."

The principle of relativity also failed to consider this, resulting in the erroneous notion that "the relative velocity of an object is absolutely relative and can be arbitrarily assumed."

Einstein had already recognized this in general relativity, such as the distortion of space and quantum fluctuations in empty space, which are all manifestations of the motion of empty space, but this error can no longer be corrected, because the constancy of the speed of light and the principle of relativity are the two fundamental pillars of special relativity, and the errors in these two principles have made it impossible to revise special relativity

Relativity lost its rational foundation.

(4) Error in the logic of simultaneity judgment

There is an obvious logical problem in Einstein's determination of simultaneity. He believed that different inertial frames would judge the simultaneity of the same event differently due to their own motion, which is the fundamental logical basis of his discussion in special relativity.

But here he confuses an inertial frame with an observer within that inertial frame.
We know that an inertial frame is a complete reference frame, whose scope can be infinitely expanded, even encompassing the entire universe. Therefore, there can be not just one but an infinite number of observers within this reference frame. We can consider that at every spatial point in the universe, there exists an observer of this reference frame. Thus, whenever two events occur, we can guarantee that there is an observer at zero distance from both locations who can instantly determine the time of the two events and thus ascertain whether they are simultaneous.

In this case, regardless of how many inertial frames there are and regardless of their motion, they will all be able to confirm the simultaneity of these two events, and their judgments will remain consistent without any differences due to the different inertial frames.

Step back, even if each inertial frame has only one human observer, we do not rely solely on our own eyes to observe things. We can have countless sensors and detectors, which can also be deployed anywhere within an inertial frame. Therefore, we can also accurately determine the time when these two events occur and confirm their simultaneity.

Therefore, Einstein believed that different inertial frames in motion would determine the simultaneity of the same event differently, which clearly falls into a logical error.
(5) The erroneous conclusion of special relativity regarding time dilation and length contraction
Special relativity is merely a mathematical means of describing the objective physical reality, and this mathematical description method

The segment distorts the understanding of the objective world and is not based on objective reality.
At the same time, Einstein made an error in deriving the time dilation and length contraction effect. Even according to the principle of the constancy of the speed of light, the conclusion should be that an object in high-speed motion is observed to have slower clocks, expanded lengths, and increased mass. It should be length expansion rather than length contraction.

If according to the time dilation and length contraction effect, then the observer on the moving spaceship would see themselves traveling a distance L (which can be assumed to be one times the length of the spaceship) in time t, with a speed of

$$V = \frac{L}{t}$$

and according to the time dilation and length contraction effects of special relativity, the Earth-based observer sees the spaceship traveling a distance of L_1 in a time of t_1, resulting in an observed speed of

$$\mathrm{V}_1 = \frac{\mathrm{L}_1}{\mathrm{t}_1} = \frac{\mathrm{L}\sqrt{1-\frac{\mathrm{V}^2}{\mathrm{C}^2}}}{\frac{\mathrm{t}}{\sqrt{1-\frac{\mathrm{V}^2}{\mathrm{C}^2}}}} = \frac{\mathrm{L}}{\mathrm{t}}\left(1-\frac{\mathrm{V}^2}{\mathrm{C}^2}\right) = \mathrm{V}\left(1-\frac{\mathrm{V}^2}{\mathrm{C}^2}\right)$$

Therefore:

$$\mathrm{V}_1 = \mathrm{V}\left(1-\frac{\mathrm{V}^2}{\mathrm{C}^2}\right)$$

The observed speed of the spaceship by the two observers is inconsistent. The speed of the spaceship observed by the Earth's stationary observer is less than the actual speed of the spaceship, even when the spaceship's speed approaches the speed of light, the ground observer sees its speed decrease rapidly to zero, which is clearly unreasonable.

More importantly, the error Einstein made here also led to the incompatibility of the special theory of relativity with the gravitational formula, forcing him to continue developing the general theory of relativity and moving further away.

The correct conclusion should be that clocks run slow and rulers expand, in which case the aforementioned problems would not occur, and both parties would observe the same speed of motion:

$$V_1 = \frac{L_1}{t_1} = \frac{\frac{L}{\sqrt{1-\frac{V^2}{C^2}}}}{\frac{t}{\sqrt{1-\frac{V^2}{C^2}}}} = \frac{L}{t} = V$$

It is also compatible with the gravitational formula:

$$F_1 = G\frac{M_1 m_1}{r_1^2} = G\frac{Mm/\left(\sqrt{1-\frac{V^2}{C^2}}\right)^2}{\left(r/\sqrt{1-\frac{V^2}{C^2}}\right)^2} = G\frac{Mm}{r^2} = F$$

Where M、m、r、F represents the mass, distance, and gravity observed by a stationary observer between two mass bodies; M_1、m_1、r_1、F_1 represents the mass, distance, and gravity observed by a moving observer between two mass bodies;

V is the velocity of the moving observer relative to the two mass bodies.
The incompatibility with the gravitational formula is the biggest problem of special relativity.

(6) Misunderstanding of the mass-energy equation in special relativity

Many theoretical scholars have two misunderstandings about the mass-energy equation:
(1) The mass-energy equation $E = mC^2$ can not only be derived from special relativity but also obtained through other non-relativistic mathematical calculation systems

Including Einstein himself, he once derived this conclusion using non-relativistic methods
(2) We have analyzed before that this mass-energy equation is not an equation describing the conversion of mass into energy, but merely describes the energy of a photon.

Just like the well-known kinetic energy formula $E = 1/2mV^2$, this formula only describes an object's kinetic energy, not the "mass-energy equation" that converts mass into energy.

Einstein's special relativity gives an incorrect interpretation here, confusing the original meaning of this formula

The significance, as shown in Figure 12-1.

Kinetic energy of matter: $E = \frac{1}{2}mv^2$

Photon energy: $E = mc^2$

Figure 12-1 Kinetic Energy and Energy of Matter

We know that light exhibits wave-particle duality. Assuming the photon has mass m, its motion as a particle at the speed of light would give it kinetic energy $E = 1/2mC^2$, but at the same time, the photon is also a wave, moving in a wave form

Therefore, it also has potential energy. Light waves are transverse waves, and the kinetic energy and potential energy of transverse waves are equal
So photons also have potential energy, thus we obtain the photon's energy as the sum of potential and kinetic energy, that is, $E = mC^2$, as shown in Figure 12—2.

Wave-Particle Duality of Light

Photons are particles

Therefore, photons have kinetic energy: $E_1 = \frac{1}{2}mc^2$
Light travels as a wave. It is a transverse wave
Therefore, photons have potential energy:
(Transverse wave kinetic energy equals potential energy)

$$E_2 = \frac{1}{2}mc^2$$

So the total energy of a photon is:

$$E = mc^2$$

Figure 12-2: Photon Energy and Mass-Energy Equation

It can be said that if stones and other objects in our macroscopic world also oscillate in the transverse wave mode, they would also have equal kinetic and potential energy

That is, the total energy is $\mathrm{E} = mV^2$, but we only consider their kinetic energy
It's only $\mathrm{E} = 1/2\mathrm{mV}^2$

So, this is truly what the "mass-energy equation" $\mathrm{E} = \mathrm{mC}^2$ means.
This formula merely indicates that the energy of a photon is the sum of its kinetic and potential energy, arising from its wave-particle duality nature,

and is not what is referred to as the "mass-energy equation" where mass is supposedly converted into energy.

(7) Paradox of the Spaceship's Superluminal and Light Speed Limit Hypothesis

According to the relativistic equation of mass and velocity in special relativity, when an object is in motion, an observer at rest outside will observe that its mass increases with velocity. As the velocity approaches the speed of light, the mass also approaches infinity, resulting in acceleration approaching zero. Therefore, relativity theory holds that a spaceship can never exceed the speed of light no matter what.

However, we further consider that when a spaceship approaches the speed of light, the mass of the fuel it carries will also increase in sync with the spaceship's mass, approaching infinity. The mass of fuel burned per second by the engine, the energy released by the fuel combustion, and the thrust will also increase in the same proportion and in sync, so the spaceship can still maintain normal acceleration. As long as it carries enough fuel, it can continue to accelerate until it reaches and even exceeds the speed of light.

On the other hand, from the perspective of the personnel inside the spaceship, since they are at rest relative to the spaceship, they observe that the spaceship's mass does not increase, the fuel mass does not increase, and the energy released by the chemical reaction of fuel combustion remains unchanged. Therefore, the thrust provided by the propulsion system will not decrease, and the spaceship can continue to maintain normal acceleration, achieving and even surpassing the speed of light.

Whether the ground-based observer or the observer inside the spaceship, they should observe the same speed and acceleration of the spaceship. Therefore, regardless of the observation perspective, the spaceship can reach or even exceed the speed of light. This contradicts the assumption of the speed limit in special relativity.

This is also one of the many paradoxes generated by the mathematical calculation system of special relativity.

(8) Misuse of the Equivalence Principle in General Relativity

Similarly to the constancy of the speed of light and the principle of relativity in Special Relativity, the Equivalence Principle is the foundation upon which Einstein built General Relativity. In simple terms, the Equivalence Principle can be described as "the effect of acceleration and a gravitational field on an object are equivalent."

In response to this, many scholars have raised the famous "charge paradox" as a challenge. That is, a charge at rest in a gravitational field does not emit electromagnetic waves, but if it is in accelerated motion, it does emit electromagnetic waves. Here, acceleration and a gravitational field are not equivalent.
The "charge" paradox also directly breaks the theoretical foundation of general relativity, causing the entire theory to fall into error. Here, the equivalence principle is just as much a conclusion based on subjective speculation as the constancy of the speed of light in special relativity and the principle of relativity. People derive various physical laws from this conclusion, even forcing the physical world's laws to conform to it, which is the fundamental cause of the distortions and errors in both general and special relativity.

Although supporters, like those of special relativity, have provided various explanations from different angles regarding the charge paradox, its essence is the same as that of special relativity—forced interpretation. They are all mathematical description systems that distort the physical world, rather than true objective reality.

Einstein's Boomerang: Inherent Contradictions Between General and Special Relativity

(1) Time Dilation Effect
In special relativity, Einstein proposed the effects of time dilation, length contraction, and mass increase for moving objects, as well as the famous Lorentz transformation.

According to this theory, the time experienced by the spaceship observer A is shorter than the time experienced by the ground observer B, making it seem as if the time inside the spaceship is running faster, which is the phenomenon of a moving clock running slower.

However, when Einstein introduced general relativity, the situation changed again. Based on his equivalence principle demonstrated through the elevator thought experiment, the time experienced by the spaceship observer A is longer than the time experienced by the ground observer B, making it seem as if the time inside the spaceship is running slower.

So, this contradicts the time dilation effect in special relativity.
In special relativity, moving clocks run slower, but according to the equivalence principle in general relativity, moving clocks actually run faster (a detailed analysis of this is provided in the section "2.1 Misuse of the constancy of the speed of light principle" in "Chapter 2: Some Issues with Current Theories," which can be referred to).

The boomerang Einstein threw in special relativity came back to hit him in general relativity.

In fact, a slight analysis will reveal that the fundamental issue here lies in the constancy of the speed of light hypothesis. As analyzed in our video earlier, the biggest problem with the constancy of the speed of light hypothesis is that it forcibly assumes the speed of light is always constant at C relative to all different observers.

This assumption has made relativity distort the description of objective physical reality, turning it into a distorted mathematical system and leading to numerous distorted, counterintuitive conclusions, which are the famous various paradoxes and contradictions.

The speed of light is constant at C in a uniform vacuum space, which is not problematic. However, claiming that the speed of light is also constant at C for different observers in motion is a fallacy. As we discussed earlier, this is also one of the four major errors in special relativity, which confuses the distinction between physical laws and observed phenomena.

In the end, as we have discussed before, the theory of special relativity does not describe objective physical reality or the phenomena observed by observers; it is merely a system of mathematical calculations, and this system distorts the description of objective physical reality, making it a distorted and even incorrect mathematical framework.

Of course, in these hypothetical experiments, we have not considered the motion of vacuum space or the dragging effect.

Its influence on the speed of light, if these factors are considered, makes the situation more complex.

(2) Regarding the Ether

Throughout the history of physics, the ether has always been a very special entity. Every change in people's understanding of it has profoundly influenced the direction of physics development.

In the 17th century, Descartes proposed that all forces between objects must be transmitted through a medium called the ether, first introducing the ether into science and assigning it mechanical properties. This view provided an important ideological foundation for later physics research.

In the 19th century, Thomas Young discovered the phenomenon of light interference and polarization, making the wave theory the dominant explanation of light's nature. Physicists believed that there must be some medium in space to carry the propagation of light, and ether once again became the focus of attention in the physics community.

Later, mainstream scientists such as Poincaré and Lorentz all acknowledged the existence of ether.
But by the end of the 19th century, the experimental results of Michelson and Morley failed to detect the ether wind, causing confusion in the physics community. Einstein built upon this and developed the special theory of relativity, completely negating the existence of ether. He argued that the propagation of light and electromagnetic forces does not require such a medium, stating that "the reference to an ether will prove to be superfluous."

This severely challenged the classical mechanics theory and caused a great stir in the physics community.
Scientific skepticism about this has never ceased; many have pointed out various paradoxes caused by special relativity, but they have all been forcibly explained by its supporters using various mathematical methods, making the special relativity that denies the existence of the ether the mainstream of physics.

However, because special relativity was incompatible with the gravitational formula, Einstein had to further develop general relativity. At this time, the ghostly concept of "ether" reappeared.

In 1920, Einstein pointed out in a lecture at the National University of Leiden in the Netherlands that, according to general relativity, space was given a physical property, so in this sense, there exists an ether.

According to general relativity, a space without ether is unimaginable. Because in such a space, not only can light not propagate, but also measuring rods and clocks would not exist, and thus there would be no physical meaning of space-time intervals.

Although he emphasized that the concept of motion cannot be applied to the ether, he formally acknowledged its existence.

The boomerang Einstein threw in special relativity came back to him in general relativity.

The concept of ether has played a positive and important role in the history of physics, although there have been misunderstandings and misdefinitions in its historical understanding, namely, that it was considered absolutely stationary and uniformly distributed throughout the universe.

Even Einstein made this mistake, believing that "the concept of motion cannot be applied to ether."

But in fact, as we analyzed earlier, if we consider the void space as the ether, then the ether should be an objective reality. As we analyzed, the essence of the void space is matter, not a virtual concept, and matter must necessarily be in motion.

Modern quantum mechanics has confirmed that there are endless quantum fluctuations in empty space, which proves that empty space exists in motion.

Einstein's own general theory of relativity also points out that the space around a mass is distorted, which itself shows that space exists in motion. The change from a flat and uniform state to a distorted state is motion.

More importantly, the speed of light is constant in uniform empty space, but empty space is local and can be distorted and dragged by massive objects. Therefore, different regions of empty space may exist in relative motion, which will directly affect the speed of light propagation relative to different regions of empty space.

The Michelson-Morley experiment failed to detect any anisotropy in the speed of light relative to the ether, which is why it did not

Taking this factor into account.

(3) Principle of Relativity

The principle of relativity is one of the two fundamental pillars upon which Einstein built the theory of special relativity, and it is the most important of these two pillars. Einstein even defined relativity as the essence of the theory of relativity.

In 1948, in the entry he wrote for the "Encyclopedia of the American People," Einstein once again explained the principle of relativity:

Motion is always displayed as the relative motion of one object with respect to another. Motion can never be observed as "motion relative to space," or so-called "absolute motion." "The principle of relativity" in its broadest sense is contained in the following statement: all physical phenomena have the characteristic that they do not provide any evidence for the concept of "absolute motion"; or, more briefly but less precisely: there is no absolute motion.

Einstein specifically mentioned that motion is absolutely impossible as "motion through space," because in special relativity, the ether and the objective reality of space were completely denied. (This was already defined when Einstein created the special theory of relativity in 1905, and here it is cited from an article he wrote in 1948 for detailed explanation.)

But when discussing general relativity, Einstein changed his view.
In 1920, Einstein spoke at a lecture at the National University of Leiden in the Netherlands:
More precise examination shows that special relativity does not necessarily require the denial of the ether. It is possible to assume the existence of the ether, but it must no longer be considered as having a definite state of motion... This idea has been confirmed by the results of general relativity.
"...Denying the existence of the ether ultimately means acknowledging that empty space has absolutely no physical properties, which contradicts the fundamental facts of mechanics."

He concluded that:
According to general relativity, space has been given physical properties, so in this sense, there exists an ether. According to general relativity, a space without ether is inconceivable. Because in such a space, not only can light not propagate, but measuring rods and clocks would also be impossible, and thus there would be no physical space-time interval.

Here, Einstein acknowledged the objective reality of space, which is no longer just a concept but an objective existence with physical properties.

In fact, quantum theory has shown that electric fields, magnetic fields, gravitational fields, and even endless quantum fluctuations exist in space, and they themselves are forms of objective reality, even existing in motion.

In general relativity, space can be distorted by massive objects, which itself demonstrates that space is an objective reality and even exists in motion.

Light is deflected due to the distortion of space, indicating that the motion of light is relative to space, directly contradicting his conclusion in special relativity that "motion can never be considered as 'motion relative to space'."

The concepts of "space" and "motion relative to space" are another boomerang of Einstein's, thrown out in special relativity and later returned to him in general relativity.

In fact, the hypothesis of the constancy of the speed of light in special relativity already implies the objective reality of space and the meaning of light's motion relative to space, so it is clearly inappropriate for Einstein to deny this.

We have already discussed in our previous video that when examining the relative motion of elements such as light, electromagnetic fields, and gravity, it is necessary to consider the虚空 space in which the system resides. The motion of an object relative to this虚空 space can be considered its "absolute" motion within this虚空 space. Moreover, the虚空 space itself also exists in motion, such as the space distortion around massive objects and the endless quantum fluctuations, all of which demonstrate this.

Einstein acknowledged the objective reality of space in general relativity and the rationality of the ether, yet he still emphasized that "the concept of motion cannot be applied to the ether," clearly indicating he had not yet escaped from this line of thinking.

This is the fundamental reason why there are so many controversies between special and general relativity.
From the time dilation and length contraction effects of moving objects, to the ether, space, and relativity, the numerous "comebacks" of special relativity that appear in general relativity have already thoroughly shaken the foundation of special relativity, and even that of general relativity. In such a situation, the academic community's failure to notice these issues, or even continuing to force explanations, will only lead the theory of physics, which has been misled for a century, further into 困境.

(4) Drag effect of inertial frames

In the 19th century, scientists generally believed that light propagated through a medium called "ether." French physicist Fresnel proposed the "partial dragging hypothesis," which stated that the ether in a vacuum was absolutely stationary, and that light could be partially dragged by transparent objects in motion.

British physicist George Stokes provided a supplement to Fresnel's hypothesis, proposing the "complete dragging hypothesis": at the Earth's surface, the ether moved with the Earth at the same speed, meaning the Earth completely dragged the ether, and only at some height above the Earth's surface could the ether be considered stationary.

The ether and dragging theories well explained many phenomena in physics at that time, including the aberration of light, Arago's glass refraction experiment, Fizeau's river experiment, and the results of Einstein's observation of "aberration of light" through a filled telescope. All of these supported the correctness of the ether and dragging theories.

However, with the emergence of the Michelson-Morley experiment, since no direct observation of the anisotropy of light's speed relative to the ether was found, Einstein proposed the special theory of relativity, directly denying the ether, the objectivity of space, and also the dragging theory.

However, the boomerang appeared again during the creation and development of general relativity.
British physicist Mark Hadley from the University of Warwick discovered that when calculating our galaxy, the rotation of a galaxy exerts a dragging effect on the surrounding spacetime. According to Einstein's general theory of relativity, the speed and angular momentum of a massive celestial body during its rotation cause distortion in the spacetime around it. This effect is considered a "strong" experimental verification of general relativity and is also known as the Lense-Thirring effect

The "frame dragging effect" of general relativity.
The Gravity Probe B, launched by NASA in 2004, has confirmed this effect.
In fact, this so-called "inertial frame dragging effect" is the ether and dragging theory that was rejected together by special relativity in history.

Einstein's boomerang thrown in special relativity came back to him in general relativity.

Although the descriptions and focuses of the two drag theories are different, their essence is actually the same.
Of course, the ether we are discussing here is the void space we discussed earlier, an objective reality that exists and undergoes relative motion, differing from the ether in historical contexts, including the concept of ether in general relativity.

In our previous video, we explored that mass objects not only distort the surrounding void space but also exert a dragging effect on it during their motion. The deflection of light around celestial bodies and the deflection of light at the edges of transparent media are manifestations of this distortion effect; the influence of flowing media on the speed of light and the "frame-dragging effect" in general relativity are also manifestations of this dragging effect.

Here, the facts once again confirm the correctness of the ether theory and the dragging theory. In this case, should we think about whether the special theory of relativity, which was originally completely negated the mainstream ether and dragging theories, still has any meaning?

12.3 Overview of the Time Principle

(1) Can time flow backward?
Can time flow backward?
This is a question that everyone wonders about in their hearts.
There are many science fiction movies that describe people from the present traveling back to the past to change history or correct mistakes, make up for regrets, but find themselves trapped in a cycle of history, even seeing their younger selves, and getting stuck in a loop.

Time paradox. Or perhaps people from the future come to the present, traveling back and forth between the future and the present, even creating various strange time loops that leave the audience amazed and bewildered.

So, can time really flow backward? If time can indeed flow backward, would those amazing time paradoxes really occur? For example, the famous grandfather paradox: Tom goes back in time and kills his grandfather, which means he would never be born, and thus would never go back in time to kill his grandfather. How can such a paradox be resolved? Etc.

To understand all these issues, we first need to understand the nature and characteristics of time, and know what time really is.

People often use the metaphor of a long river to describe time, saying time flows like a river, ceaselessly and irreversibly, "逝者如斯夫，不舍昼夜" (passing away like a river, never ceasing day or night).

We can guide rivers and change their direction, for flowing water is a tangible thing that we can see and touch.

But time is invisible and intangible, without size or mass, formless and weightless. How can we guide it, let alone reverse its flow?

What exactly is time?
The ancients said, "flowers can bloom again, but youth comes only once," and that is time.
Ancient people also said that time is like water in a sponge; as long as you squeeze, there will always be some. This is also time.
In fact, the two times mentioned here correspond exactly to the two concepts of time: the former refers to a moment, and the latter refers to a period.

A period refers to a duration of time, such as one hour, two hours, etc.
But this is merely a measurement of time, and it does not clarify what the essence of time really is.
In fact, throughout history, countless predecessors have pondered the mystery of time.
The Zhuangzi says: "Emergence has no origin, and people have no aperture. What has reality but no place is the universe; what has duration but no beginning or end is the time."

Space exists in reality but has no fixed place to seek. Time keeps extending but has no beginning or end.
The Huainanzi says: "From past antiquity to present day is called Zhòu, and the four directions and up and down are called Yuˇ."
The three-dimensionality of space and the arrow of time are clearly defined.
These are the primitive thoughts and understandings of ancient people about time and space.
Now, with the development of society, we can recognize and understand time from a deeper level, more accurately grasping its essence.

Many scholars have also clearly recognized that essentially, time is merely a human sensation, a measure of the continuity of material motion.

Whether it is a moment (such as 1 o'clock, 2 o'clock) or a period (such as an hour, two hours), it is a concept of measurement defined by humans, not an objective reality.

For example: we can say something has 100 units, 200 units, or weighs one kilogram, two kilograms, etc. Here, quantity and weight are also not objective realities; they are merely concepts of measurement, attributes used to describe the world.

Time is the same in this regard. It is not an objective existence. There is no such thing as time in the world. Like length and mass, it is merely an attribute of space and matter.

In many of our current theories and discussions, the universe and our world are considered four-dimensional, with time regarded as one dimension and space as three. This leads people to believe that time is an objective reality like matter, but this is actually a misunderstanding.

Time, like length, mass, etc., is an attribute of space.
Therefore, some people suggest: according to this view, the dimension of the universe should be at least nine-dimensional, including three-dimensional length, three-dimensional time, three-dimensional mass, rather than four-dimensional. The vectorization of several attributes such as time, length, and mass should be synchronized; however, in reality, we usually combine the dimensions of time and mass with those of space, without treating them separately.

In summary, time is merely an attribute of space, or rather, a concept created to describe the motion of matter and the continuity of the world's operation. It is not an independent objective reality. From another perspective, one could also say that time is simply a human sensation. As long as we have sensations, thoughts, and motion, time exists. Motion continues, time continues.

At this point, everyone should have understood the essence and truth of what is called "time."
Can we reverse time if time is not an objective existence like rivers and lakes, and there is no such thing as "time" in the world?

How can we reverse time?
(2) The Secret of Time: How Can Time Flow Backwards?
Can time really stop, or even flow backwards?
Do we, as humans, have the ability to return to the past?
Earlier, we discussed that time is a measure of the continuity of matter's motion. So, it's simple— if we can stop matter from moving, wouldn't time also stop?

For example, we have a basketball that is moving, and the wall clock is ticking, time is passing. However, if we place the basketball there and stop it from moving, the wall clock still ticks, time still passes, and there is no reversal.

For instance, if we break an egg, the wall clock moves forward by two seconds, indicating that time has advanced by two seconds.

Then, we restore the egg to its original state (assuming we have the technology and capability to do so), exactly as it was before it was broken, which seems as if we have returned to the past, as if time has reversed.

However, we observe the clock on the wall and notice that it has moved forward by two more seconds, indicating that time has advanced by two seconds. The egg breaking and being restored takes a total of four seconds.

So, even if the broken egg is restored to its original state intact, time has not flowed backward

There is a stop, yet it still flows forward.
This seems different from what we imagine. In our understanding, as depicted in many science fiction movies and literature, matter remains the same, so wouldn't time just reverse? The scenes in movies where matter's motion reverses and time flows backward are incredibly spectacular, yet they don't match reality.

What's really wrong here?
We return once again to the starting point: time is defined as a measure of the continuity of matter's motion.
So, the key question in the above two examples is: what kind of motion would actually affect the flow of time, even reverse the direction of time?

What characteristics does time itself have that we have not considered?
The broken egg is completely restored, and the movement of matter is entirely reversed, yet time does not stop; it continues to flow forward.

Why is this the case?
In fact, this involves a secret of time: locality.
In Newton's "Mathematical Principles of Natural Philosophy" published in 1687, it is stated: "Absolute, true, mathematical time, in and of itself, and from its own nature, flows equably without relation to anything external." This is the classical mechanics' view of absolute spacetime, in which time is uniform.

In Einstein's special theory of relativity, the effect of motion on the speed of time flow was proposed: the time of an object in uniform motion passes more slowly compared to the time of an object that is relatively at rest.

In general relativity, Einstein further proposed the effect of mass on time: the closer one is to a gravitational source, the slower time passes.

This is the spatial inhomogeneity of time, which is what we call the locality of time.
Time is a manifestation of the continuity of material motion, and the motion of matter influences even determines the speed of time flow. The speed of material motion is inconsistent in different spaces and regions, which also determines the differences in these spaces and regions

The speed of time flow will be different, which is the locality of time.
That is to say, the speed of time flow is inconsistent in different spatial regions. Even the direction of time flow might be inconsistent.

In other words, time is uneven in space.
Let's take an example.
Suppose there are two twin brothers, A and B, who are placed on two different planets from birth. The planet where A is located has the same speed of matter movement as Earth, and his body's metabolism, nerve signal transmission, etc., are also the same as on Earth, so he perceives the flow of time as the same as on Earth. The planet where B is located has a speed of matter movement half of that on Earth, and his body's metabolism, nerve signal transmission, etc., are also half of

those on Earth. Therefore, for B, the speed of time flow is only half of that on Earth. From the perspective of Earth, it seems as if everything on B's planet is playing back at half speed.

A is also 100 years old and has become an elderly man with white hair, while B is only 50 and in much better health. The time and years they have experienced show a very noticeable difference.

This is the locality of time, where different spatial regions may experience varying rates of time flow, leading to different changes.

In ancient myths and legends, there are many miraculous stories that describe how a day in heaven can be a thousand years on earth. In the heavenly realm, one might simply stroll around with the immortals, only to return to the real world and find that the world has changed beyond recognition, with things and people no longer the same. This seems to be exactly the scenario described in the previous example. But were these ancient myths and legends merely stories, or did our ancestors in ancient times actually discover something significant?
(3) Time Travel: How to Stop Time?
Can time stop? How can time be stopped?
The previous chapters have already explained that time is a manifestation of the continuity of material motion, so to stop time is to stop the continuity of material motion

Then, let all matter come to a complete stop.
The complete cessation of motion refers to making matter completely stationary at the atomic level, even at the electronic level.

For example, the case we mentioned earlier: a basketball, when we place it on the ground and leave it still, its time hasn't stopped, because the atoms and molecules inside are still moving, and it continues to age.

In the case of a person, it's his metabolism that hasn't stopped, and he is still aging.
We just can't perceive this aging and aging with our naked eyes.
To make the basketball completely still, all the molecules, atoms, and matter inside it would need to stop moving entirely.

At this moment, if we place a clock inside a basketball, because the molecules and atoms have completely stopped moving, the clock will no longer function. The basketball will also remain in that state forever, without aging.

For a person, it would be as if their metabolism had completely stopped, and they would never age.
At this moment, time really stops for both humans and basketball. It stops right at that moment.
Some may ask, what does it mean by complete stillness? We all know that the motion of matter is absolute, while stillness is relative. For example, when each of us stands still on the ground, we are still moving with the Earth's rotation and revolution, and even moving rapidly around the galaxy with the solar system.

Therefore, rest is impossible to be at rest.
Here, we say that the basketball or a person being at rest, is completely at rest relative to the local space they are in. This is the characteristic of the locality of time, or the inhomogeneity of time in space.

We can describe this more accurately using entropy. That is, when the rate of entropy change in all parts and regions of a closed system is constant at zero, the flow of time in the closed system stops.

了下来。
Therefore, as long as the entropy of all parts and regions in the system no longer changes, the flow of time within the system can be considered to have stopped, regardless of whether there is movement in the system. For the sake of understanding, we will only discuss the case where all motion has completely ceased.

Time stops here, but only within the local space of this basketball and this person.
Motion stops within the local space, and the rate of entropy change is zero, so time stops in the local space.
For the world outside this local space, time continues to flow without being affected, because the outside world is not stationary and is still in motion.

The people in the basketball court continue to cheer, the basketball players are still full of energy, and no one will notice the basketball "sealed" in the corner.

This is the locality of the speed of time.
If the time in a spatial region gradually stops, the metabolism and even the thoughts of the people inside will slow down and eventually stop, making them appear completely frozen in that moment.

Meanwhile, time in other places continues normally, and everyone lives their lives as usual. After a hundred, a thousand, or even ten thousand years, the world has changed beyond recognition, but for those in the time-stopped spatial region, this period of time feels like just a moment, or even less than a blink. Neither they nor the surrounding objects have changed at all.

This is the legendary "A Moment is an Eternity."
A moment in a local space or even in a world could be an eternity for other worlds.

In ancient Chinese mythology, the legends that a day in heaven equals a thousand years on earth may just be a reflection of this very situation.

By this method of local space-time stopping, we can "travel" to the future. Of course, this

The "time travel" mentioned here is not the real time travel in fantasy novels. It simply refers to slowing down our own time flow to wait for the world to develop into the future.

Just like the mythological saying "a day in heaven is a thousand years on earth," those who return from the heavenly realm to the world seem to have traveled to a thousand years into the future.

Some hospitals now actually freeze patients, waiting for decades or even centuries for the world to develop, and then thawing them to treat their illnesses with advanced technology, which is a real attempt at this kind of "time travel."

Therefore, by utilizing the essence of time as the continuity of material motion and the locality of time, we can control time to stop, or even control "traveling to the future." Then, can we use the same method to return to the past? How can we return to the past?
(4) Controlling Time: How to Return to the Past?
By utilizing the nature and locality of time, making time stop, or even "traveling to the future," these topics have all been discussed in the previous chapters.

Can we use the same principle to reverse time and return to the past?
Some people say that within the light cone, everything is fate, believing that going beyond the light cone would mean returning to the past. This is actually a misunderstanding of the nature of time; the flow of time has nothing to do with the light cone. Even if you go beyond the light cone, time will not go back to the past.

Earlier, we gave an example of a broken egg being restored to its original state before it broke, meaning that for this egg, it has returned to its past two seconds ago.

If there is a clock inside the egg, the time recorded by the clock also goes back to two seconds ago, the moment before the egg broke.

If there is a chick inside, the chick also goes back to two seconds ago.
This is time traveling backwards.

But why does the clock on the wall outside the egg continue to move forward in time, without going back?
This is because of the characteristics of time we mentioned earlier: locality, or in other words, the non-uniformity of time in space.

The flow speed, even the direction, of time in different regions of space varies due to the different motion states of matter in each region. If the motion state of matter in a certain region changes, it will affect the flow speed of time in that region, or even cause time to flow backward. However, this only occurs within that specific region of space and has no effect on other regions. The flow speed of time in other regions is determined solely by the motion state of matter in those regions.

For example, in this case, the time reversal occurs only within this egg, because we have only restored this single egg. All the matter inside the egg, from molecules to atoms, reverses its motion. We can call this entropy reversal. The entropy reversal causes the time in this region of the egg to reverse as well, flowing back to two seconds ago.

And outside the egg, the movement of matter does not reverse, not even affected, so time continues to flow normally, and the clock keeps moving forward.

In this case, the passage or even reversal of time is a perception of time caused by changes in life and the state of matter. In other words, it's just a feeling of people. Essentially, the reverse change of matter inside the egg is also a form of matter movement, so time is still fundamentally moving forward. It's only because the egg has completely reverted to its state two seconds ago that people feel as if time has reversed.

For the outside big world, people would have no sense at all. The world runs normally, time flows normally, and there would be no abnormal changes.

Just like a storm in a teapot, the time reversal inside the egg does not affect the normal operation of the outside world. Therefore, the clock on the wall outside continues to move forward as usual.

What if we were to change our target from an egg to our entire Earth world?
It would still be the same situation.
If we were to restore the entire Earth world exactly as it was one year ago, for everyone on Earth, the whole world would revert to the past, back to one year earlier.

But for the solar system, the galaxy, and even deeper into the universe beyond Earth, time would not be affected, nor would it reverse; it would continue moving forward as usual.

For example, if in 2023 we spent a day to bring the Earth's world back to where it was one year ago, then for people on Earth, the current time would be 2022, but for those outside Earth within the solar system, the current time would still be

2023, just with an extra day.

At this moment, if there were an astronaut on a spaceship outside Earth, he would be surprised to find that he himself and the spaceship had not changed, but Earth seemed to have returned to where it was a year ago. He would feel as if he had magically traveled back to a year earlier. All the events that happened during that year would be forgotten by humans on Earth, except for him who would remember them. If this spaceship was launched in 2023, then the people on Earth would completely not recognize it, because they are now in 2022 and have not yet launched this spaceship.

At this time, humans on Earth would be surprised to discover a mysterious spaceship from an unknown source flying around the Earth's near space, causing panic across the entire human world. Because at this moment, they have not yet launched this spaceship. This spaceship seems to come from the future.

It seems as if a scene from science fiction has appeared. This is the truth behind the miraculous time travel.
Many people will wonder again, would such time travel create a new timeline of the world? Would it cause the paradoxes mentioned in science fiction works? In the world of humans on Earth in 2022, are there still astronauts on this spaceship?
(5) Is time travel going to a parallel world?
Is time travel going to a parallel world?

There are a lot of time travel issues in many movies, literature works, and even scientific discussions, including many paradoxes, such as the grandfather paradox, the Hitler paradox, and the Borde-Guth-Vilenkin paradox, which are described in a very mysterious way, making people confused and dizzy, even burning their brains.

But would time travel really create these paradoxes?
Not only ordinary people, but even many senior scholars have fallen into a misconception here, desperately seeking solutions.

For example, many scholars believe that time travel would lead to the failure of causality, which is considered one of the most fundamental laws of the universe and cannot be broken, so world travel is impossible.

Some people directly argue, "Since it hasn't been seen, it's impossible," which has already moved beyond the realm of scientific exploration and become a subjective assumption.

Some people have proposed the idea of "parallel universes," suggesting that time travel actually involves moving to a parallel universe. These are all subjective assumptions without factual or theoretical support, as they lack solid theoretical reasoning. They cannot explain how one travels back in time or how one ends up in a parallel world, among other fundamental arguments.

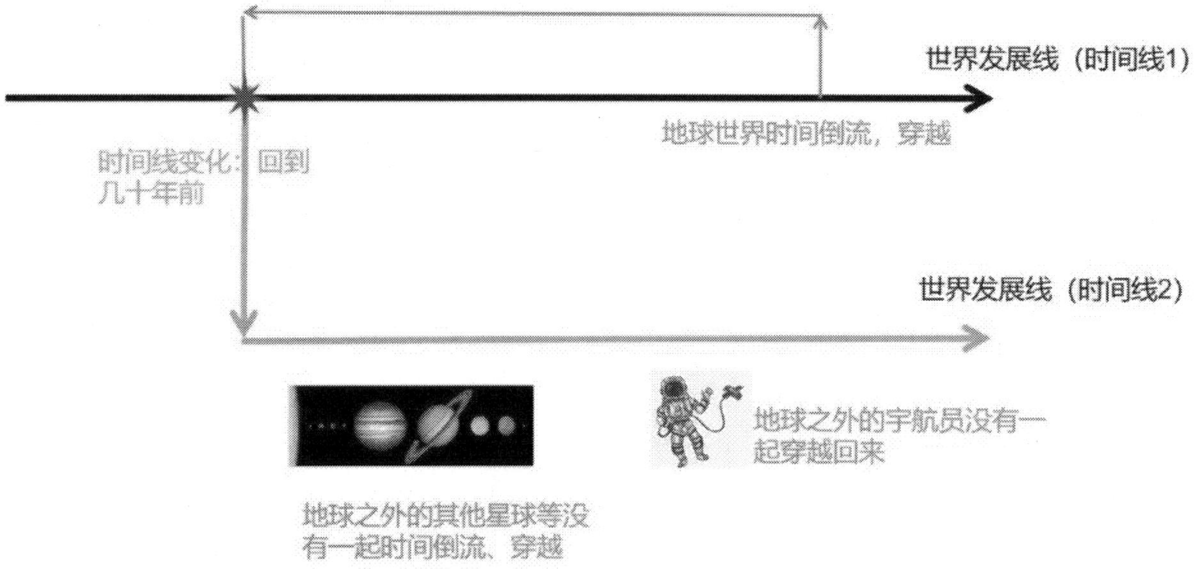

Figure 80: Figure 12-3 "Time travel" causing changes in the timeline

It is precisely due to the lack of a deep understanding of the nature of time and its local characteristics that so many confusing ideas have emerged, leading to many so-called "unsolvable" time travel paradoxes, causing many people to fall into confusion.

To fully understand these issues, we need to first clarify the core questions among these paradoxes: would time travel create a new timeline, would it allow travel to a parallel universe, and would it violate causality?

We have already discussed in the previous chapters how to achieve time travel and return to the past, as well as clarified the local nature of time.

From these discussions, we can know that if Earth and the human world were to reverse time and return to a year ago, this time reversal would be local, meaning that time in other parts of the solar system outside Earth would remain unchanged and continue to flow normally, with only Earth's time being reset to a year prior.

If there are other humans, such as astronauts, outside the Earth, they could witness the changes on Earth directly. They did not return to a year ago, but continued to live according to normal time, while the humans on Earth would completely forget about that year.

So, are those humans outside the Earth time-traveling back to a year ago? This can be metaphorically understood, but are they traveling to another timeline? Of course not.

This is simply a normal physical change. The universe remains the same universe, the solar system remains the same solar system; it's just that the matter on Earth has been reorganized and returned to its state from a year ago, as shown in Figure 12 - 3.

In this context, Earth is still the same Earth, and the world is still the same world. Earth has not crossed over to another timeline, nor has it traveled to any supposed parallel universe or multiple worlds. These are all just fantasies and not scientific facts.

Such fantasies are popular, and ordinary people do not understand such profound concepts, or have not thought deeply about them, so they mistakenly believe that different time lines or parallel universes really exist.

Of course, due to factors such as astronauts not being on Earth, when the Earth's timeline flows backward and resets to a previous year, it will have slight differences from the original scenario of that year, and this will lead the subsequent development of the world down a different path. If we understand the timeline as such different developmental paths, it's not entirely incorrect. However, this is not the same as having several time lines existing simultaneously. In the actual world, development always follows only one real timeline at any given moment.

Then, would these time reversals violate the causality principle as some people worry?
For example, if an egg falls to the ground, breaks, and then is perfectly restored to its original state on the table two seconds later, does this break the law of causality? Can a dead person be resurrected through time reversal?
Will time travel backwards violate the law of causality?
If time were to flow backwards, would that violate the law of causality?
There are numerous discussions in many movies, literary works, and scientific debates about time travel paradoxes. Many scientists are struggling to figure out how to make time travel avoid breaking the law of causality. Some scholars propose the theory of parallel worlds, while others suggest the principle of avoiding paradoxes, that is, paradoxes are forbidden to occur. Even some scholars, in order to prevent the violation of the law of causality, directly argue that time travel is impossible, not considering how to solve the problem, but instead making the problem impossible from the beginning, thus solving it once and for all.

All sorts of such arguments.
As the "Novikov Self-Consistency Principle" points out, people can travel back in time, but they cannot alter the course of history. American scholar Lloyd said: "It allows people to return to the past, but prohibits any actions that might lead to paradoxes in the future."

Etc., these views have actually violated basic physical principles and fallen into fantasy. Because of the butterfly effect in chaotic systems, even the slightest change in conditions will have a huge impact on the world's subsequent development, leading to significant changes in the evolution of the world.

As we discussed in our previous chapters, these are all results of not fully understanding the nature, characteristics, and underlying principles of time travel. Time travel will not create a new world, nor does it mean traveling to a parallel universe.

Then, does time travel violate the principle of causality? How should we understand those paradoxes that make all scholars uneasy?

We also take the example of a broken egg:
An egg falls to the ground, breaks, and then is perfectly restored, returning to the table and back to its intact state two seconds ago. In this case, is the law of causality violated?

The egg falling to the ground is the cause, and the breaking of the egg is the effect, which is the causal law we normally recognize. However, when the egg, from a broken state, starts to restore itself and even escapes gravity to return to the table, this seems to completely reverse causality, turning the breaking of the egg on the ground into the cause and its restoration on the table into the effect?

Does this violate the causality law?
In fact, if you think about it carefully, the egg's restoration and returning to the table is actually the result of external force (a particular technique and ability), not a natural occurrence. Therefore, this restoration is the outcome of our external intervention. In other words, our use of external force is the cause, and the egg's restoration to its intact state two seconds ago is the effect.

This is still normal causality and the usual way matter changes. There is no violation of causality as some people worry.

What if we don't use eggs, but instead take the example of the entire Earth's time flowing backward, returning to a year earlier, as we mentioned before?

The result is the same.
We bring the Earth human world back to one year prior from 2023 through a special kind of force or technology, which

is the result of external intervention. To astronauts in space, outside the world, seeing this may be surprising, but it is all normal material operation that conforms to physical laws and causality. Of course, if astronauts do not understand the underlying reasons, they might think the Earth's situation is abnormal, violating causality, but in reality, there is no problem.

For people within the world, they are reset back to one year prior at the level of fundamental particles, including their physical condition, everything around them, and even memories. As a result, there will be no time travel or any memories from this year. To them, the world is just arriving in 2022.

Therefore, reversing material motion to enable time travel is just as much a result of physical laws as normal material motion enabling the normal flow of time. Neither of these actions violates causality. The concerns about violating causality stem from a lack of understanding of the nature of time and the principles behind time travel.

As we mentioned earlier, fundamentally, this change that makes time flow backward and the world return to 2022 is still the normal movement of matter, so time still flows normally forward and does not reverse. It is just that for humans on Earth and astronauts in space, they feel as if they have returned to 2022, as if time has reversed. In reality, if they observe the stars and other celestial phenomena in the outside universe, they would find that time should be in 2023, because the galaxy and the universe outside Earth have not been reset and are still in the state of 2023.

Then, if time really flows backward, what about those so-called "time paradoxes," such as the grandfather paradox, the Hitler paradox, and the Borde-Guth-Vilenkin paradox, which have trapped many scholars and even famous scientists, making it difficult for them to find a good solution? How should we understand these?
(7) Time Travel: How to Understand the Grandfather Paradox
If time travel occurs, would it really create time paradoxes?
There are numerous issues of time travel paradoxes in many film and literary works, as well as scientific discussions, such as the grandfather paradox, the Hitler paradox, and the Borde-Guth-Vilenkin paradox, which are described in a mysterious way, leaving people puzzled

Are confused in the clouds, even many senior scholars have fallen into a misconception in this area, desperately seeking solutions.

In fact, as we analyzed earlier, even if time travel really occurred, there should not be any "paradoxes."

How should we understand the so-called time paradoxes and time loops in science fiction films?
The first is the grandfather paradox, also known as the grandmother paradox, which is a puzzle that has puzzled many scientists.

The grandfather paradox states: if a person named Tom travels back decades and kills his grandfather at that time, it would directly lead to a consequence where his grandfather does not marry and have Tom's father, so Tom's father would not be born, and therefore Tom would not exist. According to this result, Tom would not be able to travel back to kill his grandfather in the future. This creates a paradox: can Tom really travel back in time to kill his grandfather? As shown in Figure 12-4.

How can this paradox be understood?

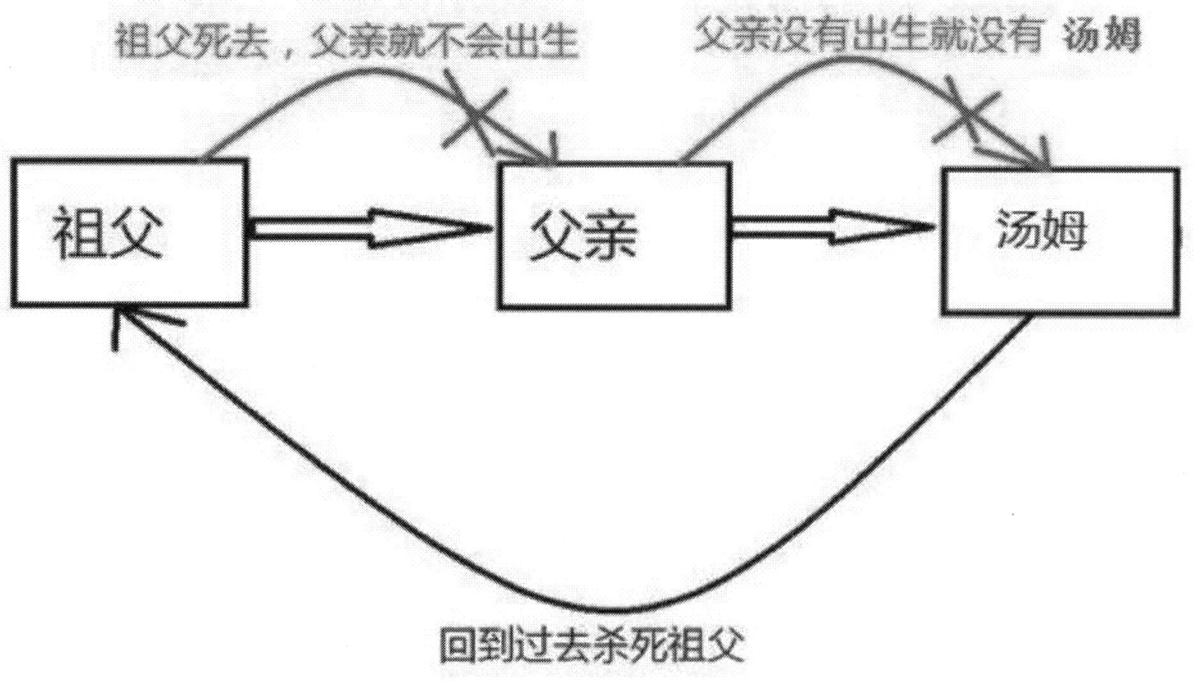

Figure 81: Figure 12-4 Grandfather's Paradox

In fact, according to what we discussed earlier regarding the nature of time and its locality, the only way for Tom to return to several decades ago is to restore the world outside of himself exactly as it was decades ago, while keeping himself in his current form.

At this moment, the world, since Tom did not follow the device to restore, would have some differences from the original world of several decades ago. Assuming we ignore these differences, if Tom really met his grandfather from several decades ago and killed him, what would happen? What strange things would occur in the world?

Would Tom disappear immediately because his grandfather died? Or would this world create parallel worlds and new time lines? Would there still be a Tom in the future world?

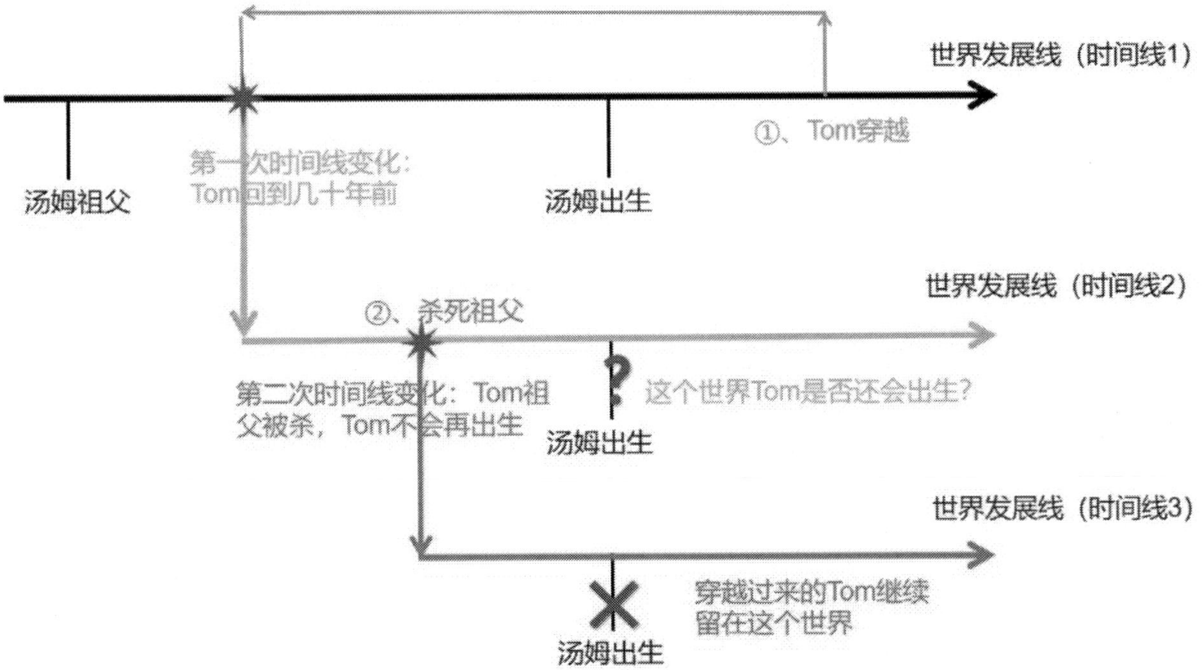

Figure 82: Figure 12-5 Grandfather Paradox World Development Line

In reality, nothing strange actually happens. Everyone can look at the diagram of this timeline: Tom's time travel and killing his grandfather, each event will affect the world's evolutionary development line. If Tom kills his grandfather, then his grandfather dies, that's all. It's like an alien coming to this world and killing one person; it won't affect the alien, and the world will still continue to develop normally, just like all other murders in the world. The slight difference is that the world after this will be a little different, because compared to the "previous world," there is one less person: Tom's grandfather, as shown in Figure 12-5.

The world will continue to develop without producing a new Tom, but this does not affect Tom now.

Tom is now a completely different person to this world, like an alien. "Beyond the three realms, not part of the five elements."

There will also not be a situation where Tom suddenly disappears in the future world, because at this moment, there is no future world in the universe. The world only exists in the present, and it will take decades of development to reach the original future time.

Of course, there is no such thing as different time lines or parallel worlds. Those are all pseudo-scientific concepts caused by a lack of understanding of time's nature. The world is still just this one, but due to this murder, the timeline of its development will diverge from the original one.

Other time paradoxes, such as the Hitler paradox, involve a person traveling back in time to kill Hitler and thus prevent World War II. However, if World War II did not occur because Hitler was killed, then the act of traveling back in time to kill him would never have happened in the first place, creating another paradox. Now that we understand the grandfather paradox, we can see that this is also a similar case. In reality, there is no paradox; it's just a misunderstanding caused by a lack of clarity about the nature of time and the principles of time travel.

We will discuss another more profound time paradox that has puzzled many scholars, to help everyone better understand.

Here's a thought question for everyone: If Tom travels back to the world decades ago and does not kill his grandfather, would the young Tom in this world still be born (as shown in the third timeline in Figure 12-5)?

(8) Time Travel: Understanding the Polchinski Paradox

Whether time can flow backward and how to make time flow backward have all been discussed in the previous chapters, and everyone has already understood and reached a clear conclusion. You now understand the essence of time and the underlying principles of changing the speed of time flow or even making time flow backward.

Regarding the so-called "time paradoxes" caused by time travel, such as the grandfather paradox and the Hitler paradox.

Those who have read the previous chapters should already understand that these所谓的 paradoxes are not actually paradoxes. They are simply misunderstandings caused by a lack of clarity about the nature of time and the principles of time reversal.

In addition to the above paradoxes, there is another well-known and influential time travel paradox among many: the Borde-Guth-Vilenkin paradox.

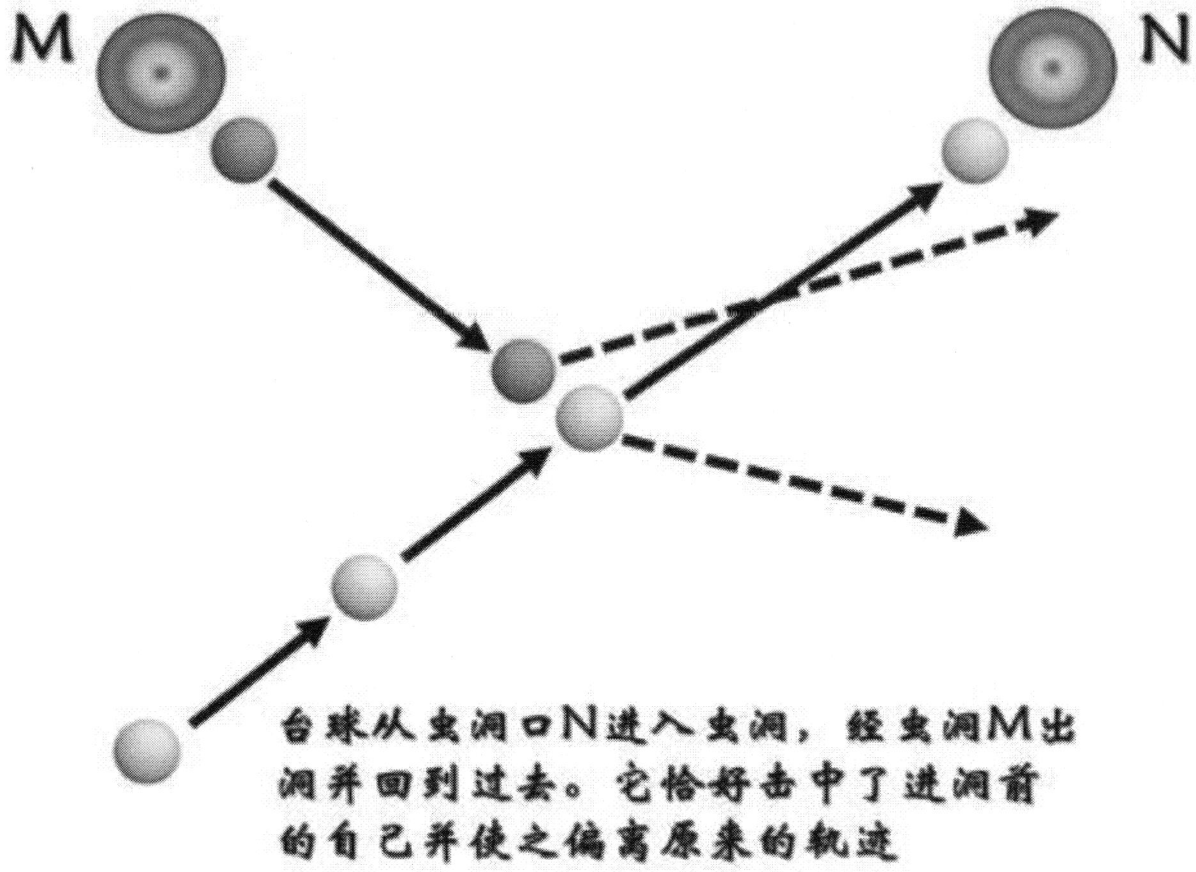

Figure 83: Figure 12-6 Time Paradox - The Polchinski Paradox

As shown in Figure 12-6, the Borde-Guth-Vilenkin paradox states that in space, there is a wormhole with an entrance N and an exit M. The wormhole has the property that entering it triggers a time reset mechanism, resetting the world to a time in the past. Outside the wormhole, assume there is a billiard ball (represented as white in the figure), which starts from an appropriate position and speed, enters the wormhole through entrance N, and exits through exit M (blue). Due to the time reset and reversal mechanism inside the wormhole, when the billiard ball exits, the time outside the wormhole is earlier than the time it entered. At this moment, the white ball, which is the younger version of itself, has just started to move and has not yet entered the wormhole. If this exiting billiard ball happens to hit the white ball just before it enters the wormhole and changes its direction, preventing it from reaching entrance N, the consequence is that the white ball would not enter the wormhole and thus would not exit through M. Therefore, it would not hit the white ball that is about to enter the wormhole.

So this creates a paradox: can this white ball actually enter a wormhole?
This paradox has caused panic among many scientists and people, and a lot of them are researching how to solve it

A paradox that required much thinking and proposed many solutions.
How can we understand this paradox?
First, let's assume that during the journey through the wormhole, some mechanism is triggered, activating a force that sends the world outside the wormhole back to a moment before, thus achieving the time travel scenario described by the paradox. What would be the result?

The ball first enters the wormhole, comes out, and returns to the moment before it entered. It just happens to hit the ball that is about to enter the wormhole, knocking it away so that it cannot enter the wormhole.

At this point, there would be no new ball flying out of the wormhole, and the world would not be sent back to that earlier moment. To the people in this world, they would see: before they throw a ball into the wormhole, it is hit by a ball that somehow comes out of the wormhole, causing the ball to fail entering the wormhole.

Then, there was nothing else.
Because the new small ball did not enter the wormhole, the time of the outside world was not reset, so the world continued to move forward normally, and there would be no paradox.

The essence of this event is actually the same as the grandfather paradox, the Hitler paradox, and others. The difference is that in the past, the time traveler was a person, but now it's this small ball. To humans, the small ball that suddenly came out of the wormhole had no memory, because their entire world, including their memories, had been reset back to a time in the past. If they carefully studied it, they might be surprised to find that the small ball that suddenly emerged was exactly the same as the one they had launched, as if it had come from the future to prevent their actions.

Of course, if people try to throw a small ball again, it would merely be repeating the same process.

So as we mentioned earlier, this so-called Polchinski paradox is not actually a paradox. The so-called paradox is merely a misunderstanding caused by people's lack of understanding of the nature and characteristics of time.

In the study of theoretical physics, it is essential to follow the correct materialist dialectical worldview and methodology to avoid falling into the trap of idealist metaphysics.

In current physical theory research, there are relatively obvious cases of this kind, which are also one of the reasons why fundamental theories in some fields have not seen major developments or breakthroughs in the past century. We will continue to discuss these issues later.

(9) Time Travel: The Crisis Implied by the Borde-Guth-Vilenkin Paradox

The Borde-Guth-Vilenkin paradox, which has puzzled many scholars and scientists, as well as other time travel paradoxes, is not actually a paradox. It is simply a misunderstanding caused by people's lack of comprehension of the nature and characteristics of time. We have already discussed these in previous chapters, and everyone should already be aware of them.

In fact, broadly speaking, all places where paradoxes occur are due to deviations in people's cognition or theoretical explanations.

However, after reading the previous chapter, careful readers may have noticed that there is actually another significant crisis hidden within this Polchinski paradox (as shown in Figure 12-7).

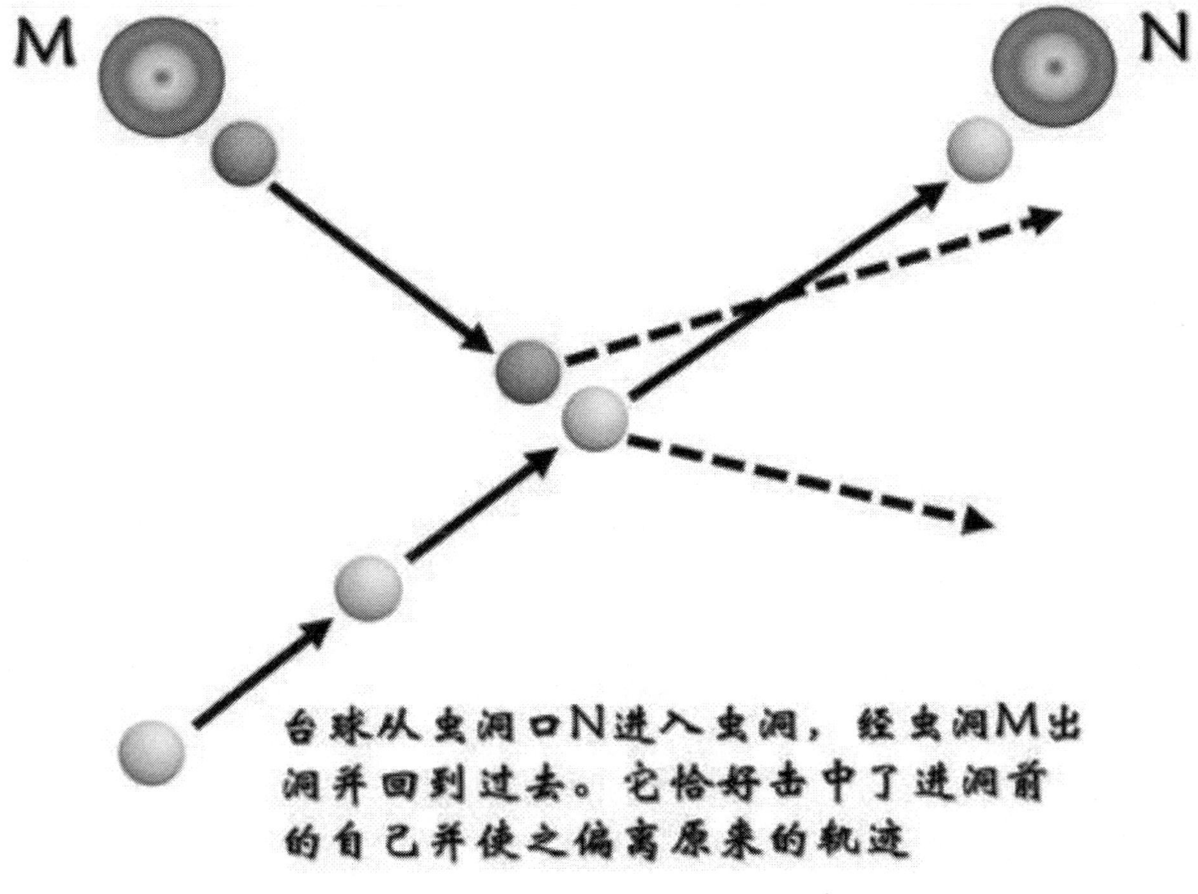

Figure 84: Figure 12-7 Polchinski Paradox

We imagine that when we launch a small ball, which we can call Ball 1, Ball 1 moves from the entrance

N entered the wormhole, triggering a mechanism that sent the outside world back to ten minutes earlier. At this time, people had just launched a small ball, which we can call Ball 2. Suppose Ball 1, coming out of the wormhole exit M, hit Ball 2 before it had entered the wormhole, preventing Ball 2 from entering and thus avoiding another time reset. In this case, there would be no problem and no paradox.

However, if the 1st small ball does not hit the 2nd small ball, allowing the 2nd small ball to continue into the wormhole and sending the world back to ten minutes ago, what would happen in that case?

This will lead to the repetition of the entire process repeatedly.
The ball triggers time reversal, resetting the world back to ten minutes ago, and the ball is fired again, causing the world to reset once more.

In other words, the world will repeatedly start from ten minutes ago and repeat until the ten minutes end, falling into a historical loop.

The entire world will infinitely repeat within this ten-minute range.
Many novels and movies you have seen also describe this time loop phenomenon, where a person endlessly cycles through the same day, which can be considered a true reflection of such situations.

Perhaps in the future, when we enter the era of cosmic exploration, we may witness the amazing sight of a civilization trapped in a time loop in a distant galaxy.

Then, how can we break this cycle and save the world from such a轮回?
There are several methods, one is that the time reset rule of a wormhole changes, making the time reset ineffective, and another is that a force outside the scope affected by the time reset can intervene to break this cycle.

Of course, as we mentioned earlier, traveling through a wormhole to change time is just a hypothesis and a fantasy without scientific evidence. Here, we continue the discussion following this paradox.

There is another case to mention. To achieve a perfect time reset, it is first necessary to ensure the conservation of mass and energy throughout the entire reset area. For the time travel example we mentioned earlier, if an astronaut is outside the world and does not undergo the same time reset, then the reset world would not conserve mass and energy. As a result, the world at the same moment in the original history would have slight differences. It is difficult to predict what changes these differences might cause. After all, in a chaotic system, the flap of a butterfly's wings might cause a storm on the other side of the world. Any minor difference in initial conditions could lead to significant subsequent changes.

Step back, even if energy and matter can be borrowed from other areas during time reset to ensure the world perfectly resets back to several decades ago, other planets, stars, and even the entire universe have not been synchronized to reset, and have undergone significant changes compared to several decades ago. These could also be factors that introduce variability into the evolution of the world.

In the previous chapters, we left you with a thought question: In the grandfather paradox, if Tom time-travels back several decades and even if he doesn't kill his grandfather, will young Tom still be born? Now, I believe you have a deeper understanding.

Returning to the case of a dead loop in the world we discussed earlier, when the world resets back to ten minutes prior after a small ball is thrown, this reset causes a discrepancy because a ball is missing, leading to non-conservation of energy and mass. As a result, the world after the reset will have slight differences from the world ten minutes ago. These differences accumulate over time and eventually influence the evolution of the world. At that point, the people in the world might decide not to throw the ball anymore or to delay throwing it, and so on. In any case, they would be free from that ten-minute dead loop.

Another important point is that all the logic behind time travel causing the world to fall into a time loop is based on the philosophical view of determinism and its logical foundation. The philosophical view of determinism itself is rooted in the theoretical foundation of Newtonian mechanics. However, after the rise of quantum mechanics, it has already caused a significant impact on this deterministic philosophical view.

The debate between these two theories and the philosophical and worldviews they have generated has yet to reach a perfect answer.

Which of these two schools will ultimately prevail? Before the heavens and earth are set in order, who will be the dark horse?
(10) Reincarnation: Poincaré Recurrence, Another Explanation of Time Reversal
Is there reincarnation in the world? Could the world possibly return to a previous moment?
In various countries around the world, there are many different legends and stories about reincarnation. Some of these stories talk about the reincarnation of individuals, where a person is reborn as another person and starts life anew. Others speak of the reincarnation of the world, suggesting that one day the world will return to an original moment and start a new cycle.

Many of these have been labeled as idealistic or superstitious, just legends, and not credible.
For scientific research, denying the possibility of something should be based on scientific and dialectical principles, not on hasty decisions. It is not acceptable to simply dismiss something as impossible just because it has never been heard of, seen, or believed, or because it goes beyond common sense.

There are many things in the world that go beyond people's imagination, which ordinary individuals may never encounter in their lifetime. In fact, many scientific developments themselves involve overturning people's common sense. Therefore, relying only on the reasons mentioned above is insufficient to deny the possibility of something.

To deny something, there must be evidence, either factual or theoretical.
Let's take the example of the cycle of the world to discuss whether this event might occur.

Will the world possibly exist in cycles?
There is a very famous theory in academia called the Poincaré theorem, also known as Poincaré recurrence, as shown in Figure 12-8.

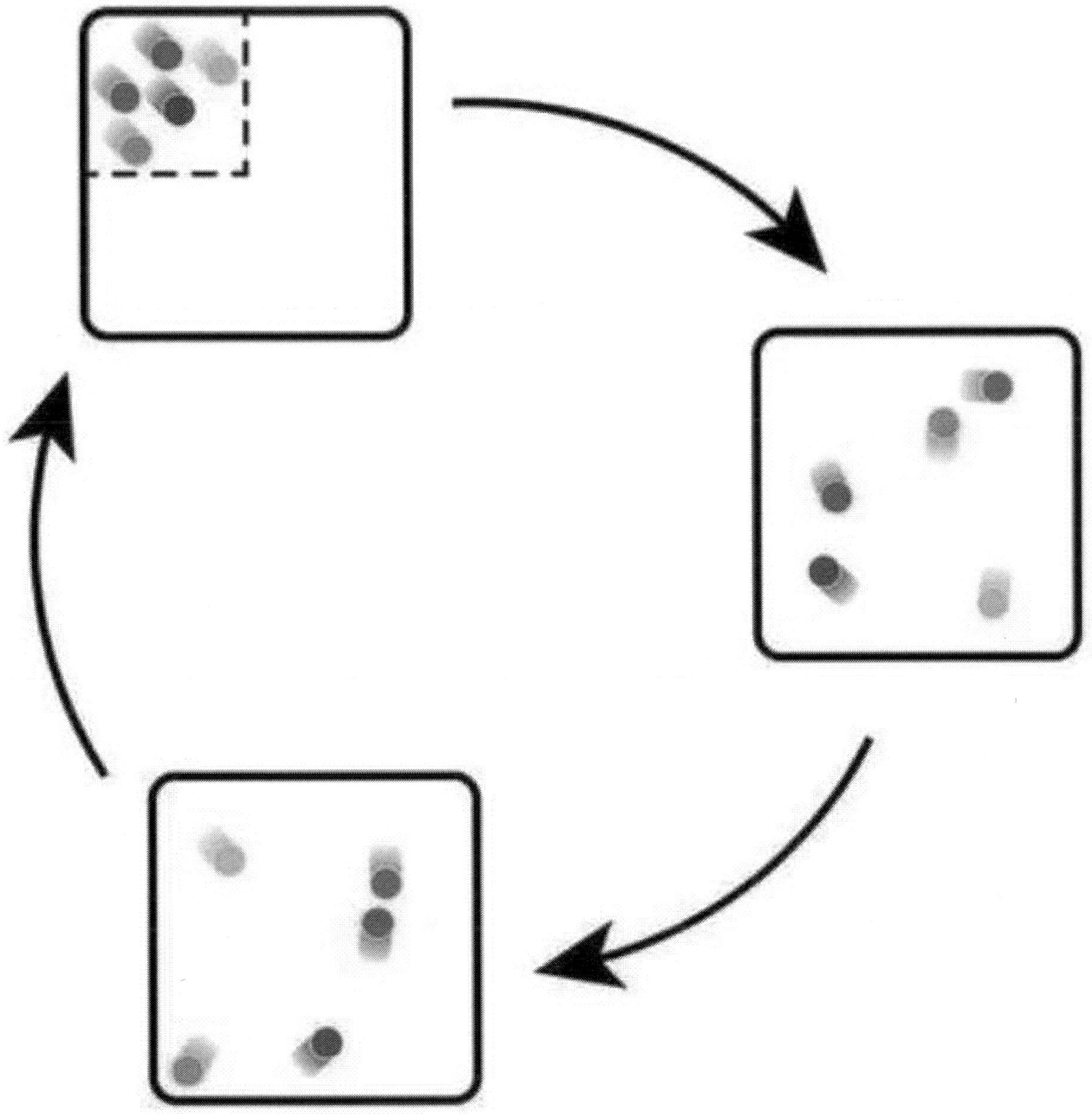

Figure 85: Figure 12-8 Poincaré recurrence of a closed thermal system

This theory states that, within a closed system, any particle will inevitably return to a position infinitely close to its initial position after a sufficiently long time, although the length of this time far exceeds our imagination. Such a cycle is called a Poincaré recurrence.

The world can be considered as a system composed of countless particles. According to this theorem, after a long period of time, all particles in the world will inevitably return to a state close to their initial positions. Here, the initial position is not only the moment of the beginning of the world, but can also be considered as any moment in the evolution of the world, which can serve as the starting point of the system's development.

At this moment, the entire world returns to its corresponding initial moment. Everything in the world, including all living and non-living entities composed of particles, also recovers to the state of that moment.

At this moment, it's as if the dead have come back to life, the vanished dynasty has reappeared, and the distant legend is once again being replayed; for this world, it's as though a brand-new cycle has begun.

According to Poincaré's theorem, this is entirely possible.
Of course, this cycle is merely an infinite approximation, not a complete repetition.
It might take almost an infinite amount of time to reach this moment.
But in the previous chapters, we have already discussed that time is local, meaning it is specific to a world.

An infinitely long time may be merely a moment, or even an instant, to another world.
Therefore, to observers from other worlds, this world may cycle through rebirths after some time, or even be in an infinite cycle of rebirth.

This cycle can be seen as another form of "time reversal," but in this case, time continues to flow forward according to the theory of chaotic systems, and the world continues to develop forward, yet returns to the past. It is purely natural and occurs without any external intervention, which is the most astonishing and unimaginable phenomenon.

For example, our world has evolved for countless years, and one day, it might return to the historical moment of the Qin Dynasty, or to the Tang or Ming Dynasties. The Qin King would still be Ying Zheng, and other countries and races in the world would remain largely the same. There might only be minor differences in geography, population numbers, livestock counts, etc. Of course, these differences could lead to significant variations in the world's development compared to the original historical course. History is not just a simple repetition; it may evolve into another history and future beyond our imagination. However, after countless years, it might once again magically return to this historical moment of the Qin Dynasty.

This can be considered another more magical time reversal.
This kind of regression is different from the time reversal we discussed earlier; this return occurs naturally without any external human intervention.

However, there is another point. Poincaré recurrence refers to particles in a chaotic system, but chaotic systems can also exhibit self-organization phenomena. For example, the emergence of humans and intelligent life is the result of such self-organization.

After the emergence of intelligent life, it would naturally intervene or even dominate the system's evolution, shaping the world in the direction it desires.

Would Poincaré recurrence still occur in such a scenario?
Will this theorem still hold?
Is this within the scope of the Poincaré recurrence theorem?

Additionally, according to the second law of thermodynamics in physics, also known as the entropy increase law, the entropy of an isolated system always tends to increase, becoming more disordered. However, the Poincaré recurrence theorem suggests that the world might return to a previous state.

How should we understand the conflicts that exist here?

(11) The Arrow of Time: Is the World Deterministic or Indeterminate

Is the world deterministic or indeterminate? This is an ancient and complex philosophical question. Numerous academic schools have emerged throughout history, with a lot of debate and even conflict, because this question directly influences human worldview.

The current main theoretical foundation of determinism is the classical theoretical system represented by Newtonian mechanics. In this classical mechanics framework, everything in the world operates deterministically. The entire world, including the development of human society, even the movement of brain signals and the functioning of thoughts, is akin to mechanics, where previous states determine future developments. Everything seems to function like a clock, predictable through precise mechanical calculations, as if everything is bound by an inevitable fate. Laplace even extended this principle to the universe, believing that the universe also follows this law. According to him, by knowing the complete state of the universe at a particular moment, one could determine its state at any time in the past or future. This is known as Laplace's determinism.

World determinism thus gained worldwide recognition, but it brought about a fatalistic philosophical view: everything you have, including your life, was determined from the moment you were born, even from the moment the universe was created. Your efforts are meaningless, your choices cannot change anything, and everything is predestined. You think it's your own effort and your own choice, but in reality, it's all decided by Newtonian mechanics. This is undoubtedly a source of despair for everyone, as if they were shackled by fate.

Many of the time travel cases that led the world into time loops we discussed earlier are based on this classical philosophical view and logical foundation: if the world returns to a state from ten years ago, it will continue to evolve forward along the original trajectory, because the development of the world is deterministic, and classical mechanics determines everything.

However, with the rise of quantum mechanics, the concepts of probability and randomness emerged. Quantum theory suggests that the microscopic origin of the world is probabilistic, random, and uncertain, so the development of the world and human society is also uncertain. This directly undermines the foundation of determinism in philosophy, breaks the chains of fatalism, and gives the philosophy of indeterminacy great success, allowing many people to regain a little hope that "my fate is determined by me, not by heaven," and to find reasons for striving.

However, since the development of quantum mechanics, the academic community has always been greatly skeptical and debated the implications of this theory regarding the nature of the universe and the probabilistic worldview, as it directly challenges our intuitive understanding of the world.

For example, Einstein argued that the probability and uncertainty revealed by quantum mechanics must have underlying reasons. Finding those reasons would lead to answers and restore determinism. Marxist materialism holds that the world is knowable, so his view essentially asserts that the world is deterministic, not indeterminate or unknowable. Etc.

The two schools of thought have been engaged in intense debate on this issue. For instance, Einstein firmly believed that he would never accept the idea that God decides the world's operation by rolling dice. However, Bohr of the quantum school countered by asking, why should ordinary people assume the right to decide how God operates the world? Perhaps God indeed does so in this way? Etc. This debate remains unresolved to this day.

It can be said that everyone has both love and hate for these two theoretical and philosophical systems.
My love for classical mechanics lies in its belief that the world is deterministic, which gives people a sense of psychological stability. However, I hate the implication that our fate is also predetermined, that everything is fated, and that all efforts and choices are meaningless.

My love for quantum mechanics lies in its view that the world is uncertain, that our efforts are meaningful, and that we can strive, that our destiny is in our own hands. Yet, people also hate this uncertainty, as it suggests the fundamental nature of the world is probabilistic and random, completely contradicting our understanding of the world.

Which of these two theoretical frameworks and philosophical views is correct? Is the world deterministic or indeterminate?

Is there a more comprehensive framework that can integrate these two theories and provide us with a better answer?

(12) The Arrow of Time: Understanding the Entropy Increase Law

In discussions about time, the second law of thermodynamics is always mentioned. Many people consider the second law of thermodynamics as a fundamental principle of physics, something that everything must follow, including the passage of time.

That is to say, the thermodynamic arrow of time, time flow cannot violate the law of entropy increase, so time cannot be reversed.

Is this really the case? How should we understand this?
We know that the entropy increase law is Clausius's interpretation of the second law of thermodynamics, stating that in an isolated system, entropy always tends to increase until it reaches its maximum state, which is the most disordered and chaotic state of the system.

Entropy, in other words, is simply a measure of "disorder." The more chaotic and disordered a system is, the higher its entropy value, and the opposite is also true.

The second law of thermodynamics implicitly defines a time direction through the change in order of an isolated system.

This is the thermodynamic arrow of time, indicating the direction of time's flow, that is, over time, isolated systems will become increasingly disordered and chaotic.

On this basis, scholars have proposed the cosmic time arrow: In 1862, Kelvin published an article titled "On the Possible Duration of the Sun's Heat," in which he suggested that if the universe is finite and follows the current laws (the Second Law of Thermodynamics), it will inevitably reach a state of stasis and death.

That is to say, as time passes, the entire universe will also become increasingly chaotic and disordered, eventually evolving into a state of complete stillness. This is known as the heat death of the universe, which is an extension of the second law of thermodynamics across the entire cosmos.

This theory, once proposed, immediately caused a great stir, and even after more than a hundred years, the debate continues.

The debate has not ceased.
Because if the universe really ends like this, it would be a despair for everyone. At that time, the entire universe would have no activity, no life, complete silence, and all our efforts would be meaningless.

But will the facts really be like this?
Can the Second Law of Thermodynamics, or the law of increasing entropy, really determine the fate of the universe?
We mentioned in the previous chapters that according to the Poincaré recurrence theorem, the arrow of time can be reversed. Given enough time, all the particles in a system will eventually return to a moment that is "close" to their initial positions. As long as time is sufficiently long, the universe may re-cycle. This directly contradicts the so-called second law of thermodynamics, which states that a system will always evolve toward greater disorder and will never spontaneously return to a more ordered state.

On the other hand, we have also mentioned earlier that self-organization phenomena can occur in chaotic systems, including the emergence of intelligence in large systems, leading to the creation of intelligent life, such as humans today or more advanced artificial intelligences in the future.

These intelligent civilizations will transform the world according to their will, making it more regular and orderly.
For example, our human civilization has built numerous cities, roads, vehicles, and equipment, all of which contribute to a more orderly world. In other words, order can spontaneously emerge from chaos, thereby reducing chaos and decreasing the entropy of the system. This concept is fundamentally in contradiction with the second law of thermodynamics and the theory of the heat death of the universe.

Actually, since the proposal of the heat death of the universe, the academic community has had many criticisms:
As Maxwell, the founder of classical electrodynamics, proposed: the second law of thermodynamics is only correct when we can only deal with large objects and cannot see or process the molecules that constitute the separation of objects. From this, he suggested that the scope of application of the second law of thermodynamics should be restricted, thereby breaking the control of the entropy increase law over the fate of the universe.

Boltzmann, another founder of thermodynamics and statistical physics, proposed a statistical explanation of the essence of entropy, suggesting that there are occasional "fluctuations" near the thermal equilibrium state, which do not follow the second law of thermodynamics. When extended to the universe, such "fluctuations" would also exist, so the universe would not reach "heat death."

Engels also proposed: this theory violates the basic principles of dialectical materialism—the principle of the imperishability of motion (also known as the first law of thermodynamics, the law of conservation and transformation of energy), and considers Clausius to have slid from metaphysics to idealism like Newton (historically, some also attribute the theory of universal heat death to Clausius).

Other scientists have proposed concepts such as "dissipative structures" or other cosmological models, aiming to rescue the universe from this terrifying fate. However, these theories lack actual evidence and experimental support, and they have many imperfections, making it impossible to eliminate all doubts and allowing the theory of universal heat death to remain a major unresolved issue worldwide.

With the rise of the Big Bang theory in the twentieth century, especially supported by evidence such as Hubble's redshift, helium abundance, and the 3 K microwave background radiation, people once again felt hopeful, believing they had found the weapon to break the cosmic fate's chains. However, they soon became disheartened to discover that even according to the Big Bang theory, the ultimate fate of the universe seemed bleak: either it continues to expand, eventually turning into a heat death, or it begins to contract, collapsing back into a single point. In either case, the fate of the universe is tragic. Moreover, the Big Bang theory itself has many unresolved issues.

In summary, there are still many debates in academia regarding the second law of thermodynamics, that is, the law of increasing entropy. However, one thing can be confirmed: the description of this thermodynamic law is not yet rigorous enough, and its scope needs strict limitations. In fact, the so-called law of increasing entropy itself is merely a statistical theory and cannot completely rule out the possibility of entropy decrease, and so on.

Moreover, the emergence and development of intelligence, consciousness, and civilization, similar to that of humans, are not covered by thermodynamic statistics.

Within the scope of the laws and the theory of the heat death of the universe, in the development history of our civilization, starting from many mythological stories such as Hou Yi shooting down the sun, Kuafu chasing the sun, Yu Gong moving mountains, and Jingwei filling the sea, the spirit of fighting against heaven and believing in man's ability to conquer nature has already been engraved in our blood. These will not fall within the range of thermodynamic laws. We have every reason to believe that the future of the universe should be bright, and it will certainly be bright.

12.4 Introduction to Telepathy

(1) Definition of Telepathy

Telepathy, more accurately termed as the remote excitation and response phenomenon of brainwave circuits between minds. For humans, it refers to the remote excitation and response phenomenon of brainwave circuits between individuals, which is the direct remote excitation and sensing between two brains without the use of instruments.

In fact, telepathy is not merely an interaction between brains. The human heart, as the center of both the physical body and the spiritual entity, plays a very important role in mental operations, including telepathy. That is why telepathy is often referred to as telepathic communication.

It can be said that the human brain and heart, and even more bodily organs and the nervous system, together form a complex system of telepathic signal loops that can be remotely stimulated and sensed.

Here, the term "heart" refers more accurately to the position in the human chest or heart area, which may not be the actual heart organ. Further accurate research and experimental testing are needed to determine this.

Currently, research and mind control products in society mainly intercept and translate brainwave signals using instruments to control other electronic devices. They do not involve direct interaction between brains, as shown in Figure 12-9.

Figure 12-9: Mature mind control products in society

(2) The Principle of Telepathy

We know that all electronic instruments based on electromagnetic effects can respond to external electromagnetic waves, can be triggered, interfered with, or even controlled. For example, our television sets can be remotely triggered and controlled by remote controls. Our radio and communication devices can be interfered with by other electromagnetic waves. Any instrument operating through electromagnetic waves or electrical signals will be affected or interfered with by external electromagnetic waves and electrical signals.

Human brain thinking and mental signals also operate based on electrical signals. The transmission of electrical signals within the brain's neural network brings about various thoughts, feelings, memories, and the brain's control over the body is also achieved through electrical signals from the nervous system, directing the body's various precise and complex movements.

It can be said that the human body itself is a highly precise and complex instrument, with the brain and nervous system serving as its command and control system. The operation of this system is also based on the transmission of electrical signals.

Any instrument that operates based on electrical signals will inevitably be affected or interfered with by external electromagnetic waves or electrical signals.

Various military radio monitoring and decryption activities follow similar principles.

The same principle applies to the human brain and animal brains, which are systems that operate through electrical signals. They must also adhere to the same rules, meaning that brainwave circuits can be remotely triggered and responded to.

This principle.

There have been extensive studies in the field of telepathic induction both domestically and internationally, achieving significant progress. However, most of these studies still focus primarily on the brain-computer interface area, as shown in Figure 12 - 10.

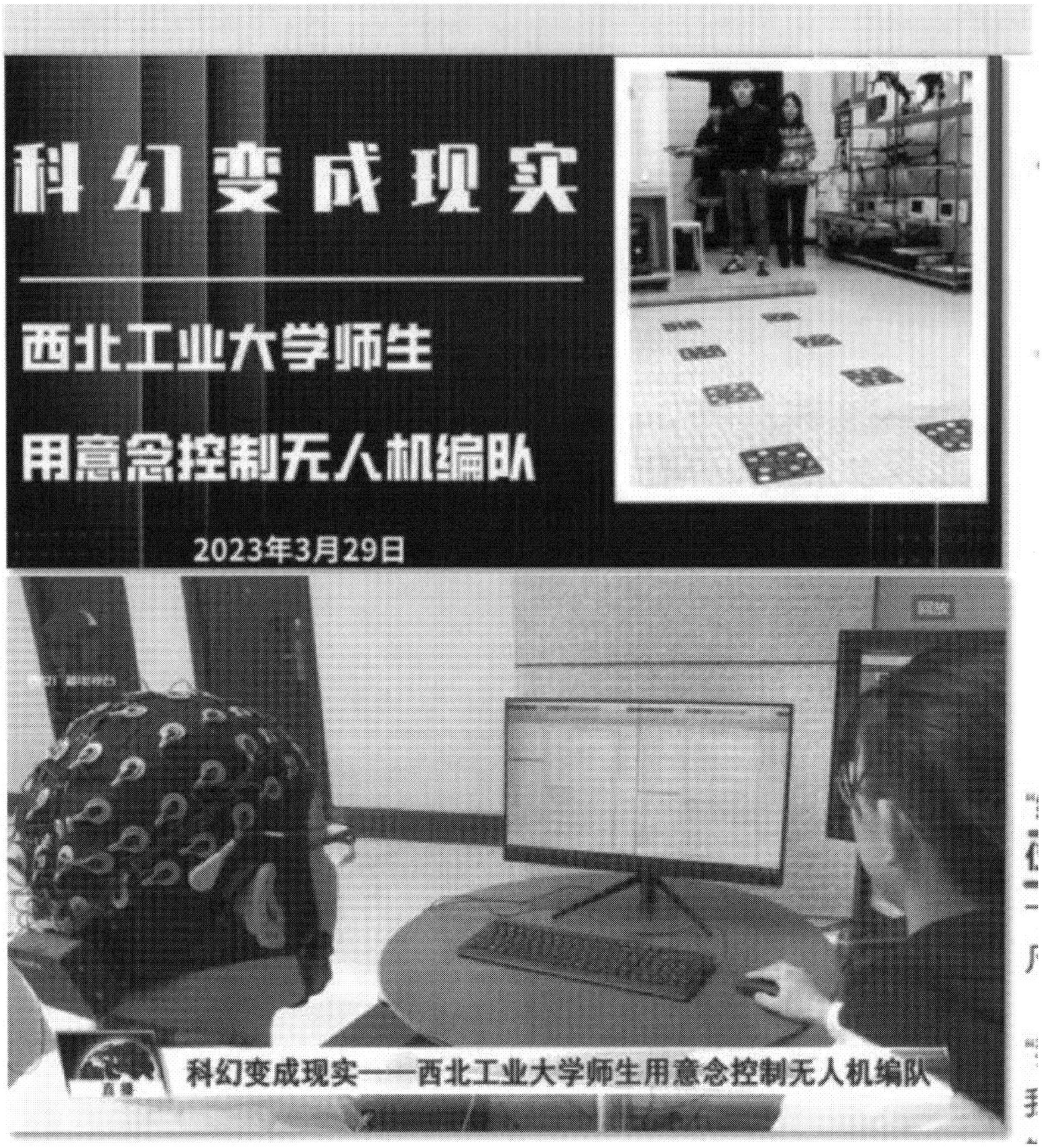

Figure 86: a) Xi'an University of Aeronautics and Astronautics mind control drone

In the aerospace field, in 2016, during the Tiangong-2 and Shenzhou-11 manned flight mission, two astronauts conducted a highly sci-fi and forward-looking experiment: the first ever space brain-machine interaction, enabling silent communication. "Human-machine integration will inevitably become a trend in future aerospace medicine and human factors engineering. Brain-computer interfaces provide a 'third hand' for special groups such as astronauts under physical constraints and enable adaptive automation collaboration at the neural function level, helping them accomplish more complex and challenging tasks," said Xu Minpeng.

Figure 87: "One day, brain-computer interface technology will revolutionize our current human-computer interaction methods. Communication between people will no longer be limited to..."

It's not just about voice or text; it can also be through the mind. "Xu Minpeng said, "In the future, maybe just by thinking, I can send a message directly into your brain. That's telepathy. The sci-fi blockbusters we watch, such as Avatar, Iron Man, Battle Angel Alita, Ready Player One, and even Inception, might all become possible."
I think in a few years, brain-computer interface devices will enter our real life like consumer electronics. We can control a "third hand" through brain-computer interfaces, turning into an eight-armed bodhisattva or a dry-hand Guanyin. By integrating brain-computer intelligence, we can become千里眼 (a mythical figure with eyes that can see far) and顺风耳 (a mythical figure with ears that can hear from afar).

b) The mind interaction experiment conducted by China on Tiangong-2

Figure 12-10 Research on mind sensing by the Chinese scientific community

(3) Basic characteristics of telepathy
(1) Inducer Type
Due to differences in people's physical, mental structures, and states, there are differences in their telepathic abilities. Those with stronger telepathic abilities can more easily generate telepathic signals actively, and we can call them A-type individuals. Others are more capable of receiving telepathic signals passively, and we can call them B-type individuals.

A-type individuals who have strong active telepathic abilities tend to have relatively weaker passive reception capabilities for the mental signals of the target. Therefore, they find it difficult to obtain the mental information of the target. On the other hand, B-type individuals who generate telepathic signals passively have stronger passive reception capabilities for the mental signals of the target, and can obtain more telepathic information.

These two situations may correspond to the difference in the abilities of a person's left and right brain, from another perspective.

There may be a difference on the surface between rational thinking ability and emotional thinking ability; those with strong rational thinking are good at stimulating, while those with strong emotional thinking can perceive more spiritual information from the other party.

(2) Directionality

The initiation of telepathy acts on specific individuals, meaning it has a directional effect on different people, indicating that telepathic actions are directional.

When an A-type simultaneously generates or possesses telepathy with multiple B-types, it can direct the telepathic target, or when all telepathic individuals actively initiate telepathic actions, such as sensing or interfering with other telepaths, they can control the direction of the telepathy, meaning the action is only directed at a specific target. This directionality is controlled through different organs and mental systems such as the brain and heart (chest).

(3) Real-time at a distance

After practical verification, this telepathic communication has a very long range and almost complete real-time capability, allowing for real-time sensing and interaction without obstacles over thousands of kilometers.

(4) Strong Penetration Ability

After practical verification, this telepathic penetration ability is extremely strong, capable of easily passing through floors, rooms, elevators, even sealed elevators, underground subways, and magnetically shielded rooms in ordinary laboratories. It is similar to the interaction of quantum holomorphic current fields between quantum soliton systems (see Chapter 9: Introduction to Spatial Soliton Dynamics), with the penetration force resembling that of neutrinos and significantly stronger than electromagnetic waves and photons.

(5) Generation and Cutoff

Telepathy is mainly generated in two ways: unconscious passive generation and active generation.
Unconscious passive telepathy mainly occurs between very close relatives and friends, as a result of long-term cohabitation, empathy, etc., where mental signals may accidentally resonate and cause telepathic communication. However, this ability is very weak, and people usually do not notice or feel it, even if it happens.

There will be a sense, but it will be very subtle. It is only during times when the mind is extremely calm, such as when sleeping, or when there is a sudden and intense fluctuation in someone's mental signals, that the sensation becomes more noticeable. For example, when a close family member suddenly experiences an unexpected event, people often feel a sense of unease or restlessness in their hearts, or receive some information in their dreams, which are all manifestations of this phenomenon.

In contrast, active spiritual perception is initiated by the person who is the sender of the spiritual perception, mainly Type A individuals. These people actively send out spiritual perceptions to others. The effect of this type of spiritual perception is very obvious, and the perceivers can actively control the action, intensity, and duration of the spiritual perception.

The two principles of producing telepathy are consistent, but actively generated telepathy has stronger ability, intensity, and types, even capable of achieving attack, murder, and complete control.

A third method might emerge later: a third party could use means to create telepathic connections between two people, or even among multiple individuals.

Telepathy once generated can exist in a very stable state. It can continue to persist even without active mental transmission, lasting for years or even decades without disappearing. It can also gradually deepen with increased active mental transmission, leading to greater abilities. This telepathic state can only be terminated by actively cutting it off or by the death of one of the parties involved. The method of active termination still requires further research.

(6) Possible Carriers

If telepathy is entirely formed through electromagnetic waves, these waves would have to be in the extremely low frequency range, because telepathy can operate over very long distances, even across the entire Earth. Therefore, only extremely low frequency ultra-long waves could possibly achieve such an effect.

At the same time, this telepathic penetration ability is also very strong, so it may not be solely due to electromagnetic waves. If it is not only electromagnetic waves, it might also involve something similar to quantum entanglement states, otherwise it would not be possible to achieve telepathy over long distances, with high precision, and in real time.

The existence and occurrence of telepathy may be related to the subconscious areas of the human brain or mental system.

The information of thoughts and other data is processed by the subconscious before it reaches the conscious mind, allowing a person to feel or think about it. However, when part of the subconscious information from person A is transmitted through telepathy to person B, it might have already been sensed by person B before it appears in person A's conscious mind. Therefore, this form of telepathic communication can be extremely real-time, even anticipatory.

The mental world of Inducer A is like a virtual world to Inducer B. The inducer can actively influence or even control what the other person feels or does within this virtual world, and can also affect the real world (i.e., the other person's body). This ability can be of great value if used with good intentions, but can lead to terrible consequences if misused.

The specific mechanism of action still requires further research and verification.
(4) Influence and Manifestation of Telepathy
All human sensations, control, thinking, memory, and so on, are based on the nervous system within the body, including the brain's nervous system, and the operation of the nervous system is entirely based on the transmission of electrical signals.

The transmission and operation of electrical signals can be influenced, interfered with, or even controlled by external electromagnetic signals. Therefore, all human sensations, control, thinking, memory, and so on, can also be affected, interfered with, and controlled by external electromagnetic signals. When this influence is carried out explicitly through external media signals such as light, sound, and electricity, it is called brainwashing. When this influence is caused by brainwave signals emitted by other people, it is referred to as telepathy.

Similar to the resonance of two brain signals, the seizure of control over the mental system.
Therefore, the effects of remote mental influence can include but are not limited to:
(1) Affecting perception
Including vision, touch, taste, hearing, and others, all of a person's senses are different electrical signals to the nervous system. These electrical signals generate specific electromagnetic waves, which can be detected and even decoded from outside the brain. Using machines to detect and decode these signals is known as a brain-computer interface, while being directly decoded by another brain is referred to as brain-to-brain interface.

Interception and decryption are telepathy.
When another brain's EEG circuitry becomes telepathically connected to this person, it can receive all sensory signals such as vision, hearing, smell, and touch from this person.

When telepathic ability can reach sufficient depth, it can influence, interfere, or even distort the sensory experiences of the target, causing the target to experience auditory and visual hallucinations, leading to incorrect perception of information.
(2) Affecting emotions
Human emotions are also the result of the operation of neural电信号, so they can also be remotely intercepted, decoded, or even interfered with and controlled.

When it reaches a certain level, it can cause emotional outbursts and actions that exceed the original state, such as anger, sadness, happiness, or even some extreme emotions. However, for the brain's electrical circuits, this is normal electrical activity, so the brain cannot determine whether the emotions or actions are generated by itself or caused by interference and influence.
(3) Affecting Thought
Human emotions are based on emotional thinking, which can be interfered with. Therefore, normal rational thinking, such as regular thought processes, can also be affected.
(4) Affecting Memory
Human memory is also stored as electrical signals, and the process of recalling is the reading and operation of these signals. Therefore, it can also be remotely intercepted or even remotely interfered with.
(5) Affecting the Body
All control over the human body, including limbs, torso, even breathing, blinking, and basic functions like eating and excretion, is achieved through electrical signals from the nervous system. These signals generate electromagnetic waves, which can be remotely interfered with.

Therefore, it can also be remotely intercepted, decrypted, influenced, interfered with, and even controlled.

Once this deep influence and control is achieved, the telepath can have a significant impact on the behavior of the person being telepathically influenced, even leading to uncontrollable consequences, including but not limited to:

1) It can interfere with the person being sensed' breathing, causing difficulty in breathing or even suffocation
2) Severe interference with muscle activity, even causing muscle damage and tearing
3) Damage to the brain's nervous system, affecting memory, thinking, language, and behavior
4) Interfering and damaging the digestive system, leading to injury and inability to defecate, causing disruption to the physiological systems and trauma to body functions and health.

5) Interfering with behavior, causing confusion, disarray, or even temporary loss of consciousness and resulting in uncontrollable actions, including dangerous movements and self-harm.

6) More seriously, it may interfere with or control the other person's actions that harm others or endanger society.
Etc.
In short, any action that a person can control with their own body, such as behavior or movement, can be achieved through mental induction, and it can even be more severe.
(6) Affecting dreams
The effect of mental signals on the brain's mental activity can also be manifested during sleep.
During sleep, the brain responds to distant mental signals, which may manifest as dream images or influence the formation of dreams.

In fact, aside from the active strong telepathy we are discussing now, most people actually experience a simple and vague form of telepathy, though it is very subtle and difficult for ordinary people to notice. This instinctive telepathy often occurs with those close to us, such as family members. For example, when a family member experiences some kind of trouble, one might suddenly feel very uneasy or irritable. These feelings are caused by the negative spirit of the family member affecting oneself, creating a similar sensation.

In the dreams of ordinary people, sometimes they can have a certain premonition about the future due to telepathy,

This telepathy is also an instinctive and vague sensation. The mental signals in dreams are very calm, making it easier to experience such sensations, thus seeing, hearing, or feeling something in the dream. These experiences are actually mental signals from people or events related to oneself, which resonate with one's own mind and influence the dream. In such cases, interpreting dreams might have some logic.

In short, this kind of "prophetic" dream is usually caused by the sleeper's brainwave signals resonating with those of others when they fall asleep. Therefore, such dreams are not solely one's own dream, but may be shared with others at the same time. Some people remember them clearly, while others have only faint memories or don't remember them at all. Alternatively, others might not even be asleep, but simply thinking or discussing issues related to the dreamer, unintentionally creating a resonance.

Of course, the probability of such telepathic phenomena occurring is relatively low. But many people may have had similar experiences.

(7) Other Influences

Telepathy arises from remote excitation and response between physical and mental states, brain signals, and so on, involving two or more parties. This kind of telepathy can be active or passive, conscious or unconscious, and even subconscious. It is not only about the sensing of consciousness and mental information, but also includes the sensing of physical signals.

For example, if a person's body organs develop diseases, the body actually emits mental signals at that time. However, ordinary people are usually unaware of these signals, including the person who is ill themselves. But when the mental state becomes calm, it is easier to perceive these signals.

As recorded in the "Zuo Zhuan • Chenggong Shínián" (The Zuo Commentary on the Spring and Autumn Annals, Year of Chenggong 10), the story of "Er Jian Wei Nue" (Two Stubborn Ones Causing Harm) tells of how Duke Jing of Jin had a dream in which two small figures discussed his illness, and the details were exactly the same as those later diagnosed by the physician. Apart from the element of storytelling, it also reveals the phenomenon that one can more clearly perceive the signals emitted by the body during dreams.

Not only the patients themselves, but sometimes close relatives of the patients, can also feel these remote physical signals in their dreams, and thus see some ominous signs, even though at this time the disease may still take one or two years to fully develop and manifest, and the patient may not even be aware of it. People with stronger sensitivity are more likely to experience this kind of sensing.

Telepathy is a very complex phenomenon, and its effects on the body, mind, and even more aspects need to be better understood as research continues. It is also hoped that this will draw attention from the industry and encourage participation in this process.

Telepathy is a very complex phenomenon, and its effects on the body, mind, and even more aspects need to be better understood as research continues. It is also hoped that this will draw attention from the industry and encourage participation in this process.

(5) The Significance of Telepathy Research

(1) Military and National Security

Telepathy has long been considered as idealistic, even superstitious. However, recent scientific developments, especially in the field of brain-computer interfaces, have confirmed that brain signals can be acquired, decoded, even interfered with and controlled. The understanding, decoding, and even control of the human brain have become a crucial area of intense competition among countries, as it is not only a matter of scientific progress, but also relates to the struggle for the high ground in the next generation of technological and military revolutions.

Military experts both domestically and internationally have also repeatedly issued warnings that human warfare has begun to enter the era of brain control. The competition for control over the brains of soldiers and civilians will deeply affect future great power security competition, becoming an increasingly important aspect of national security, as shown in Figure 12 - 11.

a)中国国防部网站文章 b)解放军报文章1 c)解放军报文章2

Figure 12-11 Military conflicts will enter an era of competition for brain control

These are all caused by interference and control over the brain through instruments, as well as direct sensing between brains.

Influence and control can be even more powerful than instruments, and will have a significant impact in military security, social security, and even the security of national leaders.

(2) The next generation of technological revolution

A mind-based internet, based on telepathy, will also be the next technological revolution that disrupts the current social model.

The next generation of the internet should be a quantum internet, and recent studies suggest that the generation and transmission of human brain thoughts and consciousness signals may be related to quantum signals.

Therefore, the quantum internet will be a revolutionary network that allows for direct connection through human thoughts. If this is validated, it will not only be a revolution in network technology but also a leap forward in the form of human civilization.

All of this requires us to pay more attention to this field and invest in research and development as soon as possible. Once we fall behind, the consequences would be unimaginable.

12.5 Variation of Vacuum Space Density and Its Effects

The superfluid in the vacuum space permeates the entire universe. As a fluid, its properties, including density, will change differently according to different movements, generally conforming to various effects of fluid mechanics. This is also an important methodological approach in the theory of space relativity.

(1) The Relationship between Space Density and Dielectric Constant

The dielectric constant is an electromagnetic property of the vacuum space, playing a fundamental role in electromagnetism. Changes in the density of the vacuum space will affect its dielectric constant properties. Simply put, the higher the density of the vacuum space, the smaller the corresponding dielectric constant, and vice versa, the lower the density of the vacuum space, the larger its dielectric constant. That is, the dielectric constant is inversely proportional to the density of the vacuum space (Equation 3.9):

$$\varepsilon_v \propto \frac{1}{\rho_v}$$

Where ε_v is the permittivity of the vacuum space in which light propagates at a speed of V , and ρ_v is the corresponding vacuum space density.

(2) The relationship between vacuum space density and the speed of light

As a medium for light propagation, changes in its density also affect changes in the speed of light. According to the speed of light formula, the speed of light is inversely proportional to the square root of the permittivity, which is essentially due to the influence of vacuum space density.

The greater the vacuum space density, the greater the speed of light propagation; the smaller the vacuum space density, the smaller the speed of light propagation. This is consistent with the change in the speed of sound (longitudinal wave speed) in fluids such as air and water, where the speed is proportional to the square root of the fluid's density when only considering the density of the vacuum space.

$$C \propto \sqrt{\rho}$$

Where $ C $ is the speed of light, and $ ρ $ is the density of the empty space through which light propagates. See Chapter "8.10 Principles of Electromagnetic Wave Generation and Propagation: (4) Principle of the Speed of Electromagnetic Waves".

(3) Relationship between vacuum space density and mass bodies

All particles, matter, and other mass bodies are always in various states of motion, which can be uniformly considered as thermal motion of the mass bodies. This thermal motion interacts with the surrounding vacuum space, exciting the corresponding thermal motion in the vacuum space.

According to fluid dynamics analysis, the greater the thermal motion such as fluid waves, or the faster the movement, the smaller the density. Therefore, the density of the vacuum space around a mass body increases from inside to the surface to the surrounding area, gradually approaching the standard density. That is, the density gradient of the vacuum space around a mass body decreases as it approaches the surface, until the density of the vacuum space inside the mass body reaches its minimum. In other words, the density of the vacuum space around a mass body is proportional to the square of the distance from the mass body and inversely proportional to its mass.

$$\rho_r \propto \frac{r^2}{M}$$

Moreover:

$\rho_r < \rho_c$

Among them: M is the mass of the mass body, r is the distance from the surface of the mass body, ρ_r is the density of the void space at a distance r from the surface of the mass body, and ρ_c is the standard void space density.

The density gradient in the surrounding vacuum space caused by thermal motion is also the reason for the gravitational effects between mass bodies, as detailed in the chapter "5.3 Unified Source of Force: (4) Gravity."

It is the same for stars.
Temperature is a direct indicator of the thermal motion intensity of a mass body, such as the temperature around the Sun, which first briefly decreases after leaving the solar surface, only reaching a few thousand degrees, then rapidly increases up to the corona, reaching millions of degrees, and then gradually decreases again to nearly absolute zero in deep space. During this process, the density of the surrounding empty space also undergoes a nonlinear change: it first increases from the solar surface, then decreases to its lowest point in the corona, and gradually increases again as the temperature (and thus thermal motion) decreases, eventually restoring to the standard empty space density in deep space (the density of empty space when the speed of light is C). See section "5.12 Mercury's Perihelion Precession" for details.

That is, the density of empty space is inversely proportional to temperature:

$$\rho \propto \frac{1}{\mathrm{T}}$$

Where ρ is the density of the vacuum space, and T is the temperature.
(4) Relationship between the density of empty space and refraction, redshift
In different density media, the density of the vacuum space also varies. The greater the density of a substance, the more intense the overall thermal motion of its internal particles can be considered, so the density of the vacuum space inside it will be smaller, causing the speed of light to be slower. When light enters a medium of different density, it will slow down due to the change in vacuum space density.

Refraction occurs due to changes in the medium, whether it's the refraction of stars in the macroscopic universe, the refraction through lenses in real life, or the refraction of particles in the microscopic world.

In this process, the frequency of light remains unchanged, while the change in speed causes the wavelength to vary, which is the redshift or blueshift of light waves that we observe. For example, light emitted from the surface of a star will have its wavelength lengthened due to the increase in space density and the acceleration of light speed, resulting in a redshift. This is the familiar gravitational redshift. Etc., see section "5.9 Redshift and Blueshift of Light".
(5) Thermal Motion of Void Space and Cosmic Microwave Background Radiation
All fluids have thermal motion, and void space superfluid is no exception. The overall thermal motion of the superfluid void space in the universe manifests as the well-known cosmic 3 K background radiation. This background radiation indicates that the superfluid void space in the universe has an overall basic thermal motion equivalent to a temperature of 3 K. See section "5.10 On the Big Bang Theory" for more details.

Many aspects of spatial relativity are related to changes in the density of empty space, and here we will summarize them uniformly.

12.6 The Paradox of Relative Observation: The Achilles' Heel of Special Relativity

In 1905, Einstein proposed the special theory of relativity, introducing the famous principle of the constancy of the speed of light, and brought about the most renowned Lorentz transformation and time dilation effect in special relativity, which states that time slows down for moving objects.

In 1911, the French physicist Larmor proposed the famous twin paradox experiment to challenge the time dilation effect of special relativity, which is also known as the twin paradox.

The Twin Paradox states that if there are two twin brothers, one staying on Earth and the other traveling in a spaceship through space, the traveling twin will experience time dilation according to the special theory of relativity. When the traveling twin returns to Earth at a speed close to the speed of light and meets his brother, he will be younger than his brother.

According to the time dilation theory of special relativity, time slows down for the moving twin.
However, if we consider the relativity of motion, we can also argue that the twin on the spaceship is stationary, while the brother on Earth is traveling at a speed close to the speed of light and returns. In this case, the brother on Earth would be younger than the traveling twin.

Therefore, this creates a contradiction. When the two twins meet again, which one is actually younger, the traveling twin or the one who stayed on Earth?

Figure 88: Figure 12-12 Twin Paradox Diagram

This hypothetical experiment has sparked widespread debate.
The supporters of special relativity provided explanations stating that special relativity is a theory applicable to inertial frames, and in this twin paradox experiment, the brother's journey of traveling in a spaceship away from Earth and then returning involves acceleration, which is not part of an inertial frame. Therefore, it is not suitable to directly apply the time transformation formula.

However, this explanation is very forced and unsatisfactory.
In fact, the so-called Lorentz transformation and relativistic effects such as time dilation and length contraction are merely mathematical calculation patterns. To ensure the speed of light remains constant at C, time and length must change, making the mathematics self-consistent. This mathematical description is not the true physical essence, but rather a distortion of the physical reality

Practical understanding.
We can slightly modify the twin paradox experiment, assuming a simpler relative observation experiment to illustrate the concept.

Suppose there are two inertial reference frames A and B, moving relative to each other at a constant speed of 0.9 C. Each inertial frame has two observers, A_1、A_2 and B_1、B_2, respectively. Since they are each in their own inertial frame, their clocks are synchronized. That is, A_1 and A_2 clocks are synchronized, and B_1 and B_2 clocks are synchronized, as shown in Figure 12 - 13(a).

At a certain moment when A_1 and B_1 meet, A_2 also meets B_2 at the same time, because clocks in the same reference frame are synchronized, so A_1 and A_2、B_1 and B_2 can respectively confirm that their meetings are simultaneous.

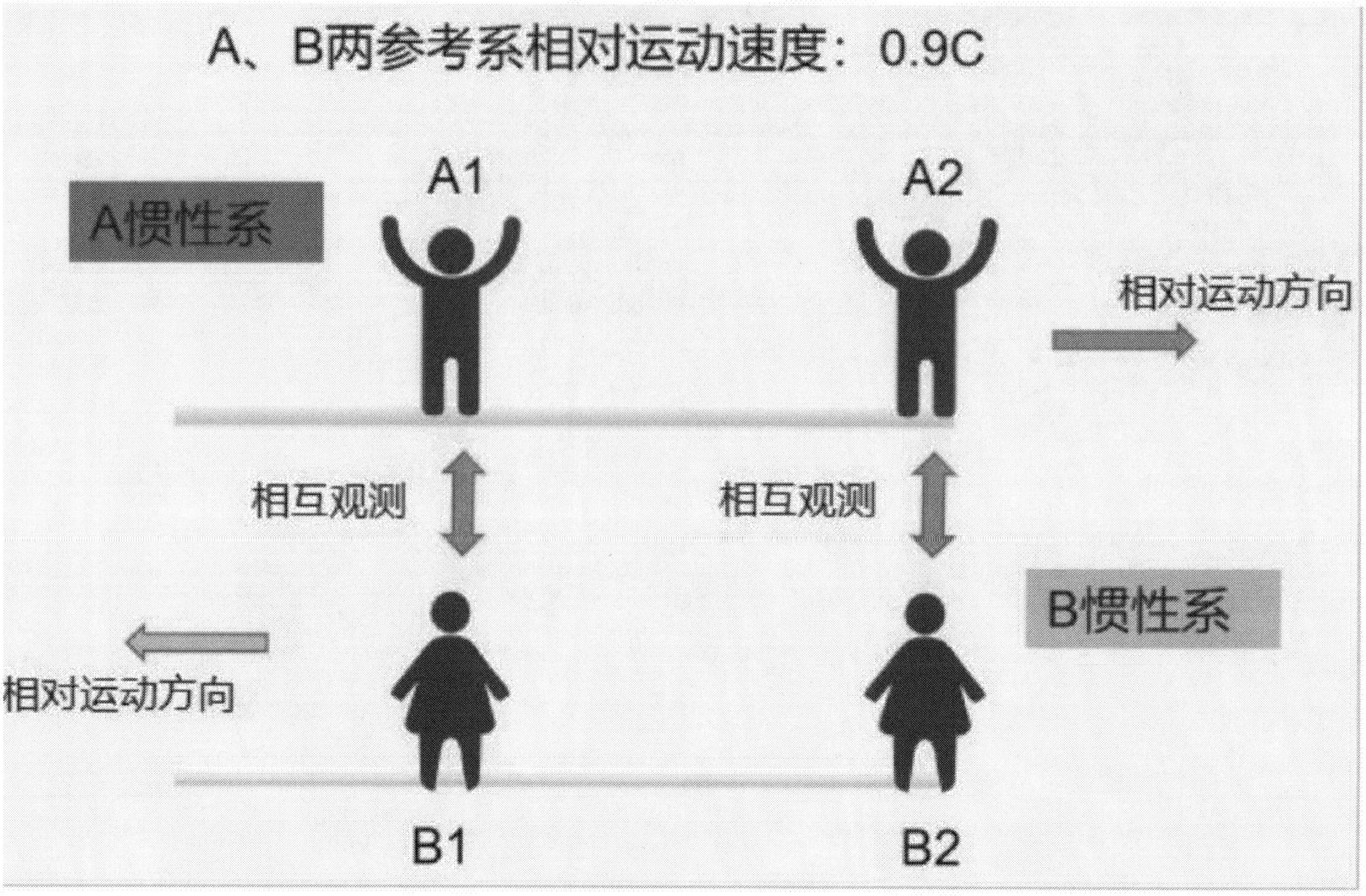

When A1 and B1, A2 and B2 positions coincide, they interact with each other
Record the observed age of the other party.

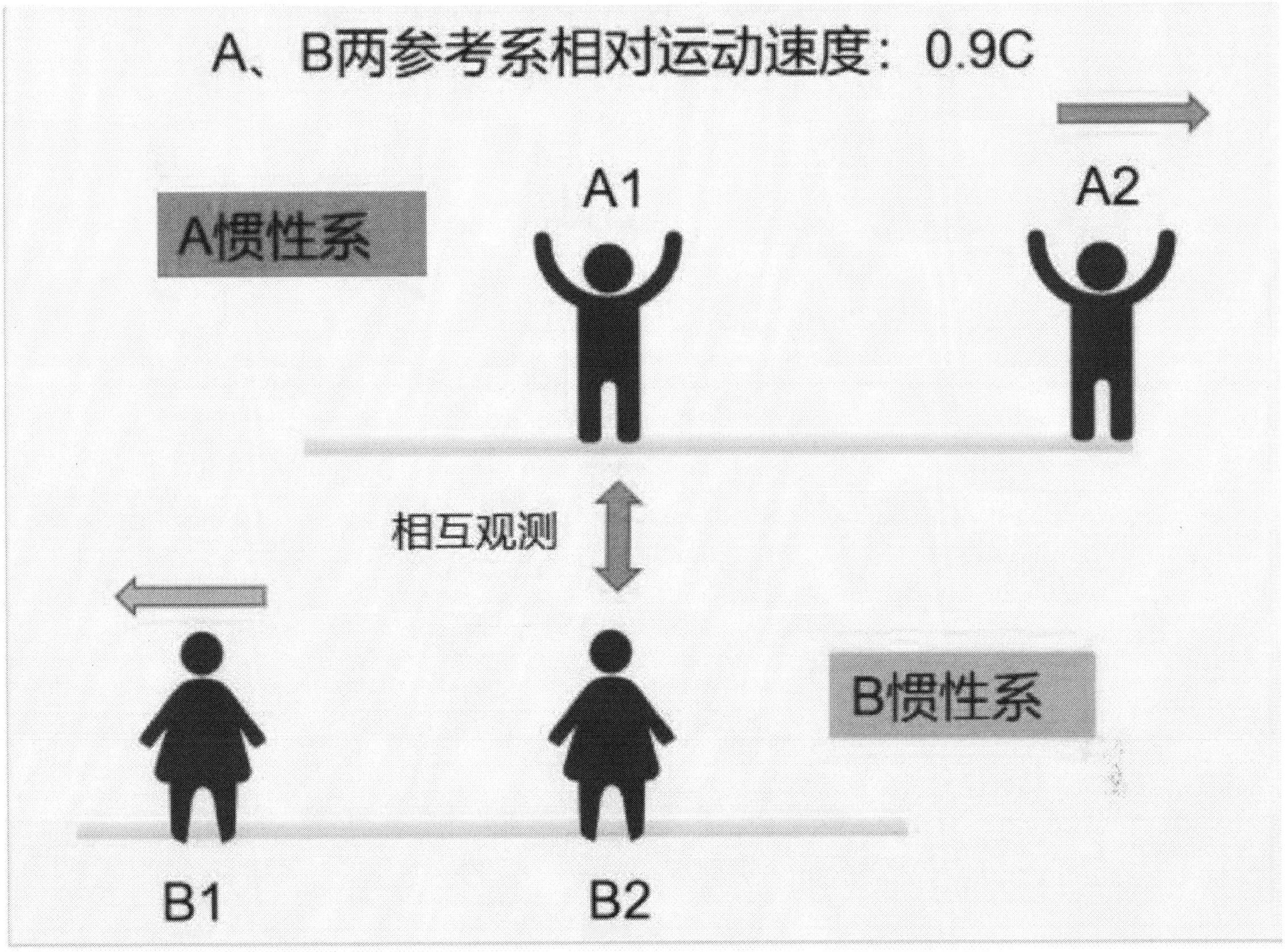

b) After 40 years in a reference frame, the positions of A1 and B2 coincide Synchronization, where each observes and records the other's age.

Fig. 12-13 Observers' clocks are synchronized in two inertial frames

At this moment, they measured each other's ages and found that both were 20 years old, that is: A_1 and B_1、 A_2 and B_2 measured each other's ages, and all four were 20 years old, both sides recorded it.

Then, the two inertial frames continued moving towards each other. In the A reference frame, after 40 years, A_1 and B_2 met, as shown in Figure 12-13(b). At this time, how much older did A_1 and B_2 observe each other to be?

Assuming they are moving relative to each other at 0.9 times the speed of light, if we assume the A reference frame is stationary and the B reference frame is moving, then the moving clock runs slower. Therefore, after 40 years in the A reference frame, the B reference frame actually experienced only about 17 years.

Year.
Then, at this time, the age of A_1 is 60 years old, and the age of B_2 is 37 years old.

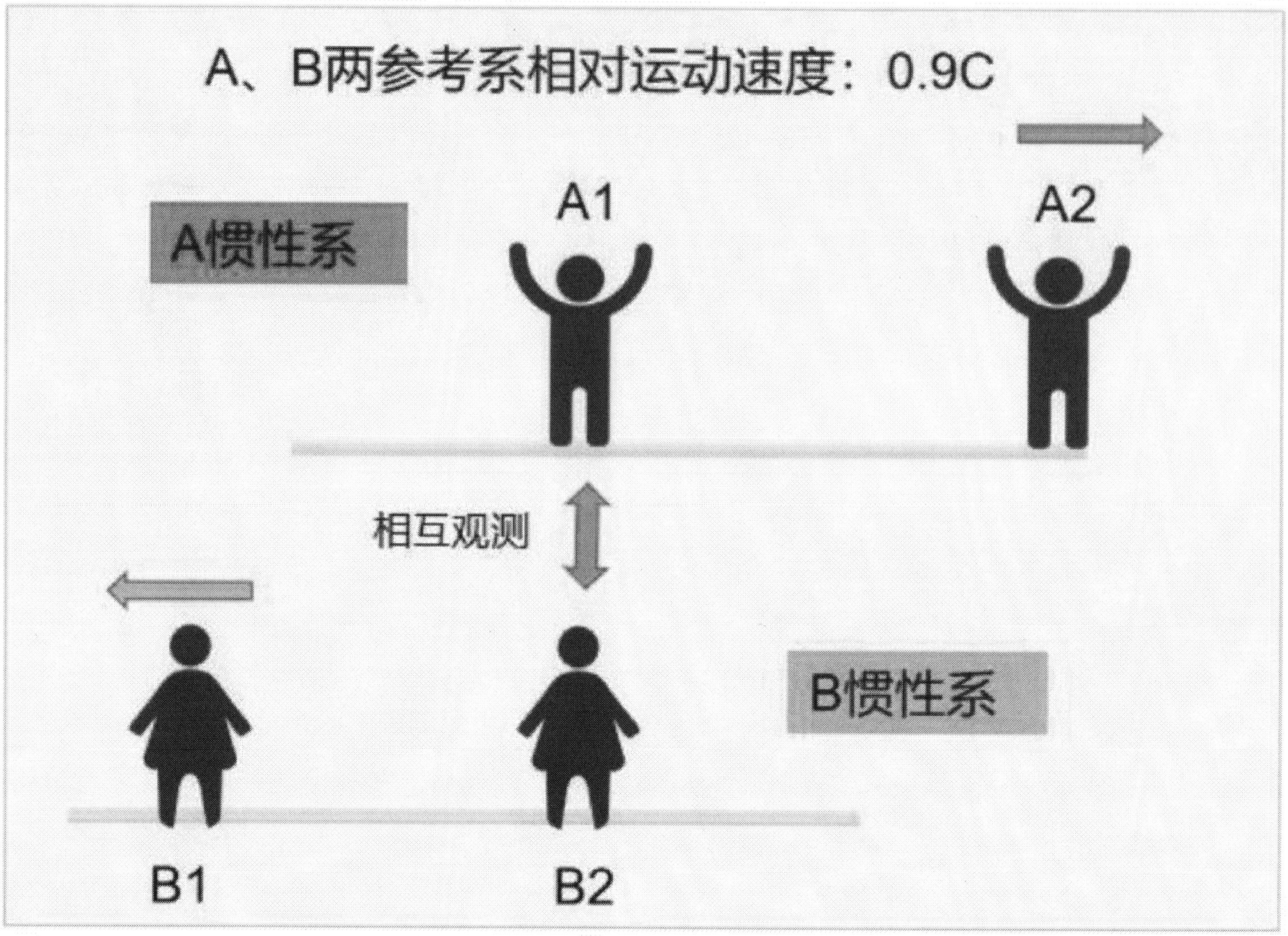

The time dilation formula of special relativity
t: time in the moving inertial frame
Time in the stationary inertial frame

1) Assume A is stationary, B is moving
A has passed 40 years, B has passed about 17 years, so:
A1: 60 years old, B2: 37 years old

$$t_1 = \frac{t}{\sqrt{1 - \frac{V^2}{C^2}}}$$

$$t_B = t_A\sqrt{1 - \frac{V^2}{C^2}} = 40 * \sqrt{1 - 0.9^2}$$

$$t_B \cong 17$$

2) Assuming A is moving and B is stationary
A has passed 40 years, B has passed about 92 years, so:
A1: 60 years old, B2: 112 years old

$$t_B = \frac{t_A}{\sqrt{1 - \frac{V^2}{C^2}}} = \frac{40}{\sqrt{1 - 0.9^2}}$$

$$t_B \cong 92$$

Figure 12-14 Time Transformation in Relative Motion Between Two Inertial Systems

But according to the principle of relativity of motion in special relativity, we can also assume that reference frame A is moving and reference frame B is stationary. In this case, A_1 is 60 years old, and his clock slows down due to motion, so the clock in reference frame B should be faster. Calculating accordingly, reference frame A has passed 40 years, and reference frame B has passed approximately 92 years, meaning B_2 should be 112 years old, as shown in Figure 12 - 11.

Similarly, assuming various different velocities for the two inertial frames while keeping their relative velocity constant, according to the time dilation and length contraction effects in special relativity, each will observe the other's age to have changed differently.

But the actual result should be only one.
So, in this case, how old would each of them observe the other to be?
Throughout the entire process, neither of the two inertial frames undergoes an acceleration phase; they are always moving towards each other at a constant velocity. Both are inertial frames, so at this point, which one should experience time dilation—the A inertial frame or the B inertial frame?

This is a paradox that special relativity cannot resolve, stemming from errors in the fundamental principles of special relativity, including

Misunderstandings of the principle of relativity and the invariance of the speed of light, as well as errors in the logic of determining simultaneity (see sections "2.1 Misuse of the Invariance of the Speed of Light Hypothesis Principle" and "2.2 Misuse of the Principle of Relativity" and "2.4 Misunderstanding of the Logic of Simultaneity Determination" in chapter "2 Some Issues with the Current Theory").

It is precisely these errors that lead to the appearance of an age paradox in this thought experiment, which cannot be resolved by the special theory of relativity.

The reason for this is that the Lorentz transformation and the time dilation effect it causes in special relativity are actually just a mathematical calculation system: in order to keep the speed of light constant, parameters such as length and time must change, ensuring consistency within the mathematical framework.

This mathematical system distorts the description of physical phenomena and our understanding of the essence of the physical world.

By removing the assumption of the constancy of the speed of light and allowing it to vary relative to different observers, and returning to the normal Galilean transformation, we can truly describe the physical essence and real results in this situation.

Therefore, the twin paradox is not really a paradox, but a contradiction arising from the theory of special relativity, stemming from the incorrect understanding and application of the two fundamental principles of special relativity: the invariance of the speed of light and the principle of relativity, as well as the flawed logic of simultaneity judgment, leading to a distortion in the mathematical description of light and motion in the physical world.

Instead of addressing this issue directly, attempting to explain it through more complex Minkowski geometry or spacetime distortion is like trying to fish for carp in a tree, completely off track.

Special relativity and the invariance of the speed of light, due to their immense influence, seem like a shackle that restricts people's thinking, distorts their understanding of the physical world, and has affected the development of theoretical physics in related fields.

We must recognize the problem and confront it directly in order to break free from the constraints and guide the development of theoretical physics back onto the right path.

References

[1] Einstein, et al. Principles of Relativity (Collection of Classic Papers on Special and General Relativity). Beijing: Science Press, 1980
[2] Dirac, Principles of Quantum Mechanics. Beijing: Science Press, 2008
[3] Ruan Xiaogang. Why Does Spacetime Curve in Einstein's General Relativity? [J]. Journal of Beijing University of Technology, 2023, 49(2): 103-178. doi:10.11936/bjutxb2022040007
[4] Isaac Newton. Mathematical Principles of Natural Philosophy. Beijing: Commercial Press. 2006
[5] Zhou Shixun. Quantum Mechanics Tutorial. Shanghai: Shanghai Science and Technology Press, 1961
[6] Zhang Yuanzhong. Experimental Foundations of Special Relativity. Beijing: Science Press, 1994
[7] Liu Liao, Fei Baojun, Zhang Yunzhong. Special Relativity. Beijing: Science Press, 2008
[8] Lee Zongwei, Xiao Xinghua. Astrophysics. Beijing: Higher Education Press, 2000
[9] Liu Liao, Zhao Feng. General Relativity. Beijing: Higher Education Press, 2004
[10] Liang Canbin. Electromagnetism (Third Edition). Beijing: Higher Education Press, 2012
[11] Ni Yunsun, Wei Rongjue. Solitary Waves in a Water Channel. Shanghai: Shanghai Science and Technology Education Press, 1997
[12] Liu Shida, Liu Shi Shi. Solitary Waves and Turbulence. Shanghai: Shanghai Science and Technology Education Press, 1994
[13] Lin Sen. Study on the Characteristics of Optical Current Sensors Based on the Faraday Magneto-Optic Effect [D]. Chongqing University, 2010
[14] Griffiths, David J. , Introduction to Quantum Mechanics (2nd ed.) , Pren - tice Hall, 2004
[15] Sangster K, Hinds E A, Barnett S M, et al. Aharonov - Casher phase in an atomic system [J] . Physical Review A, 1995, 51 (3) : 1776 - 1786. DOI: 10. 1103 / PhysRe - vA. 51. 1776.
[16] Witze A. These six distant galaxies captured by JWST are wowing astrono - mers [J] . Nature, 2023, 619

(7968) : 16 - 17.
[17] Yang Juan, Zhao Tong, Song Yujia, Gou Lijun. Review of the Hot Topics in Astronomy in 2023. Science and Technology Trends. 2024, 42(01): 72-86
[18] Gupta R P. JWST early Universe observations andΛCDM cosmology [J] . Monthly Notices of the Royal Astronomical Society, . 2023, 524 (3) : 3385 - 3395
[19] Lee Zongwei, Xiao Xinghua. Astrophysics. Beijing: Higher Education Press, 2000
[20] Zhang Minghui, Teng Guirong. Engineering Fluid Mechanics. Beijing: Machinery Industry Press, 2018
[21] Zhou Bangrong. Quantum Field Theory. Beijing: Higher Education Press, 2008
[22] Lee Zhengdao. A Brief Introduction to Particle Physics and Field Theory. Beijing: Science Press, 1984
[23] Hu Jinghuan, Wang Zhonglie, Liu Yuhua. Atomic Physics [M]. Beijing: Beijing Normal University Press, 1989
[24] Zhu Hongyuan. Quantum Field Theory. Beijing: Science Press, 1960
[25] Wu Daxian. Quantum Mechanics (Two Volumes: Part A and Part B). Beijing: Science Press, 1984
[26] Ji Jiangyan. Lecture Notes on Quantum Mechanics. Beijing: University of Science and Technology Beijing, 2003
[27] John W. M. Bush, Wang Zhan. Guiding Wave Hydrodynamics. Advances in Mechanics. 2021, 51(01): 155 - 177.
[28] Chen Guanrong. de Broglie and His Pilot-Wave Theory. Science and Technology Review. 2019, 37(22): 149 - 152.
[29] Shen Xingyang. Superfluidity. Beijing: Science Press, 1982
[30] Chen Shuhong, Xiong Chunyan, Liu Shuai. Dynamical Behavior of Transition from Superfluidity to Bose-Einstein Supercondensation. Hefei: University of Science and Technology of China Press, 2020
[31] Wu Wangyi. Fluid Mechanics (Upper and Lower Volumes). Beijing: Peking University Press, 1982
[32] Liu Quanzhong, Li Xiaobin. Advanced Fluid Mechanics. Harbin: Harbin Institute of Technology Press, 2017
[33] L. Ryder. Particles and Symmetries. Beijing: Science Press, 1983
[34] Luo Liao Fu, Lu Can. Fundamental Particles. Beijing: Beijing Press, 1981
[35] Maxwell. Treatise on Electricity and Magnetism. Beijing: Peking University Press, 2010
[36] M. Born. Principles of Optics: Electromagnetic Theory of Propagation, Interference and Diffraction (Volumes I and II). Beijing: Science Press, 1981
[37] Zhu Hongyu, Xie Shichong. Some Notes on Displacement Current in Vacuum, Maxwell's Equations, Sources of Electromagnetic Fields, and Jefimenko's Formula. University Physics, 2023, 22(5): 3-8+24
[38] Zhao Kaihua. On the displacement current and conduction current not exciting the magnetic field according to the same law (method) [J]. University Physics, 2001, 20(8): 29-31. DOI: 10.16854/j.cnki.1000-0712.2001.08.007
[39] Yan Zhongqiang. The Biot-Savart law in quasi-steady situations cannot be extended to rapidly changing fields [J]. University Physics, 1986, 5(7): 43. DOI: 10.16854/j.cnki.1000-0712.1986.07.017
[40] Zhao Kaihua. A simple example showing that displacement current does not excite a magnetic field [J]. University Physics, 2001(06): 40-45. DOI: 10.16854/j.cnki.1000-0712.2001.06.012
[41] Liu Jifeng. Particle Physics. Guilin: Guangxi Normal University Press, 2002
Chapter 42: Zhang Naosen. Particle Physics (Vol. 1, Vol. 2). Beijing: Science Press, 1986
Chapter 43: The U.S. Special Committee on Elementary Particles. Elementary Particle Physics. Beijing: Science Press, 1992
Chapter 44: Pauling (Linus Pauling). The Nature of the Chemical Bond. Beijing: Peking University Press, 2024
Chapter 45: Yuan Goutao. New Advances in Dark Matter Research. Modern Physics Knowledge, 1996, 8(1): 22–23
[46] Yu Kuanxin. Advanced Optics. Beijing: Beijing University of Technology Press, 2009
[47] Xie Chufang, Rao Kejin, Yang Xianqing, Zhao Jiaseng. Electromagnetic Fields and Electromagnetic Waves. Beijing: Higher Education Press

Publisher, 2019
[48] Ge Jingye, Yang Lingling, Li Qing. Exploring the Mystery of the Gravitational Constant. Physics, 2023, 52(09): 653-656
[49] Han Rusan. Physics of High-Temperature Superconductivity, Second Edition. Beijing: Peking University Press, 2014
[50] Luo Huiqian. "The Superconducting 'Little Era': The Past, Present and Future of Superconductivity." Beijing: Tsinghua University Press, 2022
[51] London F, London H. The Electromagnetic Equations of the Supraconduc - tor [JJ. Proc. Roy. Soc. , (London) , 1935, A155: 71.
[52] Min Jiayin. The Philosophical Significance of Field (Daochang) Research. Journal of Hangzhou Normal University (Social Sciences Edition), 2004(06): 26-30
[53] Lei Jinzhi, Jiang Xingliu, et al. Electrochemical Anomalies and the Theory of Torsion Field. Science and Technology Review, 2000
[54] Yu Ye. Ghost Scratch Field. Big Technology (Science Mysteries), 2006(10): 4-9
[55] Li Huaijiang, Zhou Zuzhen. Inconsistency of Relativistic Quantum Mechanics - A Re-examination of Basic Theoretical Issues in Quantum Field Theory. Journal of Yunnan Normal University, 2000, 20(6): 35-37
[56] Huang Zhixun, Jiang Rong. Does "Relativistic Quantum Mechanics" Truly Exist. Frontiers of Science, 2017, 11(04): 12-38

Author's Biography

Fang Wenxiang, male, 48 years old, founder of Datong Wanwu Biotechnology Co., Ltd.

Since 2015, I have experienced an awakening of spiritual thinking and developed a strong interest in natural sciences. After ten years of in-depth thinking, a new scientific perspective has been proposed on the materialization, quantization, and integration of space and time.

Since 2023, the brand new scientific thinking has been promoted on Tiktok, video number and other We Media platforms, and has been recognized by many people. My video number is "too easy to live". The understanding of space that I elaborated in the video is consistent with the views of Professor Su Enjie.

All physical phenomena are actually derived from the evolution of spatial superfluids, and as observers, we also perceive, describe, and define this objective spatial entity. A vacuum is not empty, and space is not a blank virtual background board in physics, but a real entity that exists. The objective material world we consider is actually just an observation, definition, and description of spatial entities.

Spatial superfluids have density and uneven distribution, and are constantly in motion due to surface polarization and spatial pressure differences. The four forces in physics are all generated and transmitted through space. Without the entity and foundation of space, we view the material world in an isolated and fragmented way of thinking, which is the biggest deficiency facing the scientific community today.

May this book enable human scientific researchers to elucidate and refine human knowledge from the perspective of materialization and integration in space. Human cognition is limited, while the universe is infinite. With limited human cognition, it is impossible to fully understand the universe. If we can understand from the perspective of spatial materialization, humanity will be one step closer to truth.

2025

Su Enjie, male, 47 years old, from Shanghai.

Since I accidentally derived the mass energy equation based on the de Broglie material wave formula in junior high school, it seems like I have an inexplicable connection with relativity. I still remember when the teacher asked me if I knew about relativity after reading the equation, I looked confused and thought the name was very strange. I recognized every word, but when put together, I didn't understand what it meant.

Later, I went all the way to Northwestern Polytechnical University for further studies, pursuing a bachelor's degree and a master's degree. After joining Huawei, I started working and have been paying close attention to relativity for more than ten years. I also feel that there are many unsatisfactory aspects to this theory. After leaving Huawei, I wanted to study it carefully and write down these questions and ideas, but I never had the time. Until now, I feel that I should sit down and seriously settle this matter, which can be considered as an explanation to myself. After several months, I have written this article.

2025

Made in United States
Orlando, FL
04 November 2026

90201060R00131